Fly Biology, Ecology, Behavior and Management

Fly Biology, Ecology, Behavior and Management

Editors

Sérgio M. Ovruski
Flávio Roberto Mello Garcia

Basel • Beijing • Wuhan • Barcelona • Belgrade • Novi Sad • Cluj • Manchester

Sérgio M. Ovruski
Planta Piloto de Procesos
Industriales Microbiológicos
y Biotecnología
(PROIMI-CONICET)
San Miguel de Tucumán
Argentina

Flávio Roberto Mello Garcia
Departamento de Ecologia,
Zoologia e Genética
Universidade Federal de Pelotas
Pelotas
Brazil

Editorial Office
MDPI AG
Grosspeteranlage 5
4052 Basel, Switzerland

This is a reprint of the Special Issue, published open access by the journal *Insects* (ISSN 2075-4450), freely accessible at: https://www.mdpi.com/journal/insects/special_issues/fly_management.

For citation purposes, cite each article independently as indicated on the article page online and using the guide below:

Lastname, A.A.; Lastname, B.B. Article Title. *Journal Name* **Year**, *Volume Number*, Page Range.

ISBN 978-3-7258-1575-3 (Hbk)
ISBN 978-3-7258-1576-0 (PDF)
https://doi.org/10.3390/books978-3-7258-1576-0

Contents

About the Editors

Sergio M. Ovruski

Dr. Sérgio M. Ovruski earned his PhD in Biological Sciences from the National University of Tucumán (Argentina) in 1995 and has been working with economically important fruit flies and their hymenopterous parasitoids since 1998 as researcher of the National Council of Science and Technology of Argentina (CONICET). He is currently a CONICET Senior Researcher at the Pilot Plant of Industrial Microbiological Processes and Biotechnology (PROIMI in Spanish) in Tucumán, Argentina. He is the head of the Laboratory for Bioecological Research on Fruit Flies and their Natural Enemies at the Pest Biological Control Department. His research interests include biology, ecology, behavior, and management via eco-friendly methods, with a focus on biological control. He has been the supervisor and co-supervisor of numerous PhD theses, doctoral and postdoctoral fellows, junior researchers of CONICET, national and international research projects, and technological development and research agreements between CONICET and provincial governments. His h-index is currently 22.

Flávio Roberto Mello Garcia

Dr. Flavio Roberto Mello Garcia is a professor at the Universidade Federal de Pelotas (UFPel), Brazil. He achieved his master's degree and a PhD in Zoology (Entomology) at the Pontifical Catholic University of Rio Grande do Sul (PUCRS), Brazil. He carried out postdoctoral training at the University of Florida (UF) and trained at the US Department of Agriculture (USDA). He is the author of more than 240 articles and chapters and 11 books. He was the Project Coordinator for the evaluation of the use of sterile insects and pupae parasitoids for the management of *Drosophila suzukii* in greenhouses, which was approved and encouraged by the International Atomic Energy Agency (IAEA). He is an advisor for postdoctoral, doctorate, and masters' students and a scientific project initiator. He has coordinated more than 34 research projects in Brazil, Africa, Uruguay, and the United States.

Preface

Diptera, known as the "true" flies, is one of the most numerous and diverse orders of insects, comprising about 125,000 species widely distributed throughout the world. These insects play an ecologically relevant role and bear significant importance to humans. Just as certain flies act as pollinators, decomposers, predators, and parasitoids, others are vectors of human diseases and represent serious pests of agriculture, stables, and barnyards. Several dipterans are involved in forensic legal investigations. This Special Issue collects the original research articles and reviews to deepen the biological, ecological, and behavioral knowledge of this diversified insect group and to cover management strategies for both pest and disease vector species. This involves a wide range of studies, such as life history, physiological aspects, reproduction, demographic traits, the ecology and evolution of behavioral patterns, population fluctuation and dynamics, diversity, abundance, trophic relationships, host range and status, natural enemies, integrated pest management, and control tactics.

Sérgio M. Ovruski and Flávio Roberto Mello Garcia
Editors

 *insects*

Article

The Population Dynamics and Parasitism Rates of *Ceratitis capitata*, *Anastrepha fraterculus*, and *Drosophila suzukii* in Non-Crop Hosts: Implications for the Management of Pest Fruit Flies

María Josefina Buonocore-Biancheri [1], Xingeng Wang [2,*], Segundo Ricardo Núñez-Campero [3,4], Lorena Suárez [5,6], Pablo Schliserman [7], Marcos Darío Ponssa [1], Daniel Santiago Kirschbaum [8,9], Flávio Roberto Mello Garcia [10] and Sergio Marcelo Ovruski [1]

[1] Planta Piloto de Procesos Industriales Microbiológicos y Biotecnología (PROIMI-CONICET), División Control Biológico, Avda. Belgrano y Pje. Caseros, San Miguel de Tucumán T4001MVB, Argentina; mjbuonocoreb@hotmail.com (M.J.B.-B.); mdpwolf@gmail.com (M.D.P.); sovruski@conicet.gov.ar (S.M.O.)

[2] USDA-ARS Beneficial Insects Introduction Research Unit, Newark, DE 19713, USA

[3] Centro Regional de Investigaciones Científicas y Transferencia Tecnológica de La Rioja (CRILAR-CONICET), Entre Ríos y Mendoza s/n, Anillaco, La Rioja 5301, Argentina; segundo.nc@conicet.gov.ar

[4] Departamento de Ciencias Exactas, Físicas y Naturales, Instituto de Biología de la Conservación y Paleobiología, Universidad Nacional de La Rioja (UNLaR), Av. Luis M. de la Fuente s/n., La Rioja 5300, Argentina

[5] Dirección de Sanidad Vegetal, Animal y Alimentos de San Juan (DSVAA)-Gobierno de la Provincia de San Juan, CONICET, Nazario Benavides 8000 Oeste, Rivadavia, San Juan J5413ZAD, Argentina; lorenasuarez@conicet.gov.ar

[6] CCT CONICET San Juan, Argentina Av. Libertador Gral. San Martín 1109, San Juan J5400AR, Argentina

[7] Centro Regional De Energía y Ambiente para el Desarrollo Sustentable (CREAS), CONICET-UNCA, Prado 366, de Catamarca 4700 SFV, Argentina; schliserman73@yahoo.com.ar

[8] INTA Estación Experimental Agropecuaria Famaillá, Tucumán Ruta Prov. 301, km 32, Famaillá 4132, Argentina; kirschbaum.daniel@inta.gob.ar

[9] Cátedra Horticultura, Facultad de Agronomía y Zootecnia, Universidad Nacional de Tucumán, San Miguel de Tucumán 4000, Argentina

[10] Departamento de Ecologia, Zoologia e Genética, Instituto de Biologia, Universidade Federal de Pelotas, Pelotas 96000, Rio Grande do Sul, Brazil; flavio.garcia@ufpel.edu.br

* Correspondence: xingeng.wang@usda.gov

Citation: Buonocore-Biancheri, M.J.; Wang, X.; Núñez-Campero, S.R.; Suárez, L.; Schliserman, P.; Ponssa, M.D.; Kirschbaum, D.S.; Garcia, F.R.M.; Ovruski, S.M. The Population Dynamics and Parasitism Rates of *Ceratitis capitata, Anastrepha fraterculus*, and *Drosophila suzukii* in Non-Crop Hosts: Implications for the Management of Pest Fruit Flies. *Insects* **2024**, *15*, 61. https://doi.org/10.3390/insects15010061

Academic Editor: Christos G. Athanassiou

Received: 15 December 2023
Revised: 10 January 2024
Accepted: 12 January 2024
Published: 15 January 2024

Simple Summary: Non-crop host plants inhabiting wild vegetation areas surrounding crops strongly influence the dynamics and abundance of polyphagous pest fruit flies, including *Ceratitis capitata* (*Cc*), *Drosophila suzukii* (*Ds*), and *Anastrepha fraterculus* (*Af*). The two former species are dangerous invasive pests widespread in all Argentinean fruit-producing regions, whereas the latter species, native to the Neotropics, coexists with those exotic species in northwestern Argentina. Integrated and eco-friendly management strategies are needed against those pests, targeting both crop and non-crop areas. Therefore, this study assessed the abundance of these pest dipterans, their seasonal infestation levels in five non-crop fruit species, relationships with competing saprophytic drosophilids, and natural parasitism. Fruits were surveyed in a disturbed wild habitat in northwestern Argentina over 40 months, and fruits were sampled from the tree canopies and ground. The results revealed that *Af* had the highest abundance, followed by *Cc* and *Ds*. Saprophytic drosophilids were predominant only from ground fruit samples. Spatiotemporal overlaps of different host fruit availability enabled continuous and suitable sources for pest proliferation throughout the year. The population peaks of both exotic pests coincided with the highest availability of peaches from December to January, whereas the *Af* population peaked during guava fruiting from February to April. These pest flies were attacked mainly by generalist parasitoids that could be useful in the conservation and augmentative biological control of these pests.

Abstract: Understanding the seasonal dynamics inherent to non-crop host–fruit fly–parasitoid interactions is vitally important for implementing eco-friendly pest control strategies. This study assessed the abundance and seasonal infestation levels of three pest fly species, *Ceratitis capitata* (Wiedemann), *Anastrepha fraterculus* (Wiedemann), *Drosophila suzukii* (Matsumura), as well as the related saprophytic drosophilids, and their natural parasitism in a disturbed wild habitat characterized by non-crop hosts in northwestern Argentina over 40 months. *Juglans australis* Griseb (walnut), *Citrus aurantium* L. (sour orange), *Eriobotrya japonica* (Thunb.) Lindley (loquat), *Prunus persica* (L.) Batsch (peach), and *Psydium guajava* L. (guava) were sampled throughout their fruiting seasons. Fruits were collected from both the tree canopies and the ground. The most abundant puparia was *A. fraterculus*, followed by *C. capitata* and *D. suzukii*. *Drosophila* species from the *D. melanogaster* group were highly abundant only in fallen fruits. Spatiotemporal overlaps of different host fruit availability provided suitable sources for pest proliferation throughout the year. The populations of both invasive pests peaked from December to January, and were related to the highest ripe peach availability, whereas the *A. fraterculus* population peaked from February to April, overlapping with the guava fruiting period. The three pest fly species were parasitized mainly by three generalist resident parasitoids, which are potential biocontrol agents to use within an integrated pest management approach.

Keywords: medfly; spotted-wing drosophila; South American fruit fly; seasonal infestation level; fruit fly abundance; parasitoid; non-crop host; disturbed natural habitat

1. Introduction

Landscape fragmentation plays an essential role in the establishment, dispersal, and population dynamics of invasive species in a new location [1]. Disturbance of the natural habitat strongly influences the composition and abundance of related biota [2–4]. This occurs through the competitive displacement of native species, changes in natural enemy abundance and diversity, and the capacity of the invader to occupy empty or disturbed niches, among other factors [5]. In the case of invasive fruit flies, the distribution and abundance of host plants, the structure of vegetation surrounding crops as alternative habitats, and the distribution of essential resources such as food, shelter, and oviposition substrates strongly influence the spatiotemporal dynamics, distributions, and abundances of fruit fly pests [6–8]. Representative examples of habitat-driven pest dynamics are global invasive species *Ceratitis capitata* (Wiedemann) (Diptera: Tephritidae), native to Mediterranean Africa and commonly known as medfly [9], and *Drosophila suzukii* (Matsumura) (Diptera: Drosophilidae), originally from Southeast Asia and known worldwide as the spotted-wing Drosophila [10,11]. These two exotic fruit fly species are severe pests of economically valuable fruit crops worldwide [12,13], although *D. suzukii* mainly attacks soft-skinned small fruits, such as berries and cherries [14]. Unlike most *Drosophila* species, *D. suzukii* females lay eggs in fresh, healthy, ripening fruit because it has a serrated ovipositor, which allows females to oviposit inside the fleshy mesocarp [15]. Another relevant example of the habitat-driven pest population dynamic is the Neotropical-native *Anastrepha fraterculus* (Wiedemann) (Diptera: Tephritidae). This tephritid fruit fly is the most economically important species of *Anastrepha* in South America, and it is a quarantine pest for the United States and several European and Asian countries [16]. All three dipteran species are highly polyphagous and can exploit various crop and non-crop host plants [9,17,18].

The availability of alternative hosts in non-crop habitats could play an important role in sustaining the populations of polyphagous fruit flies and dictating their local movement patterns when favorable hosts are not available in crops, and non-crop habitats could act as sinks, sources, shelters, or overwintering sites for the fly populations [7,19–21]. Thus, the effectiveness of any control measures for those polyphagous and highly mobile pests requires in-depth knowledge of their seasonal field ecology, including the role of non-crop host plants in the landscape structure for population dispersal and persistence [22]. Such information is essential to implement integrated and area-wide pest management strategies that minimize environmental impact and maximize sustainability to reduce reliance on insecticides alone [23–27]. In this context, resident natural enemies may play a unique role in reducing insect pest populations in non-crop environments that could provide reservoirs for the pest populations moving into crops after they have been treated [28–30]. Non-crop hosts can also provide various ecological services to neighboring agricultural environments, including maintaining and amplifying the numbers of beneficial insects, such as parasitoids [31]. Therefore, biological control properly used in natural environments may be a valuable option for long-term, landscape-level management of insect pests [32–35].

The subtropical mountain rainforest, locally known as Yungas, is one of the South American mountain cloud forests divided into sections along an altitudinal gradient that extends discontinuously from Venezuela to northwestern Argentina [36]. In Argentina, the Yungas lowlands have been strongly transformed into crop and pasture areas because of agricultural development and human settlement [36]. However, in the last five decades, vast sectors of croplands have been restored as nature conservation areas. Thus, the native vegetation naturally regenerated, although abundant exotic plants also grew [37]. Some feral plants have been recorded as hosts of *C. capitata*, *A. fraterculus*, and *D. suzukii*, coupled with local parasitoid assemblages [38–42]. Therefore, natural sites with high and medium disturbance levels are interesting frameworks to evaluate how non-crop hosts adjacent to fruit crops can increase the risk of infestation during the fruiting season.

The current study aimed to describe the abundances and infestation levels of *C. capitata*, *A. fraterculus*, *D. suzukii* and related saprophytic drosophilids infesting five prevalent non-crop fruit species in a highly disturbed natural habitat adjacent to commercial crops and family orchards in northwestern Argentina. Although most saprophytic drosophilids are not considered pests, they share many generalist drosophila parasitoids with *D. suzukii* and may act as alternative hosts for these parasitoids. We compared temporal variations of the infestation levels by the three pest dipteran species during the fruiting seasons of the five host fruit species and assessed natural parasitism levels. Focusing on the tri-trophic interaction (host fruit–fruit fly–parasitoid) over a long-term period in a disturbed wild area would allow a better understanding of how the three fruit fly pests use non-crop fruits based on temporal patterns of host availability. Simultaneously, it is also feasible to identify key hosts accountable for pest population increase, persistence, and the incidence of resident parasitoids in the landscape as the season progresses. This information is useful for not only the different fruit-growing regions of Argentina but also for regions of Latin America and throughout the world affected by some of those pest dipterans.

2. Materials and Methods

2.1. Study Area

The area, located in Horco Molle, Yerba Buena district, Tucumán province, northwestern Argentina, originally belonged to the Low Montane Forest sector from the southernmost end of the subtropical mountain Yungas Forest [36]. The study site belongs to the Horco Molle Experimental Reserve (HMER), a protected wildlife area. This area lies between 26°47′ S latitude and 65°18′ W longitude at 600 m and covers a total surface area of 200 ha. Adjacent to the HMER is the Sierra de San Javier Park, both managed by the National University of Tucumán. A disturbed secondary rainforest, i.e., both exotic and native plant species co-exist, characterizes the study site (Photograph, File S1). The surrounding landscape is a mosaic of various commercial citrus crops, small familiar multi-fruit orchards, scattered rural houses, and wild secondary forest patches, with the closest crops located < 0.5 km away from the study site (Scheme, File S2). The climate in this region is classified as "humid warm–temperate" with a rainy warm season from October through April, and a dry cold season from May through September, with ≈22 °C and 900 mm of average annual temperature and rainfall, respectively [43]. The variation in mean temperature and accumulated rainfall during collecting periods at the study site is shown in Figure 1A.

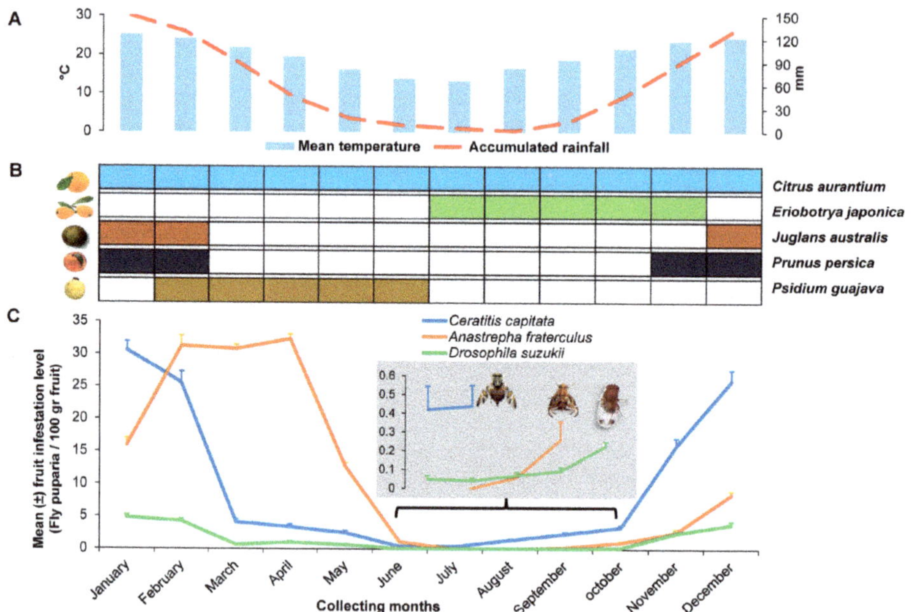

Figure 1. Mean temperature and accumulated rainfall during collecting periods (**A**), temporal patterns of availability for five host plant species (**B**), and seasonal dynamics of total fruit infestation levels (data were pooled from different host plants) of *Ceratitis capitata*, *Anastrepha fraterculus*, and *Drosophila suzukii* (**C**) at the study site (Horco Molle, Tucuman, northwestern Argentina).

2.2. Host Fruit Sampling

A total of 56, 176, 54, 64, and 72 *Juglans australis* Griseb (wild walnut) (Juglandaceae), *Citrus aurantium* L. (sour orange) (Rutaceae), *Eriobotrya japonica* (Thunb.) Lindley (loquat) (Rosaceae), *Prunus persica* (L.) Batsch (peach) (Rosaceae), and *Psydium guajava* L. (guava) (Myrtaceae) trees, respectively, were sampled according to the temporal patterns of fruit ripening throughout their fruiting seasons (Figure 1B). Two trees of each species were chosen randomly on bi-weekly sampling dates from November 2016 to March 2020. Plants were not sampled after March 2020 due to the confinement established by the Argentinian government due to the COVID-19 pandemic. These five non-crop hosts are highly abundant and widely spread throughout disturbed wildland areas of northwestern Argentina [44]. *Juglans australis* was the only native species sampled, and the remainder were feral exotic species. Sample size varied according to fruit weight and relative fruit availability per host species. Fruit samples by the collection date were 48, 16, 200, 60, and 40 wild walnuts, sour oranges, loquats, peaches, and guavas, respectively. Half of the ripe fruit in each sample was randomly collected from the tree canopies and the remaining half from the ground beneath the canopies. Fruits were separately handled to determine whether there were differences in both fly and parasitoid species composition at each level. To collect fruit located in canopies above 1.8 m high, a plastic basket attached to a 3.5 m long extendable metal pole was placed beneath the fruit, and the branch was shaken. Each fruit sample was placed individually into a 20 × 30 m (diameter × deep) cloth bag and transported in plastic crates for processing at the Pest Biological Control Department (DCBP, Spanish acronym). This department belongs to the Biotechnology and Microbiological Industrial Processes Pilot Plant (PROIMI, Spanish acronym) in San Miguel de Tucumán, Tucumán, 15 km from the study site.

2.3. Host Fruit Processing

All collected fruits from the canopies or ground were rinsed with a 30% sodium benzoate and 70% sterile water solution and weighed individually. Mean (±SE) individual fruit weight was 39.7 ± 1.1, 126.4 ± 3.5, 10.5 ± 1.0, 34.5 ± 2.3, and 49.5 ± 1.7 g for walnut, sour orange, loquat, peach, and guava, respectively. Fruit from the ground and the canopies were separately processed and kept individually. First, each fruit was placed in a 48 × 28 × 15 cm plastic crate with a slotted bottom. Then, the crate was placed over another plastic crate of the same size but with a non-perforated bottom and with a thin layer of sterilized, moistened vermiculite Intersum® (Aislater S.R.L., Cordoba, Argentina) on the bottom as a pupation medium. Both crates were tightly covered with a shiny polyester organza fabric lid. The double crate method prevented mixing sand with fruit, fungal growth, and bacterial contamination. All collected samples from the same date were grouped on shelves, which were kept in a dark room under natural environmental conditions for two weeks. Vermiculite was sifted daily to collect fly puparia. Finally, each fruit was dissected to search for larvae or puparia remaining inside the fruit.

2.4. Fly Puparia Processing and Identification

Fly puparia were identified at the DCBP's laboratory. Both *A. fraterculus* and *C. capitata* puparia were identified using external characters of everted anterior spiracles, tubes with finger-like projections [45]. *Drosophila suzukii* puparia were also differentiated from those saprophytic drosophilids by the external characteristic shape of the anterior spiracles [46]. Puparia of different saprophytic drosophilid species were not identified. The puparia of each fly species belonging to the same fruit sample were processed separately. Then, they were transferred to 200 cc translucent plastic cups filled with sterilized moist vermiculite. Each cup was covered with a shiny polyester organza fabric and tied with a rubber band. Cups were placed into 32 × 24 × 12 cm plastic containers. Each container housed the puparia of a particular fly species from the same fruit sample. The numbers of emerged flies and parasitoids were recorded weekly. Voucher adult specimens were stored at the entomological collection of the Fundación Miguel Lillo in San Miguel de Tucumán.

2.5. Data Analysis

The response variables analyzed were the monthly accumulated fruit infestation level by fly species, infestation level recorded in each fruit species by fly species, total parasitism on each fly species, and the parasitoid abundance per fly species. All variables were estimated for both fallen and canopy fruit samples. The fruit infestation level was calculated as the total number of recovered fly puparia per 100 g of fruit weight. The monthly accumulated infestation level was calculated by combining infestation values obtained from all host fruit species during a particular month and by fly species. The infestation level recorded by host fruit species was calculated by including infestation values recorded over a 40-month survey period and by fly species. The total parasitism on each fly species was calculated as the total adult parasitoid number over the total number of puparia recovered from a particular fly species throughout all collecting periods, regardless of host fruit species. The parasitoid abundance was calculated as the total number of parasitized puparia by host fly species from all fruit species collected over the 40-month survey period. The statistical analysis was performed using the software R-4.3.2 [47]. Kruskal–Wallis' rank sum tests were performed to compare fruit infestation levels and parasitoid abundance per fly species. Dunn's post hoc pairwise comparison tests were conducted to show differences between factor levels using a Bonferroni–Holm adjustment method. Mann–Whitney–Wilcoxon tests, with a Bonferroni–Holm adjustment method, were performed to compare parasitism on fly species recovered from both canopies and ground fruit samples. Violin box plots were used to show the resulting data. Violin box plots were used for the figures with statistical data. A violin plot is a mixture of a box plot and a kernel density plot, which shows peaks in the data. Figures were made with the 'grouped_ggbetweenstats' function from the 'ggstatsplot' package [48]. Each plot involves media (horizontal line inside the box), median (red dot inside the box), interquartile range Q1–Q3 (vertical line inside the box), range (minimum: Q0, maximum: Q4; both ends of the whisker on the vertical line outside the box), and raw data dispersal (colored circles). The library 'rcompanion' function was used to include letters that display the significant difference in figures.

3. Results

3.1. Fly Abundance and Infestation Levels

A total of 11,212 fruits (408.8 kg) were collected, 50% from the tree canopies and 50% from the ground during this study, which yielded 19,989 *A. fraterculus*, 19,187 *C. capitata*, 3242 *D. suzukii*, and 23,999 *Drosophila* spp. puparia (Table 1). Saprophytic *Drosophila* species were from the *Drosophila melanogaster* species group. Tephritid puparia accounted for 59% of the total recovered fly puparia, whereas the remaining 41% were drosophilid puparia, from which only 12% belonged to *D. suzukii*. Fruit infestation levels by the three pest dipteran species varied sharply across sampling months (Figure 1C). *Ceratitis capitata* yielded significantly the highest infestation levels particularly between November and February, with a peak in January, in fruits collected either from canopies ($\chi^2_{\text{kruskal-Wallis (11)}}$ = 125.75, $p < 0.0001$) (Figure 2A) or from ground ($\chi^2_{\text{kruskal-Wallis (11)}}$ = 109.75, $p < 0.0001$) (Figure 2B).

Table 1. Total numbers of *Anastrepha fraterculus* (*Af*), *Ceratitis capitata* (*Cc*), *Drosophila suzukii* (*Ds*), and *Drosophila* spp. from *D. melanogaster* group (*Dspp*) *puparia*, and emerged adult flies, recovered from *Citrus aurantium* (*Ca*), *Eriobotrya japonica* (*Ej*), *Juglans australis* (*Ja*), *Prunus persica* (*Pp*), and *Psidium guajava* (*Pg*) fruits collected from canopies and ground between November 2016 and March 2020 in Horco Molle, Tucumán, northwestern Argentina.

Fruit Origin	Fruit Species	No. of Collected Fruit (Weight, Kg)	Total Numbers							
			Af Puparia	*Af* Adults	*Cc* Puparia	*Cc* Adults	*Ds* Puparia	*Ds* Adults	*Dspp* Puparia	*Dspp* Adults
Canopy	*Ca*	692 (87.4)	17	8	514	203	0	0	0	0
	Ej	2700 (26.9)	492	245	1442	763	286	144	16	11
	Ja	672 (28.3)	2819	1437	923	493	4	2	0	0
	Pp	960 (32.6)	1122	550	6120	2948	1537	725	36	23
	Pg	580 (29.1)	6321	3059	824	358	224	86	73	47
Ground	*Ca*	92 (87.8)	108	45	767	301	0	0	7595	3158
	Ej	2700 (27.1)	291	94	895	336	148	41	2458	1015
	Ja	672 (28.1)	1974	832	724	299	61	23	3145	1209
	Pp	960 (32.3)	1195	543	6376	3009	887	374	3249	1339
	Pg	580 (29.2)	5650	2350	612	228	95	25	7427	2857

Anastrepha fraterculus showed significantly the highest infestation levels between December and May in fruits sampled either from canopies ($\chi^2_{\text{kruskal-Wallis (11)}}$ = 85.08, $p < 0.0001$) (Figure 2C) or from the ground ($\chi^2_{\text{kruskal-Wallis (11)}}$ = 130.00, $p < 0.0001$) (Figure 2D). *Drosophila suzukii* exhibited the highest infestation levels between October and May, although infestation peaked between November and January in fruits collected either from canopies ($\chi^2_{\text{kruskal-Wallis (11)}}$ = 29.59, $p < 0.0001$) (Figure 2E) or from the ground ($\chi^2_{\text{kruskal-Wallis (11)}}$ = 49.58, $p < 0.0001$) (Figure 2F). Saprophytic drosophilids had significantly similar infestation levels, <1 fly puparium/100 g fruit, in fruits collected from canopies throughout the year ($\chi^2_{\text{kruskal-Wallis (11)}}$ = 23.31, $p = 0.0160$) (Figure 2G). Infestation levels by saprophytic drosophilids were remarkably high in fallen fruits from January to April ($\chi^2_{\text{kruskal-Wallis (11)}}$ = 200.97, $p = 0.0001$) (Figure 2H). Infestation levels by the three pest dipteran species and by saprophytic *Drosophila* species showed significant differences among the different fruit species, collected either from the canopies or from the ground (Table 2). Significantly higher infestation levels by *A. fraterculus* than those of the other pest fly species were recorded from walnut (Figure 3A,B) and guava (Figure 3E,F), whereas *C. capitata* had significantly the highest infestation levels in peach (Figure 3I,J), loquat (Figure 3C,D), and sour orange (Figure 3G,H). Infestation levels by *D. suzukii* in peach were high, but similar to that of *A. fraterculus* (Figure 3I,J). Infestation levels by *Drosophila* spp. from *D. melanogaster* group were the highest in all sampled fruit species, but only in fruit samples collected from the ground (Figure 3B,D,F,H,J).

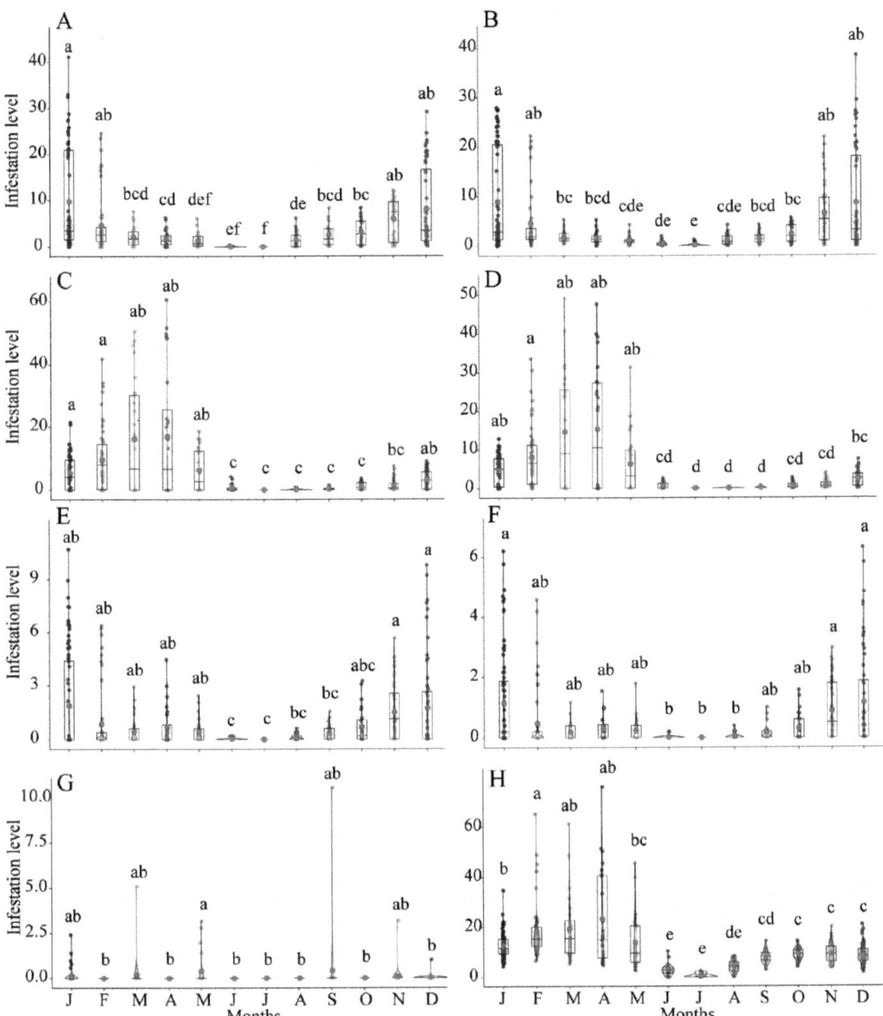

Figure 2. Monthly fruit infestation levels (fly puparia/100 g fruit) by *Ceratitis capitata* (**A**,**B**), *Anastrepha fraterculus* (**C**,**D**), *Drosophila suzukii* (**E**,**F**), and *Drosophila* spp. (*D. melanogaster* group) (**G**,**H**) recorded from fruits collected from tree canopies (left column) and from the ground (right column) at the study site (Horco Molle, Tucuman, northwestern Argentina). Different lowercase letters represent significant differences at $\alpha = 0.05$ (Dunn's Test).

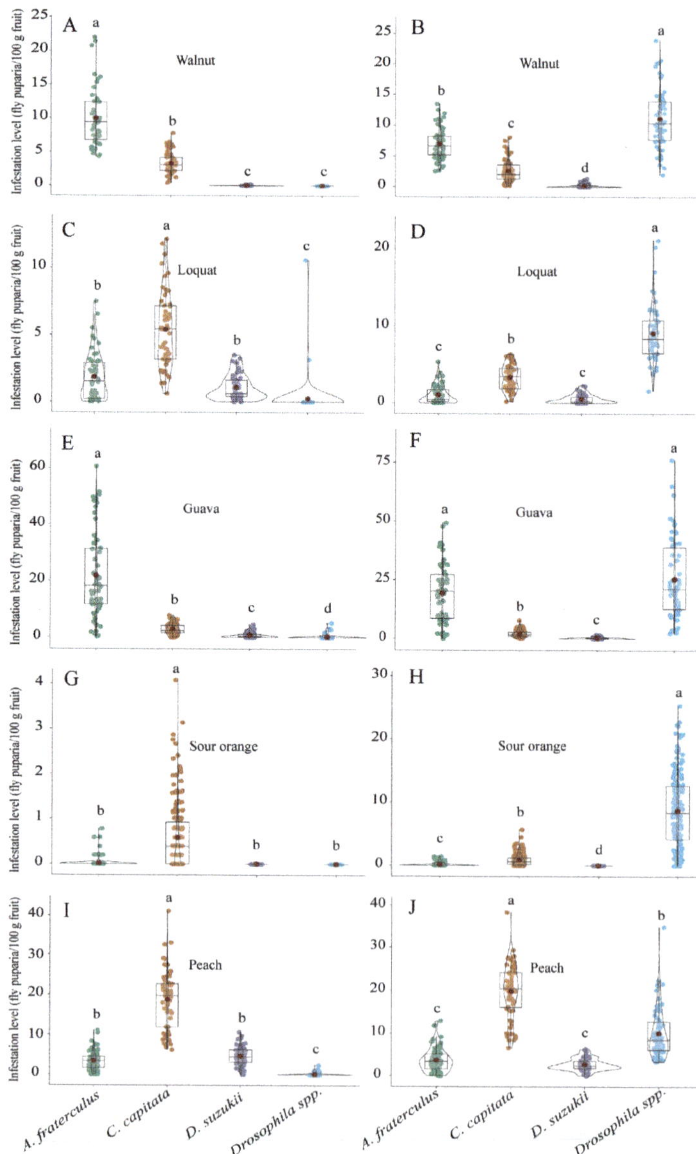

Figure 3. Infestation levels (fly puparia/100 g fruit) of *Anastrepha fraterculus*, *Ceratitis capitata*, *Drosophila suzukii*, and *Drosophila* spp. (*D. melanogaster* group) recorded from host fruit species collected from tree canopies (**left column**) and from the ground (**right column**) in Walnut (**A,B**), loquat (**C,D**), guava (**E,F**), sour orange (**G,H**), and peach (**I,J**) at the study site (Horco Molle, Tucuman, northwestern Argentina). Different lowercase letters represent significant differences at α = 0.05 (Dunn's Test).

Table 2. Summary of Kruskal–Wallis models on the infestation levels by *Ceratitis capitata*, *Anastrepha fraterculus*, *Drosophila suzukii*, and *Drosophila* spp. (*D. melanogaster* species group) on *Citrus aurantium*, *Eriobotrya japonica*, *Juglans australis*, *Prunus persica*, and *Psidium guajava* fruits collected from both canopies and the ground during fruiting seasons between November 2016 and March 2020 in Horco Molle, Tucumán, Northwestern Argentina.

Fruit Origin: Fruit Species	Statistical Results			
	df	*n*	χ^2	*p*
Canopy:				
Citrus aurantium	3	348	316.90	<0.0001
Eriobotrya japonica	3	108	145.38	<0.0001
Juglans australis	3	112	209.24	<0.0001
Prunus persica	3	128	208.25	<0.0001
Psidium guajava	3	116	164.67	<0.0001
Ground:				
Citrus aurantium	3	348	485.34	<0.0001
Eriobotrya japonica	3	108	155.39	<0.0001
Juglans australis	3	112	179.20	<0.0001
Prunus persica	3	128	181.80	<0.0001
Psidium guajava	3	116	173.21	<0.0001

3.2. Parasitoid Abundance and Parasitism Levels

A total of 7349 adult parasitoids belonging to six different species, *Ganaspis pelleranoi* (Brèthes) (28.6%) (Figitidae), *Trichopria anastrephae* Lima (28.2%) (Diapriidae), *Pachycrepoideus vindemiae* Rondani (18.1%) (Pteromalidae), *Leptopilina* sp. cf. *boulardi* (Barbotin, Carton, and Kelner-Pillault) (Figitidae) (14.9%), *Doryctobracon areolatus* (Szèpligeti) (5.7%) (Braconidae), and *Doryctobracon brasiliensis* (Szèpligeti) (4.5%) (Braconidae), were obtained from fly puparia recovered over the 40-month study. Five parasitoid species, *D. areolatus*, *D. brasiliensis*, *G. pelleranoi*, *P. vindemiae*, and *T. anastrephae*, were recovered from *A. fraterculus*, whereas only *G. pelleranoi* and *P. vindemiae* were associated with *C. capitata*, and *T. anastrephae*, *Leptopilina* sp. cf. *boulardi*, and *P. vindemiae* with both *D. suzukii* and *Drosophila* spp. The latter three parasitoid species prevailed on saprophytic drosophilids, whereas *G. pelleranoi* mostly parasitized *A. fraterculus* and to a minor extent *C. capitata* (Figure 4). The braconid species were found as associated only with *A. fraterculus* (Figure 4). The numbers of parasitized host puparia recorded in all three pest fly species and in saprophytic drosophilid species were significantly different between the ground ($\chi^2_{\text{kruskal-Wallis (3)}} = 1298.81$, $p < 0.0001$) (Figure 5A) and canopy ($\chi^2_{\text{kruskal-Wallis (3)}} = 281.66$, $p < 0.0001$) (Figure 5B) fruit samples. The highest number of parasitized host puparia was on saprophytic drosophilids recovered from fallen fruits (Figure 5A). The number of parasitized *C. capitata* puparia was significantly higher than that of *A. fraterculus* and both were significantly higher than that of *D. suzukii* (Figure 5A). The number of parasitized *A. fraterculus* puparia recorded from the canopy fruit was significantly higher than that recorded for other tested fly species (Figure 5B). Moreover, the number of parasitized *C. capitata* puparia was significantly higher than that recorded from both *D. suzukii* and saprophytic drosophilids (Figure 5B). Significant positive correlations between parasitism and infestation levels were recorded for *C. capitata* ($\tau = 0.51$, $z = 18.75$, $p < 0.0001$), *A. fraterculus* ($\tau = 0.75$, $z = 27.10$, $p < 0.0001$), *D. suzukii* ($\tau = 0.37$, $z = 11.85$, $p < 0.0001$), and *Drosophila* spp. ($\tau = 0.85$, $z = 30.34$, $p < 0.0001$). The total levels of parasitism were significantly different between the host puparia recovered from fruits still in the canopies and those from fallen fruits. Significantly, more parasitoids were recovered from puparia collected from fallen fruits than from the canopy fruits. This pattern was consistent for *A. fraterculus* ($W_{M\text{-}W} = 7.01^5$, $n = 2450$, $p < 0.0001$) (Figure 6A), *C. capitata* ($W_{M\text{-}W} = 1.59^5$, $n = 1384$, $p < 0.0001$) (Figure 6B), *Drosophila* spp. ($W_{M\text{-}W} = 7650$, $n = 1221$, $p < 0.0001$) (Figure 6C), and *D. suzukii* ($W_{M\text{-}W} = 1.01^5$, $n = 975$, $p < 0.0001$) (Figure 6D).

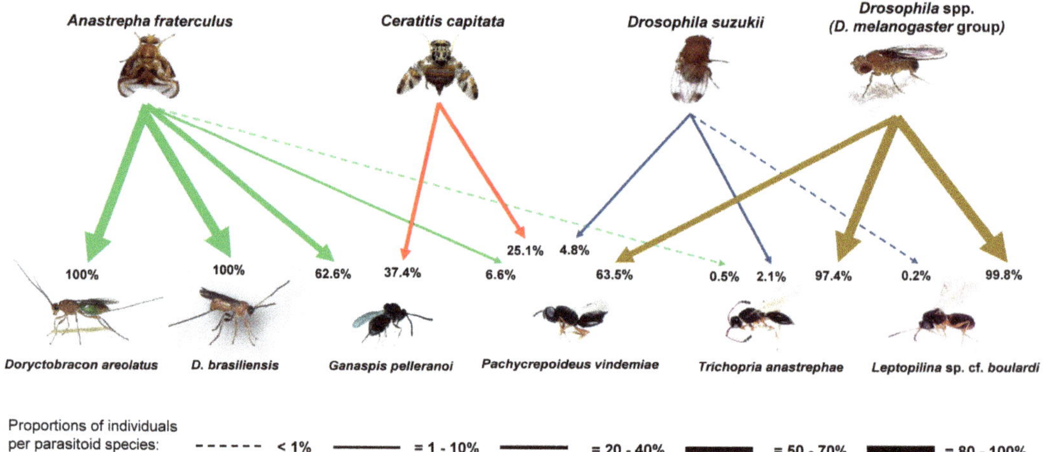

Figure 4. The proportion of individuals belonging to a particular parasitoid species over the total number of specimens from that parasitoid species recovered from each host fly species during collection periods at the study site (Horco Molle, Tucuman, northwestern Argentina).

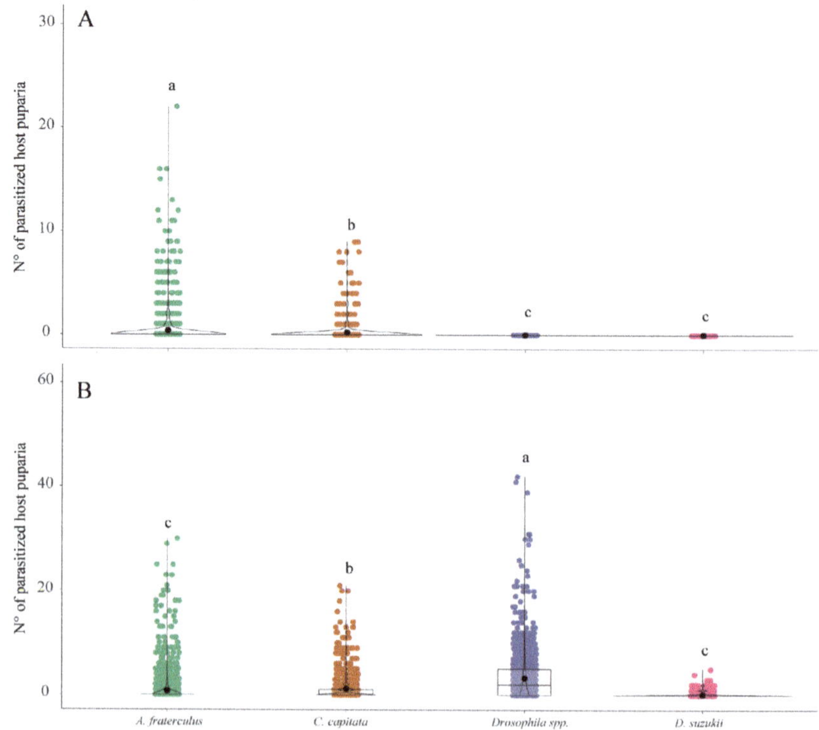

Figure 5. Comparison of the parasitized puparia number of *Anastrepha fraterculus*, *Ceratitis capitata*, *Drosophila* spp. (*D. melanogaster* group), and *Drosophila suzukii* recovered from (**A**) fallen fruits on the ground and (**B**) fruits still on the canopies at the study site (Horco Molle, Tucuman, northwestern Argentina). Different lowercase letters represent significant differences at $\alpha = 0.05$ (Dunn's Test).

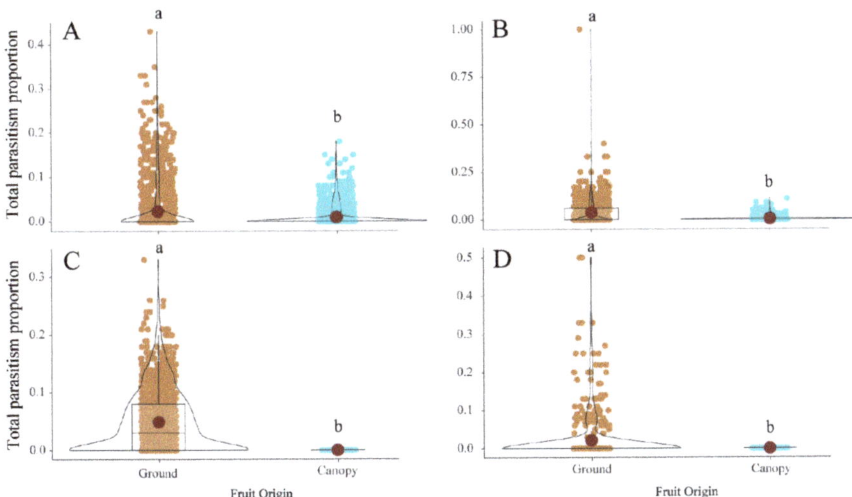

Figure 6. Total parasitism proportion recorded on (**A**) *Anastrepha fraterculus*, (**B**) *Ceratitis capitata*, (**C**) *Drosophila* spp. (*D. melanogaster* group), and (**D**) *Drosophila suzukii* from fruit samples collected from both ground and tree canopies at the study site (Horco Molle, Tucuman, northwestern Argentina). Different lowercase letters represent significant differences at $\alpha = 0.05$ (Mann-Whitney's test).

4. Discussion

The current study provides significant information needed to develop fruit fly IPM strategies that minimize environmental impact and maximize long-term sustainability. The results showed that (1) highly disturbed wild habitats adjacent to crops are suitable sites for the development and increase of the pest fruit fly species *C. capitata*, *A. fraterculus*, and *D. suzukii*; (2) non-crop host fruit species influence the relative and temporal abundances of these flies; (3) overlaps in fruiting seasons of different host species throughout the year allow these flies to access regularly resources to sustain their populations in the disturbed habitats; (4) temporal infestation levels by both invasive pest species are similar but differ from the native pest; and (5) the abundance and diversity of resident parasitoids, as well as parasitism levels, depend largely on non-crop fruit species where the larval or pupal hosts developed, dipteran host species associated with the host fruits, and fruit infestation levels.

Firstly, we found high abundance and fruit infestation levels by the three fly species in a forest regenerated from anthropogenic disturbances. Many characteristics in the disturbed habitats, such as the presence of abundant and diverse indigenous and exotic host fruit species as well as the high thermal and humidity variation may allow the occurrence and coexistence of these species. Some introduced host plants widely dispersed in this habitat such as *C. aurantium* and *E. japonica*, are uncontested or poorly contested by native fly species, thereby providing empty niches mainly exploited by *C. capitata* [38], but less so by *D. suzukii*. In addition to the high level of polyphagy of *C. capitata* and *D. suzukii*, both exotic flies have high thermal plasticity, allowing them not only to persist but also to thrive in such disturbed environments [6,9,49–52]. Although *A. fraterculus* prevails in low-disturbed environments with a high abundance of native plants, it is also usually found in association with exotic host fruits in highly disturbed environments [38,53].

Secondly, this study revealed different preferences for certain host plants among the three fly species. *Prunus persica* was the preferred host fruit for *C. capitata*, followed in decreasing order by *P. guajava*, *E. japonica*, *C. aurantium*, and *J. australis*. It was evident that *C. capitata* preferred introduced feral fruits that are usually underutilized by *A. fraterculus*, with the exception of *P. guajava*. Previous records [38,39] pointed out feral *P. persica* as the most relevant multiplying hosts in northern Argentina and one of the key hosts for

C. capitata dispersing in all Argentinian fruit-growing regions. Feral *P. guajava* was also the preferred exotic host for *A. fraterculus* and together with the native *J. australis* were the ones that mainly allowed the highest population growth of *A. fraterculus* as shown in this study. Interestingly, the presence of *P. guajava* in disturbed habitats characterized by low native plant cover and by sectors with a higher incidence of sun, increasing *A. fraterculus*'s abundance, although preferred native hosts, such as walnuts, are still present. This occurs because *P. guajava* is the most commonly recorded *A. fraterculus* host plant throughout the Neotropics [17]. Interestingly, *J. australis* had not previously been recorded as a host of *D. suzukii* in Argentina or South America. Although *D. suzukii* was occasionally abundant in that native fruit species, it is a novel host to 24 exotic and native, crop, and non-crop fruits thus far recorded for this invasive pest in Argentina [40]. *Drosophila suzukii* was preferentially more abundant in *P. persica* in the study site, followed in decreasing order by *P. guajava* and *E. japonica*. Both feral peach and guava have previously been recorded as alternative hosts to *D. suzukii* in wilderness areas in northwestern Argentina [40]. Similarly, *D. suzukii* was previously recorded infesting *E. japonica* fruits in crop areas of northwestern Argentina [54], as well as in commercial peaches in northeastern Buenos Aires (central-eastern Argentina) [55]. Data from the current study on the abundance of *D. suzukii* on *P. persica* and *E. japonica* are not surprising since Rosaceae is the plant family with the largest number of host species recorded for *D. suzukii* worldwide [21]. *Citrus aurantium* evidently is not a suitable oviposition host for *D. suzukii*. However, two *Citrus* species, *C. sinesis* (L.) Osbeck and *C. reticulata* Blanco, have been recorded as alternative reproductive hosts on damaged fruits in California (USA) [56], whereas *C. sinensis* was also recorded in Uruguay [21]. Rutaceae apparently include host species not preferred by *D. suzukii* [21]. The high abundance and high infestation levels of saprophytic drosophilids on all sampled fruit species can be mainly attributed to the fact that these dipterans are associated with a wide variety of habitats, particularly related to rotting fallen fruits [57]. Precisely, data from the current study show the highest infestation levels of these drosophilids in ripe fruits sampled only from the ground.

Thirdly, we showed overlaps of temporal availability of *P. persica* with the remaining host fruit species and a constant availability of ripe *C. aurantium* fruit throughout the year. This provides these fruit fly pest species with year-round resources for oviposition in the study site. In this context, mainly both *E. japonica* and *C. aurantium*, but also *P. guajava*, play important bridging roles during the cold-dry season, which spans from late autumn and winter to early spring. During this period of the year, *P. persica* is not available and its availability is low throughout mid- and late autumn when compared to the summer and early autumn seasons. The role of *E. japonica* as a host for the three fly pests is crucial, despite the low infestation levels recorded for this exotic, feral fruit species. This is because *E. japonica* provides an alternative host when the latest guavas are not highly available until late autumn, and when the earliest peaches ripe in late spring. This was previously recorded only for *C. capitata* and *A. fraterculus* [38]. This study also showed that *D. suzukii* used the same resource as the other two tephritid fly pests to persist at low density during a period of unfavorable climatic conditions and a shortage of primary hosts. This is new information on the ecological aspects of *D. suzukii* in northwestern Argentina, as the loquat apparently is also a reservoir host for this invasive pest, whereas *C. aurantium* is a non-host. This fact is relevant because *P. persica* is not only the main multiplying host for *C. capitata*, but also for *D. suzukii*, as *P. guajava* is for *A. fraterculus* and the second proliferating host for both *C. capitata* and *D. suzukii*. Therefore, *D. suzukii* may be found throughout the year in environments with floristic characteristics similar to the site of the current study. This is mainly due to the presence of late ripe guavas and the early ripening of the loquat during the dry cold period. Isolated *D. suzukii* adult catches in liquid traps were recorded in August (mid-winter) in blueberry-growing lowland areas of Tucumán [58]. The presence of infested feral loquats in wild vegetation areas surrounding berry crops may explain the winter catches of *D. suzukii* adults. Similarly, in southern Brazil, *D. suzukii* can still remain at low natural infestation rates in native non-crop hosts, such as *Psidium cattleianum* Sabine (strawberry guava) and *Eugenia unifora* L. (surinam cherry), and in feral loquat, even in winter [59]. Therefore, *D. suzukii* females that have overwintered in alternative non-crop host fruits are probably a source of infestations in crop fruits available during spring in northwestern Argentina. *Drosophila suzukii* has high dispersal abilities, which enable it to move freely between both non-crop and crop habitats throughout the year [22,60–63]. The same dispersal behavior between crops and patches of wild vegetation and surrounding family gardens in a heterogeneous landscape has been recorded for *C. capitata* [44,64,65] and *A. fraterculus* [53,66,67]. Structurally complex landscapes influence trophic interactions mainly because suitable resources occurring in different types of patches can support consumer species [68]. This essentially shapes the spatial and temporal dynamics of the biological communities in these landscapes [69].

Fourthly, we showed that the population dynamics of *D. suzukii* and *C. capitata* appeared to be similar but partially differ from *A. fraterculus*. Both *D. suzukii* and *C. capitata* populations gradually increased from August (cold dry winter), reaching the highest peak in January in *C. capitata*, and between December and January in *D. suzukii* (warm-humid summer) and then sharply declined in March to maintain a low abundance throughout autumn and winter. The population peaks may be associated with the highest availability of peach and, to a lesser degree with the walnut fruiting period. However, the availability of guava may have also influenced the infestation level of *D. suzukii* during February. These low populations are not only associated with the absence of preferred host fruits but also essentially due to the decrease in temperature and humidity at the end of the warm humid season as previously discussed [44,58]. Earlier studies [38] in a secondary forest of northwestern Argentina indicated two continuous population peaks for *C. capitata* in December and January, coinciding with the greatest availability of both *P. persica* and *C. aurantium*. In the current study, only one population peak was detected in January, as accumulated infestation levels recorded for *C. capitata* in December and February were very similar and lower than in January. The native host *J. australis* played a relevant role in increasing *C. capitata* population, as in January the infestation level was 2.7-fold higher than that of *C. aurantium*. As for *D. suzukii*, the current study provides first-hand information on the temporal abundance variation of this invasive pest in Argentina, because the few known studies on population fluctuation of this pest in Argentina were only carried out in berry-growing areas using trap catches of adult flies. In northern and central Argentinian fruit-producing regions, trap catches detected two adult population peaks in late spring–early summer (November and December) and in mid-autumn (April and May), respectively, with the catches being lower in the second than the first peaks and declining from late autumn onward [54,55,58,70–72]. However, in the Alto Valle de Rio Negro, northern Patagonia (i.e., in the cold and dry southern Argentina), the peak of trapped *D. suzukii* adults occurred between late summer and late autumn (March-May), coinciding with raspberry and cherry fruiting seasons [73]. Climatic conditions are probably the major factors affecting *D. suzukii* abundance [74]. The hottest and coldest months of the year in temperate and subtropical climates may reduce *D. suzukii* populations; therefore, this pest usually increases its population in late spring and mid-autumn [75]. This indeed reflects the population dynamics based on adult catches in berry-growing areas from Argentina but is not consistent with data of the current study, because the major *D. suzukii* population peak occurred during the month with high temperature and humidity. The diverse microhabitats in this environment and the phenotypic and thermal plasticity of *D. suzukii*, as well as a high availability of suitable fruits, are probably responsible for the population increase in the middle of the warm humid season. Similarly, the infestation levels of *A. fraterculus* also gradually increased from August as *C. capitata* and *D. suzukii*, but the infestation levels of *A. fraterculus* continually increased after January and reached population peaks between February and April. This coincides with the guava fruiting period and the gradually rising temperature and humidity as summer progresses [44]. Infestation levels *A. fraterculus* decreased sharply after May and remained low during late autumn and throughout winter.

Finally, the current study revealed the trophic associations among these host plants, dipteran pests, and resident parasitoid species as well as the relative abundance and diversity of parasitoids throughout the year. Although *C. capitata* was the dominant pest fly in three feral introduced fruit species, *C. aurantium*, *E. japonica*, and *P. persica*, it was parasitized only by *G. pelleranoi* and *P. vindemiae*, both generalist parasitoids [76]. Similarly, *C. capitata* was also only parasitized by these two parasitoids on *P. guajava* and *J. australis*, a major host of *A. fraterculus*. The figitid *G. pelleranoi* is one of the few Neotropical-native larval parasitoid species sympatrically associated with *Anastrepha* that can successfully develop on *C. capitata* larvae [76]. *Ganaspis pelleranoi* females frequently forage fly larvae inside fallen fruit and mainly attack the host by entering through the fissures produced in the fruit or holes produced by its jaws [77]. Faced with this behavior, physical features, such as large size, rind thickness, and pulp depth, do not limit the parasitoid's access to locate and parasitize host larvae. This was supported in the current study as 85% of the total identified *G. pelleranoi* specimens were from fallen fruit samples. The pteromalid *P. vindemiae* is a cosmopolitan species that attacks puparia of various cyclorrhaphous dipteran species, among which, *C. capitata* is a host recurrently recorded in the American continent [76]. *Pachycrepoideus vindemiae* is an abundant and widespread species in wild vegetation environments from northwestern Argentina, where it was recorded as a common pupal parasitoid on *C. capitata* [39]. In terms of parasitoid diversity and abundance, *C. capitata* was parasitized by two of the six identified species (33%), but the abundance of parasitoids associated with this invasive pest was high. The abundance of *G. pelleranoi* recovered from *C. capitata* prevailed on highly available fruits during the warm humid season, while *P. vindemiae* from *C. capitata* was more abundant (76%) than *G. pelleranoi* only on loquat. This may be because loquat is mostly available during the cool dry season, a time of the year with low *C. capitata* infestation levels, and the absence of *G. pelleranoi* in the study area. Interestingly, *P. vindemiae* was recovered from *C. capitata* puparia collected from loquat from mid-August to mid-November. Apparently, *P. vindemiae* is a parasitoid not only with high adaptability to diverse environments but also with greater thermal plasticity than native parasitoids, such as *G. pelleranoi*. The other invasive species, *D. suzukii*, also showed low parasitoid diversity. Only three species were recovered, with *P. vindemiae* as the prevalent parasitoid in the four host plant species associated with *D. suzukii*, followed by *T. anastrephae*, but only in *P. persica* and *P. guajava*. In line with the latter, both host fruit species had the highest infestation levels by *D. suzukii*. However, the abundance of *P. vindemiae* recovered from *D. suzukii* was low compared with that of the other two identified flies, but *T. anastrephae* was mainly abundant on *D. suzukii*, rare on *A. fraterculus* and absent on *C. capitata*. The South American-native *T. anastrephae* is a pupal endoparasitoid previously associated with both *A. fraterculus* and *D. suzukii* in Argentina [39] and Brazil [54]. However, *T. anastrephae* has a strong preference for parasitizing puparia of resident saprophytic drosophilid species located inside the fruit [42]. The low diversity of parasitoids associated with *D. suzukii* in the study area may correlate with the absence of host–parasitoid co-evolution and co-adaptation processes, especially for larval endoparasitoids that must overcome the hosts' immune response, and for this reason, they are highly co-evolved with their particular hosts. This also applies to the case of *C. capitata* as correlation coefficients between parasitism and fruit infestation by *D. suzukii* and *C. capitata* were between 1.5- and 2.3-fold lower than those recorded for both *A. fraterculus* and *Drosophila* spp. (*D. melanogaster* group). Although some larval parasitoid species were recovered from *D. suzukii* puparia in Argentina, such as *Dieucoila octofagella* Reche, *Ganaspis brasiliensis* (von Ihering), *Leptopilina* sp., *Hexacola* sp. [40], parasitism levels were extremely low. The figitid specimens recovered from *D. suzukii* were taxonomically similar to *L. boulardi*, a worldwide saprophytic drosophilid's parasitoid. *Leptopilina boulardi* was recently associated with *D. suzukii* in Argentina (Vanina Reche, unplublished data). In contrast to the two invasive fly species, five of the six identified parasitoid species (83%) were recovered from the native *A. fraterculus*. In addition, the highest parasitoid abundance in *P. guajava* and *J. australis* came from *A. fraterculus*. *Anastrepha fraterculus* was mostly parasitized by *G. pelleranoi*, followed by two native braconid parasitoids, *D. areolatus* and *D. brasiliensis*, whereas sporadically by the pupal parasitoids *P. vindemiae* and *T. anastrephae*. Both *Doryctobracon* species integrate an assemblage of several Neotropical-native parasitoids that co-evolved in sympatry with *A. fraterculus* in South American rainforest areas [76].

5. Conclusions

The current study improves our understanding of the temporal and spatial dynamics of these three important pest fruit flies, the utilization patterns and relative importance of non-crop hosts for these pests, as well as the trophic associations with resident parasitoids in the disturbed non-crop habitats surrounding cultivated crops. As shown in this study, the disturbed natural habitat would inevitably provide sources of the fly populations that may move into adjacent fruit crops. The three pests also showed different host preferences. Both *C. capitata* and *D. suzukii* preferred peach and loquat, their highest infestation levels thus occurred between December and February when peaches were highly available. In contrast, high levels of infestations by *A. fraterculus* occurred between February and April when guavas were highly available. Both *P. vindemiae* and *T. anastrephae* are key natural mortality factors of *D. suzukii* while *G. pelleranoi* is the main natural mortality factor of both *C. capitata* and *A. fraterculus*. Consequently, area-wide management strategies must consider reducing pest pressure in susceptible crops by reducing sources of fly populations in the non-crop habitats. In this context, biological control is highly desirable to naturally regulate the fly populations. The current study suggests that timed mass releases of these parasitoids during early or peak infestation stages of these pests in disturbed habitats may help suppress the fly populations prior to their main spread to commercial crops.

Supplementary Materials: The following supporting information can be downloaded at: https://www.mdpi.com/article/10.3390/insects15010061/s1. File S1: photograph—disturbed sector of the Yungas subtropical rainforest in Horco Molle, Tucuman, northwestern Argentina; File S2: Scheme—Schematic representation of the area showing the study site, Horco Molle Experimental Reserve from the National University of Tucumán, located in Horco Molle, Yerba Buena district, Tucumán province, northwestern Argentina.

Author Contributions: Conceptualization, M.J.B.-B., X.W., S.R.N.-C., L.S., M.D.P., P.S., D.S.K., F.R.M.G. and S.M.O.; data curation, M.J.B.-B., S.R.N.-C., L.S., M.D.P. and S.M.O.; funding acquisition, D.S.K. and S.M.O.; formal Analysis, S.R.N.-C.; investigation, M.J.B.-B., X.W., S.R.N.-C., L.S., M.D.P., P.S., D.S.K., F.R.M.G. and S.M.O.; methodology, M.J.B.-B., X.W., S.R.N.-C., L.S., M.D.P., P.S., D.S.K., F.R.M.G. and S.M.O.; project administration, D.S.K. and S.M.O.; resources, D.S.K. and S.M.O.; software, S.R.N.-C.; supervision, S.M.O.; validation, M.J.B.-B., X.W., S.R.N.-C., L.S., M.D.P., P.S., D.S.K., F.R.M.G. and S.M.O.; visualization, M.J.B.-B., S.R.N.-C., L.S., M.D.P. and S.M.O.; writing—original draft preparation; M.J.B.-B., X.W., S.R.N.-C., L.S., M.D.P., P.S., D.S.K., F.R.M.G. and S.M.O.; writing—review and editing, M.J.B.-B., X.W., S.R.N.-C., L.S., M.D.P., P.S., D.S.K., F.R.M.G. and S.M.O. All authors have read and agreed to the published version of the manuscript.

Funding: Fondo para la Investigación Científica y Tecnológica (FONCYT) from Agencia Nacional de Promoción Científica y Tecnológica de Argentina (ANPCyT) (grants PICT 2017-0512 and PICT 2020-01050).

Data Availability Statement: The dataset is available at the CONICET-Digital Institutional Repository, URI: http://hdl.handle.net/11336/221026, accessed on 8 January 2024.

Acknowledgments: The authors would like to thank the authorities of the Horco Molle Experimental Reserve and Universidad Nacional de Tucumán, Argentina for allowing us to carry out the surveys, and Patricia Colombres and Guillermo Borchia (PROIMI—CONICET) for laboratory assistance. The USDA is an equal opportunity provider and employer and does not endorse products mentioned in this publication.

Conflicts of Interest: The authors declare no conflicts of interest.

References

1. Didham, R.K.; Tylianakis, J.M.; Gemmell, N.J.; Rand, T.A.; Ewers, R.M. Interactive effects of habitat modification and species invasion on native species decline. *Trends Ecol. Evol.* **2007**, *22*, 489–496. [CrossRef] [PubMed]
2. Simberloff, D. How much information on population biology is needed to manage introduced species? *Conserv. Biol.* **2003**, *17*, 83–92. [CrossRef]
3. Vinatier, F.; Tixier, P.; Duyck, P.F.; Lescourret, F. Factors and mechanisms explaining spatial heterogeneity: A review of methods for insect populations. *Methods Ecol. Evol.* **2011**, *2*, 11–22. [CrossRef]
4. Vasseur, C.; Joannon, A.; Aviron, S.; Burel, F.; Meynard, J.-M.; Baudry, J. The cropping systems mosaic: How does the hidden heterogeneity of agricultural landscapes drive arthropod populations? *Agric. Ecosyst. Environ.* **2013**, *166*, 3–14. [CrossRef]
5. Duyck, P.F.; David, P.; Quilici, S. Can more K-selected species be better invaders? A case study of fruit flies in La reunion. *Divers. Distrib.* **2007**, *13*, 535–543. [CrossRef]
6. Duyck, P.F.; David, P.; Quilici, S. A review of relationships between interspecific competition and invasions in fruit flies (Diptera: Tephritidae). *Ecol. Entomol.* **2004**, *29*, 511–520. [CrossRef]
7. Aluja, M.; Ordano, M.; Guillen, L.; Rull, J. Understanding long-term fruit flies (Diptera: Tephritidae) population dynamics: Implications for area wide management. *J. Econ. Entomol.* **2012**, *105*, 823–836. [CrossRef]
8. Hennig, E.I.; Mazzi, D. Spotted wing Drosophila in sweet cherry orchards in relation to forest characteristics, bycatch, and resource availability. *Insects* **2018**, *9*, 118. [CrossRef]
9. Copeland, R.S.; Wharton, R.A.; Luke, Q.; De Meyer, M. Indigenous hosts of *Ceratitis capitata* (Diptera: Tephritidae) in Kenya. *Ann. Entomol. Soc. Am.* **2002**, *95*, 672–694. [CrossRef]
10. Asplen, M.K.; Anfora, G.; Biondi, A.; Choi, D.S.; Chu, D.; Daane, K.M.; Gibert, P.; Gutierrez, A.P.; Hoelmer, K.A.; Hutchison, W.D.; et al. Invasion biology of spotted-wing Drosophila (*Drosophila suzukii*): A global perspective and future priorities. *J. Pest Sci.* **2015**, *88*, 469–494. [CrossRef]
11. De la Vega, G.J.; Corley, J.C. *Drosophila suzukii* (Diptera: Drosophilidae) distribution modelling improves our understanding of pest range limits. *Int. J. Pest Manag.* **2019**, *65*, 217–227. [CrossRef]
12. Malacrida, A.R.; Gomulski, L.M.; Bonizzoni, M.; Bertin, S.; Gasperi, G.; Guglielmino, C.R. Globalization and fruit fly invasion and expansion: The medfly paradigm. *Genetica* **2007**, *131*, 1–9. [CrossRef] [PubMed]
13. Mazzi, D.; Bravin, E.; Meraner, M.; Finger, R.; Kuske, S. Economic impact of the introduction and establishment of *Drosophila suzukii* on sweet cherry production in Switzerland. *Insects* **2017**, *8*, 18. [CrossRef] [PubMed]
14. De Ros, G.; Grassi, A.; Pantezzi, T. Recent trends in the economic impact of *Drosophila suzukii*. In *Drosophila Suzukii Management*; Garcia, F.R.M., Ed.; Springer Nature: Cham, Switzerland, 2020; pp. 11–28.
15. Atallah, J.; Teixeira, L.; Salazar, R.; Zaragoza, G.; Kopp, A. The making of a pest: The evolution of a fruit-penetrating ovipositor in *Drosophila suzukii* and related species. *Proc. R. Soc. B* **2014**, *281*, 20132840. [CrossRef] [PubMed]
16. CABI, Centre for Agricultural Bioscience International. Invasive Species Compendium, *Ceratitis capitata* (Mediterranean Fruit Fly). CAB International. 2022. Available online: https://www.cabi.org./isc/datasheet/12367 (accessed on 2 October 2023).
17. Norrbom, A.L. Host Plant Database for *Anastrepha* and *Toxotrypana* (Diptera: Tephritidae: Toxotrypanini). In *Diptera Data Dissemination Disk [CD-ROM]* 2; North American Dipterist's Society: Washington, DC, USA, 2004.
18. Kenis, M.; Tonina, L.; Eschen, R.; van der Sluis, B.; Sancassani, M.; Mori, N.; Haye, T.; Helsen, H. Non-crop plants used as hosts by *Drosophila suzukii* in Europe. *J. Pest Sci.* **2016**, *89*, 735–748. [CrossRef] [PubMed]
19. Ims, R.A. Movement patterns related to spatial structures. In *Mosaic Landscapes and Ecological Processes*; Hansson, L., Fahrig, L., Merriam, G., Eds.; Chapman & Hall: London, UK, 1995; pp. 85–109.
20. Gutierrez, A.P.; Ponti, L. Assessing the invasive potential of the Mediterranean fruit fly in California and Italy. *Biol. Invasions* **2011**, *13*, 2661–2676. [CrossRef]
21. Kirschbaum, D.S.; Funes, C.F.; Buonocore-Biancheri, M.J.; Suárez, L.; Ovruski, S.M. The biology and ecology of *Drosophila suzukii* (Diptera: Drosophilidae). In *Drosophila Suzukii Management*; Garcia, F.R.M., Ed.; Springer Nature: Cham, Switzerland, 2020; Chapter 4; pp. 41–91.
22. Elsensohn, J.E.; Loeb, G.M. Non-crop host sampling yields insights into small-scale population dynamics of *Drosophila suzukii* (Matsumura). *Insects* **2018**, *9*, 5. [CrossRef]
23. Lewis, W.J.; van Lenteren, J.C.; Phatak, S.C.; Tumlinson, J.H. A total system approach to sustainable pest management. *Proc. Natl. Acad. Sci. USA* **1997**, *94*, 12243–12248. [CrossRef]
24. Aluja, M. Fruit fly (Diptera: Tephritidae) research in Latin America: Myths, realities, and dreams. *An. Soc. Entomol. Bras.* **1999**, *28*, 565–594. [CrossRef]
25. Bianchi, F.J.J.A.; Booij, C.J.H.; Tscharntke, T. Sustainable pest regulation in agricultural landscapes: A review on landscape composition, biodiversity and natural pest control. *Proc. R. Soc. B Biol. Sci.* **2006**, *273*, 1715–1727. [CrossRef]
26. Aluja, M.; Rull, J. Managing pestiferous fruit flies (Diptera: Tephritidae) through environmental manipulation. In *Biorational Tree Fruit Pest Management*; Aluja, M., Leskey, T., Vincent, C., Eds.; CAB International: Wallingford, UK, 2009; pp. 214–252.
27. Thomson, L.J.; Hoffmann, A.A. Natural enemy responses and pest control: Importance of local vegetation. *Biol. Control* **2010**, *52*, 160–166. [CrossRef]
28. Landis, D.A.; Wratten, S.D.; Gurr, G.M. Habitat management to conserve natural enemies of arthropod pests. *Annu. Rev. Entomol.* **2000**, *45*, 175–201. [CrossRef] [PubMed]

29. Veres, A.; Petit, S.; Conord, C.; Lavigne, C. Does landscape composition affect pest abundance and their control by natural enemies? A review. *Agric. Ecosyst. Environ.* **2013**, *166*, 110–117. [CrossRef]

30. Schmidt, J.M.; Whitehouse, T.S.; Green, K.; Krehenwinkel, H.; Schmidt-Jeffris, R.; Sial, A.A. Local and landscape-scale heterogeneity shape spotted wing drosophila (*Drosophila suzukii*) activity and natural enemy abundance: Implications for trophic interactions. *Agric. Ecosyst. Environ.* **2019**, *272*, 86–94. [CrossRef]

31. Aluja, M.; Sivinski, J.; van Driesche, R.; Anzures-Dadda, A.; Guillén, L. Pest management through tropical tree conservation. *Biodivers. Conserv.* **2014**, *23*, 831–853. [CrossRef]

32. Ehler, L.E. Parasitoid communities, parasitoid guilds, and biological control. In *Parasitoid Community Ecology*; Hawkins, B.A., Sheehan, W., Eds.; Oxford University Press: Oxford, UK, 1994; pp. 418–436.

33. Holland, J.M.; Bianchi, F.; Entling, M.H.; Moonen, A.C.; Smith, B.M.; Jeanneret, P. Structure, function and management of semi-natural habitats for conservation biological control: A review of European studies. *Pest Manag. Sci.* **2016**, *72*, 1638–1651. [CrossRef]

34. Markó, V.; Elek, Z.; Kovács-Kostyánszki, A.; Kőrösi, Á.; László, S.; Földesi, R.; Varga, Á.; Iván, Á.; Báldi, A. Landscapes, orchards, pesticides—Abundance of beetles (Coleoptera) in apple orchards along pesticide toxicity and landscape complexity gradients. *Agric. Ecosyst. Environ.* **2017**, *247*, 246–254. [CrossRef]

35. Vargas, G.; Rivera-Pedroza, L.-F.; Cracía, L.F.; Mundstock Jahnke, S. Conservation biological control as an important tool in the Neotropical region. *Neotrop. Entomol.* **2023**, *52*, 134–151. [CrossRef]

36. Brown, A.D.; Grau, H.R.; Malizia, L.R.; Grau, A. Argentina. In *Bosques Nublados del Neotrópico*; Kappelle, M., Brown, A.D., Eds.; Instituto Nacional de Biodiversidad: San José, Costa Rica, 2001; pp. 623–659.

37. Grau, H.R.; Aragon, R. Arboles invasores de la Sierra de San Javier, Tucumán, Argentina. In *Ecología de Árboles Exóticos en las Yungas Argentinas*; Grau, H.R., Aragon, R., Eds.; Laboratorio de Investigaciones Ecológicas de las Yungas, Universidad Nacional de Tucumán: San Miguel de Tucuman, Argentina, 2020; pp. 5–20.

38. Schliserman, P.; Aluja, M.; Rull, J.; Ovruski, S.M. Temporal diversity and abundance patterns of parasitoids of fruit-infesting Tephritidae (Diptera) in the Argentinean Yungas: Implications for biological control. *Environ. Entomol.* **2016**, *45*, 1184–1198. [CrossRef]

39. Buonocore-Biancheri, M.J.; Suárez, L.; Kirschbaum, D.S.; Garcia, F.R.M.; Funes, C.F.; Ovruski, S.M. Natural parasitism influences biological control strategies against both global invasive pests *Ceratitis capitata* (Diptera: Tephritidae) and *Drosophila suzukii* (Diptera: Drosophilidae), and the Neotropical-native pest *Anastrepha fraterculus* (Diptera: Tephritidae). *Environ. Entomol.* **2022**, *51*, 1120–1135. [CrossRef]

40. Buonocore-Biancheri, M.J.; Kirschbaum, D.S.; Suárez, L.; Ponssa, M.D.; Ovruski, S.M. *Drosophila suzukii* in Argentina: State of the art and further perspectives. *Neotrop. Entomol.* **2023**. [CrossRef] [PubMed]

41. Guillén, D.; Sánchez, R. Expansion of the national fruit fly control programme in Argentina. In *Area-Wide Control of Insect Pests: From Research to Field Implementation*; Vreysen, M.J.B., Robinson, A.S., Hendrichs, J., Eds.; Springer: Dordrecht, The Netherlands, 2007; pp. 653–660.

42. Buonocore-Biancheri, M.J.; Núñez-Campero, S.R.; Suárez, L.; Ponssa, M.D.; Kirschbaum, D.S.; Garcia, F.R.M.; Ovruski, S.M. Implications of the niche partitioning and coexistence of two resident parasitoids for *Drosophila suzukii* management in non-crop areas. *Insects* **2023**, *14*, 222. [CrossRef] [PubMed]

43. Brown, A.D. Las selvas pedemontanas de las Yungas: Manejo sustentable y conservación de la biodiversidad de un ecosistema prioritario del noroeste argentino. In *Selva Pedemontana de las Yungas: Historia Natural, Ecología y Manejo de un Ecosistema en Peligro*; Brown, A.D., Blendinger, P.G., Lomáscolo, T., García Bes, P., Eds.; Ediciones del Subtrópico, Yerba Buena: Tucumán, Argentina, 2009; pp. 13–36.

44. Schliserman, P.; Aluja, M.; Rull, J.; Ovruski, S.M. Habitat degradation and introduction of exotic plants favor persistence of invasive species and population growth of native polyphagous fruit fly pests in a Northwestern Argentinean mosaic. *Biol. Invasions* **2014**, *16*, 2599–2613. [CrossRef]

45. White, I.M.; Elson-Harris, M.M. *Fruit Flies of Economic Significance: Their Identification and Bionomics*; CAB International-ACIAR, Redwood Press Ltd.: Melksham, UK, 1992.

46. Hauser, M. A historic account of the invasion of *Drosophila suzukii* (Matsumura) (Diptera: Drosophilidae) in the continental United States, with remarks on their identification. *Pest Manag. Sci.* **2011**, *67*, 1352–1357. [CrossRef] [PubMed]

47. R Core Team. *R: A Anguage and Environment for Statistical Computing*; R. Foundat Statistical Computing: Vienna, Austria, 2023. Available online: https://www.R-project.org/ (accessed on 5 October 2023).

48. Patil, I. Visualizations with statistical details: The 'ggstatsplot' approach. *J. Open Source Softw.* **2021**, *6*, 3167. [CrossRef]

49. Diamantidis, A.D.; Carey, J.R.; Nakas, C.T.; Papadopoulos, N.T. Population-specific demography and invasion potential in medfly. *Ecol. Evol.* **2011**, *1*, 479–488. [CrossRef] [PubMed]

50. Rossi-Stacconi, M.V.; Kaur, R.; Mazzoni, V.; Ometto, L.; Grassi, A.; Gottardello, A.; Rota-Stabelli, O.; Anfora, G. Multiple lines of evidence for reproductive winter diapause in the invasive pest *Drosophila suzukii*. *J. Pest Sci.* **2016**, *89*, 689–700. [CrossRef]

51. Grassi, A.; Gottardello, A.; Dalton, D.T.; Tait, G.; Rendon, D.; Ioriatti, C.; Gibeaut, D.; Rossi-Stacconi, M.V.; Walton, V.M. Seasonal reproductive biology of *Drosophila suzukii* (Diptera: Drosophilidae) in temperate climates. *Environ. Entomol.* **2018**, *47*, 166–174. [CrossRef]

52. Stockton, D.; Wallingford, A.; Loeb, G. Phenotypic plasticity promotes overwintering survival in a globally invasive crop pest, *Drosophila suzukii*. *Insects* **2018**, *9*, 105. [CrossRef]

53. Kovaleski, A.; Sugayama, R.L.; Malavasi, A. Movement of *Anastrepha fraterculus* from native breeding sites into apple orchards in Southern Brazil. *Entomol. Exp. Appl.* **1999**, *91*, 457–467. [CrossRef]

54. Garcia, F.R.M.; Lasa, R.; Funes, C.F.; Buzzetti, K. *Drosophila suzukii* management in Latin America: Current status and perspectives. *J. Econ. Entomol.* **2022**, *115*, 1008–1023. [CrossRef] [PubMed]

55. Gallardo, F.E.; Reche, V.A.; Margaría, C.B.; Aquino, D.A.; Ansa, M.A.; Dettler, M.A.; Vazquez, F.; Barrientos, G.; Santadino, M.; Martínez, E.; et al. Survey of potential parasitoids (Hymenoptera) of *Drosophila suzukii* (Diptera: Drosophilidae) in Buenos Aires province. Argentina. *Rev. Soc. Entomol. Arg.* **2022**, *81*, 71–78. [CrossRef]

56. Wang, X.G.; Kaçar, G.; Daane, K.M. Temporal dynamics of host use by *Drosophila suzukii* in California's San Joaquin Valley: Implications for area-wide pest management. *Insects* **2019**, *10*, 206. [CrossRef]

57. Markow, T.A.; O'Grady, P. Reproductive ecology of *Drosophila*. *Funct. Ecol.* **2008**, *22*, 747–759. [CrossRef]

58. Funes, C.F.; Escobar, L.I.; Dadda, G.E.; Villagrán, M.E.; Olivera, G.I.; Gastaminza, G.G.; Kirschbaum, D.S. Occurrence and population fuctuations of *Drosophila suzukii* (Diptera: Drosophilidae) in blueberry crops of subtropical Argentina. *Acta Hortic.* **2023**, *1357*, 257–264. [CrossRef]

59. Wollmann, J.; Schlesener, D.C.H.; Ferreira, M.S.; Kruger, A.P.; Bernardi, D.; Garcia, J.A.B.; Nunes, A.M.; Garcia, M.S.; Garcia, F.R.M. Population dynamics of *Drosophila suzukii* (Diptera: Drosophilidae) in berry crops in southern Brazil. *Neotrop. Entomol.* **2019**, *48*, 699–705. [CrossRef]

60. Arnó, J.; Sola, M.; Riudavets, J.; Gabarra, R. Population dynamics, non-crop hosts, and fruit susceptibility of *Drosophila suzukii* in Northeast Spain. *J. Pest Sci.* **2016**, *89*, 713–723. [CrossRef]

61. Bal, H.K.; Adams, C.; Grieshop, M. Evaluation of off-season potential breeding sources for spotted wing Drosophila (*Drosophila suzukii* Matsumura) in Michigan. *J. Econ. Entomol.* **2017**, *110*, 2466–2470. [CrossRef]

62. Panel, A.D.C.; Zeeman, L.; van der Sluis, B.J.; van Elk, P.; Pannebakker, B.A.; Wertheim, B.; Helsen, H.H.M. Overwintered *Drosophila suzukii* are the main source for infestations of the first fruit crops of the season. *Insects* **2018**, *9*, 145. [CrossRef]

63. Drummond, F.; Ballman, E.; Collins, J. Population dynamics of spotted wing Drosophila (*Drosophila suzukii* (Matsumura)) in Maine wild blueberry (*Vaccinium angustifolium* Aiton). *Insects* **2019**, *10*, 205. [CrossRef]

64. Gavriel, S.; Gazit, Y.; Leach, A.; Mumford, J.; Yuval, B. Spatial patterns of sterile Mediterranean fruit fly dispersal. *Entomol. Exp. Appl.* **2012**, *142*, 17–26. [CrossRef]

65. De Villiers, M.; Manrakhan, A.; Addison, P.; Hattingh, V. The distribution, relative abundance, and seasonal phenology of *Ceratitis capitata*, *Ceratitis rosa*, and *Ceratitis cosyra* (Diptera: Tephritidae) in South Africa. *Environ. Entomol.* **2013**, *42*, 831–840. [CrossRef] [PubMed]

66. Araujo, E.S.; Lino, B.; Monteiro, R.S.; Nishimura, G.; Franck, P.; Lavigne, C. Impact of native forest remnants and wild host plants on the abundance of the South American fruit fly, *Anastrepha fraterculus* in Brazilian apple orchards. *Agric. Ecosyst. Environ.* **2019**, *275*, 93–99. [CrossRef]

67. Aluja, M.; Piñero, J.; Jácome, I.; Diaz-Fleischer, F.; Sivinski, J. Behavior of flies in the genus *Anastrepha* (Trypetinae: Toxotrypanini). In *Fruit Flies (Tephritidae): Phylogeny and Evolution of Behavior*; Aluja, M., Norrbom, A.L., Eds.; CRC Press LLC., Corporate Blvd.: Boca Raton, FL, USA, 2000; pp. 375–406.

68. Tscharntke, T.; Klein, A.M.; Kruess, A.; Steffan-Dewenter, I.; Thies, C. Landscape perspectives on agricultural intensification and biodiversity–ecosystem service management. *Ecol. Lett.* **2005**, *8*, 857–874. [CrossRef]

69. Thies, C.; Steffan-Dewenter, I.; Tscharntke, T. Effects of landscape context on herbivory and parasitism at different spatial scales. *Oikos* **2003**, *101*, 18–25. [CrossRef]

70. Santadino, M.V.; Riquelme-Virgala, M.B.; Ansa, M.A.; Bruno, M.; Di Silvestro, G.; Lunazzi, E.G. Primer registro de *Drosophila suzukii* (Diptera: Drosophilidae) asociado al cultivo de arándanos (*Vaccinium* spp.) de Argentina. *Rev. Soc. Entomol. Arg.* **2015**, *74*, 183–185.

71. Dagatti, C.V.; Marcucci, B.; Herrera, M.E.; Becerra, V.C. First record of *Drosophila suzukii* (Diptera: Drosophilidae) associated to blackberry in Mendoza, Argentina. *Rev. Soc. Entomol. Arg.* **2018**, *7*, 26–29. [CrossRef]

72. Dettler, M.A.; Barrientos, G.N.; Martinez, E.; Ansa, M.A.; Santadino, M.V.; Coviella, C.E.; Riquelme-Virgala, M.B. Field infestation level of *Zaprionus indianus* (Gupta) and *Drosophila suzukii* (Matsumura) (Díptera: Drosophilidae) in *Ficus carica* L.. (Rosales: Moraceae) and *Rubus idaeus* L. (Rosales: Rosaceae) in the Northeast of Buenos Aires province. *Rev. Soc. Entomol. Argent.* **2021**, *80*, 43–47. [CrossRef]

73. Cichón, L.I.; Garrido, S.; Lago, J. Drosophila suzukii: Una Nueva Plaga Presente en la Norpatagonia. *Ediciones-INTA, Instituto Nacional de Tecnología Agropecuaria, Río Negro, Argentina*. 2016. Available online: https://repositorio.inta.gob.ar/bitstream/handle/20.500.12123/3119/INTA_CRPatagoniaNorte_EEAAltoValle_Cichon_LI_Drosophila_suzukii_nueva_plaga_presente_Norpatagonia.pdf (accessed on 10 July 2023).

74. Tait, G.; Cabianca, A.; Grassi, A.; Pfab, F.; Oppedisano, T.; Puppato, S.; Mazzoni, V.; Anfora, G.; Walton, V.M. *Drosophila suzukii* daily dispersal between distinctly different habitats. *Entomol. Gen.* **2020**, *40*, 25–37. [CrossRef]

75. Wiman, N.G.; Walton, V.M.; Dalton, D.T.; Gianfranco Anfora, G.; Burrack, H.J.; Chiu, J.C.; Daane, K.M.; Grassi, A.; Miller, B.; Tochen, S.; et al. Integrating temperature-dependent life table data into a matrix projection model for *Drosophila suzukii* population estimation. *PLoS ONE* **2014**, *9*, e106909. [CrossRef]

76. Garcia, F.R.M.; Ovruski, S.M.; Suárez, L.; Cancino, J.; Liburd, O.E. Biological control of tephritid fruit flies in the Americas and Hawaii: A review of the use of parasitoids and predators. *Insects* **2020**, *11*, 662. [CrossRef] [PubMed]

77. Aluja, M.; Ovruski, S.M.; Guillén, L.; Oroño, L.E.; Sivinski, J. Comparison of the host searching and oviposition behaviors of the tephritid (Diptera) parasitoids *Aganaspis pelleranoi* and *Odontosema anastrephae* (Hymenoptera: Figitidae, Eucoilinae). *J. Insect Behav.* **2009**, *22*, 423–451. [CrossRef]

Article

The Effect of Genotype Combinations of *Wolbachia* and Its *Drosophila melanogaster* Host on Fertility, Developmental Rate and Heat Stress Resistance of Flies

Natalya V. Adonyeva [1], Vadim M. Efimov [1,2] and Nataly E. Gruntenko [1,*]

[1] Institute of Cytology and Genetics SB RAS, Novosibirsk 630090, Russia; nadon@bionet.nsc.ru (N.V.A.); efimov@bionet.nsc.ru (V.M.E.)
[2] Department of Natural Sciences, Novosibirsk State University, Novosibirsk 630090, Russia
*   Correspondence: nataly@bionet.nsc.ru; Tel.: +7-383-3634963 (ext. 3103)

Simple Summary: *Wolbachia*, the intracellular symbiont of insects, is of big interest and importance for its numerous effects on the host's life-history traits. However, the details of *Wolbachia*–host interaction are still not well studied and understood. Here, we present data on the influence of two different *Wolbachia* strains on the life-history traits of two different wild-type *D. melanogaster* lines. The results obtained allow us to assume that the effect of *Wolbachia* on the flies' life-history traits depends on the genotypes of both the host and the symbiont, but the fact of recent transfer of the symbiont to a new host could also be a factor.

Abstract: The best-known effect of the intracellular bacterium *Wolbachia* is its mostly negative influence on the reproduction of the host. However, there is evidence of a positive influence of *Wolbachia* on the host's resistance to stress, pathogens, and viruses. Here, we analyzed the effects of two *Wolbachia* strains belonging to wMel and wMelCS genotypes on *D. melanogaster* traits, such as fertility, survival under acute heat stress, and developmental rate. We found that *D. melanogaster* lines under study differ significantly in the above-mentioned characteristics, both when the natural infection was preserved, and when it was eliminated. One of *Wolbachia* strains, wMel, did not affect any of the studied traits. Another strain, wMelPlus, had a significant effect on the development time. Moreover, this effect is observed not only in the line in which it was discovered but also in the one it was transferred to. When transferred to a new line, wMelPlus also caused changes in survival under heat stress. Thus, it could be concluded that *Wolbachia*–*Drosophila* interaction depends on the genotypes of both the host and the symbiont, but some *Wolbachia* effects could depend not on the genotypes, but on the fact of recent transfer of the symbiont.

Keywords: *Wolbachia*; *Drosophila*; fertility; developmental rate; heat stress; viability

Citation: Adonyeva, N.V.; Efimov, V.M.; Gruntenko, N.E. The Effect of Genotype Combinations of *Wolbachia* and Its *Drosophila melanogaster* Host on Fertility, Developmental Rate and Heat Stress Resistance of Flies. *Insects* 2023, 14, 928. https://doi.org/10.3390/insects14120928

Academic Editors: Sérgio M. Ovruski and Flávio Roberto Mello Garcia

Received: 27 October 2023
Revised: 29 November 2023
Accepted: 4 December 2023
Published: 5 December 2023

1. Introduction

Maternally inherited alpha-proteobacterium *Wolbachia pipientis*, best known for its ability to induce cytoplasmic incompatibility and manipulate host reproduction [1], occurs in more than 40% of arthropod species, including *Drosophila melanogaster* [2,3]. According to its effect on reproductive biology of the host, *Wolbachia* has long been considered a parasite, but now a lot of data have been accumulated indicating that the host, in turn, can benefit from *Wolbachia* infection [4].

Another positive aspect of this symbiosis is the possibility of using *Wolbachia* in the control of insect pests. *Wolbachia* have two main uses in this regard: incompatible insect technique (IIT) based on cytoplasmic incompatibility (CI) caused by *Wolbachia* in many insect species [5], and pathogen blocking technique (PBT) based on the ability of *Wolbachia* to provide antiviral protection to their host and to spread in a wild population due to CI-provided fitness advantages of *Wolbachia*-infected females [6,7]. IIT achieves a suppression of insect pests' populations due to *Wolbachia*-infected males' failure to produce viable embryos by mating with wild-type uninfected females [8]. PBT blocks the spread of distinct human pathogens, including Zika, Dengue, Yellow fever and West Nile viruses, the Plasmodium parasites that cause malaria, and filarial worms in insect vectors such as *Aedes aegypti* and *Ae. albopictus* [7,9]. The combination of IIT and female sterilization with ionizing radiation was recently used for the population suppression of a fruit pest, *Drosophila suzukii* [10,11], making *D. melanogaster* a valid model for *Wolbachia* studies.

In the members of the *Drosophila* genus, some evidence concerning the role of the genetic features of both bacterium and its host in their interaction was obtained. The antiviral protection of *Wolbachia* strains transferred to the same genetic background of *D. simulans* from different *Drosophila* species depends on the *Wolbachia* genotype [12]. Various *Wolbachia* genotypes transferred to the same genetic background of *D. melanogaster* also demonstrate different effects on the host's hormonal status and survival under heat stress [13,14]. On the other hand, the positive effect of *Wolbachia* infection on *D. melanogaster* longevity [15] as well as fitness benefits caused by a *Wolbachia* infection in population cage experiments in *D. simulans* [16] depend on the fly genotype.

In *D. melanogaster*, *Wolbachia* genotypes are classified into two groups, wMel and wMelCS, based on polymorphic markers [17–19]. The wMel group is dominant all over the world and thus could be considered as giving more benefits to its host [17,20]. However, we have earlier discovered a *Wolbachia* strain, wMelPlus, which belongs to the wMelCS group of genotypes but differs from other described members of this group by a large inversion [21], and found out that this strain changes host fertility [14] and heat stress resistance [22]. The positive effect of the wMelPlus strain on *D. melanogaster* survival under acute heat stress was demonstrated on two different lines, Bi90^T and Canton S, after infection transfer from donor w153 line to them by 10–20 generations of backcross with corresponding males [22]. On the other hand, the effect of wMelPlus *Wolbachia* on its "native" host, *D. melanogaster* line w153, was never studied. At the same time, the uninfected Bi90^T line obtained by tetracycline treatment from wild type line Bi90, which carried wMel *Wolbachia* from the beginning, did not differ from it in heat stress resistance and fertility level [13].

So we could not be certain if the previously discovered effects of wMelPlus *Wolbachia* on the host's fitness depends only on the strain's characteristics and not on the effect of *Wolbachia* transfer from line to line. In order to clarify this question, we performed the present study of fertility level, developmental rate, and stress resistance in two pairs of *D. melanogaster* lines: Bi90–Bi90^T and w153–w153^T, infected and uninfected with wMel and wMelPlus *Wolbachia* genotypes, respectively. The characteristics of line Bi90^{wMelPlus} carrying the wMelPlus strain on the Bi90^T line's genetic background were also investigated when the effect of the strain on the trait under study was found in the "native" host line, w153, compared to the tetracycline-treated w153^T line.

2. Materials and Methods

2.1. Drosophila Lines and Rearing

The females of the following *D. melanogaster* lines were used in the study: isofemale lines w153, carrying *Wolbachia* infection of wMelPlus genotype, and Bi90 carrying infection of wMel genotype, established as full-sib families from a single inseminated wild-caught female from Tashkent (Uzbekistan) and Bishkek (Kyrgystan), correspondingly [19]; the corresponding uninfected lines, w153^T and Bi90^T, treated with tetracycline for three generations no less than 10 generations prior to the start of the experiments; and line Bi90^{wMelPlus}, obtained as a result of wMelPlus strain transfer to the Bi90^T line by backcrossing with Bi90^T males for 20 generations as described earlier [22].

Flies were maintained on standard food (agar-agar, 7 g/L; corn grits, 50 g/L; dry yeast, 18 g/L; sugar, 40 g/L) in a MIR-554 incubator (Sanyo, Osaka, Japan) at 25 °C under a 12:12 h light–dark cycle.

2.2. Developmental Rate Analysis

Males and females with a difference in age no more than 4–5 h since eclosion were chosen as parents; as they reached 3 days of age, flies were placed into vials (3–5 parent pairs; 10 vials per experiment group), where they laid eggs for 24 h. Eclosed progeny (imagoes) were counted every 12 h, at 9 a.m. and 9 p.m., up to the eclosion of the last descendant. Developmental rate was presented as a percentage of the total number of eclosed progeny for every period of measurement.

2.3. Fertility Analysis

To measure fertility, 3 or 5 pairs of young females and males with a difference in age no more than 4–5 h since eclosion were placed into vials (10 vials per experiment group), where they laid eggs under standard conditions. For 10 days, the flies were placed into new vials every 24 h for oviposition. Fertility was measured as the number of progeny (imagoes) eclosed from the eggs laid every 24 h per female.

2.4. Viability Analysis

To estimate viability under acute heat stress, the vials with flies of all groups (10–15 vials with 5 females and 5 males each per group) at the age of 6 days were transferred from 25 °C to 38 °C for 4 h. Then they were returned to 25 °C and the surviving females were counted 24 h later. The survival rates were calculated as the percentage of survivors in each vial.

2.5. Statistical Analyses

2.5.1. Fertility

Each culture vial was considered as a separate case, and the fertility in it per female per day was considered as a separate trait. Euclidean distance was used to estimate differences between vials by fertility over a range of days. The matrix of pairwise Euclidean distances by fertility between all vials of all considered lines was processed by the principal coordinate method (PCoA). For Euclidean distances, this method is equivalent to principal component analysis (PCA) [23]. As a rule, differences between lines manifested themselves in one of the first two principal components, which together accounted for more than half of the total variance. The significance of the differences between each pair of lines for each principal component was assessed using a two-sample *t*-test, applying the Benjamini–Hochberg P adjustment to correct for multiple testing [24].

Some inconvenience of the Benjamini–Hochberg P adjustment is that it is necessary, in addition to each sample's p_i-value, to calculate the corresponding standard pBH_i-adjustment for comparison. However, the calculation can be greatly simplified by multiplying both indicators by N/i. Then the indicator $Np_i = p_i$-value $\times N/i$ must be compared with $NpBH_i = pBH_i$-adjustment $\times N/i = i\alpha/N \times N/i = \alpha$, that is, with the standard tabular level to which everyone is accustomed. It is simple, but very convenient. For $i = 1$, this technique also works for the Bonferroni method [25].

For convenience of calculation and presentation of results, each pair of lines compared by one quantitative characteristic (for example, by the principal component) was combined into one sample and a dichotomous variable was additionally formed for it so that each value of the quantitative sample was marked 0 for one of the lines, 1 for another. The Pearson correlation coefficient r was calculated between the quantitative variable and the dichotomous one. It is known [26] that calculating the significance of this point-biserial correlation coefficient r is equivalent to calculating the significance of Student's *t*-test used to compare the means of two normal populations with equal variance. Additionally, the squared correlation coefficient (r^2) is an estimate of the currently recommended effect size [27], so we also present it in the tables.

2.5.2. Developmental Rate

Every 12 h, the number of emerging flies was recorded for each vial. After the end of the experiment, the resulting dynamics were normalized to the total number of all eclosed flies. Euclidean distance was used to assess differences in developmental rates for each pair of vials. Next, just as for fertility, the principal components were calculated by the Gower principal coordinate method and the lines were compared with each other according to the first two principal components by *t*-test, applying the Benjamini–Hochberg P adjustment [24] (see the previous section).

2.5.3. Viability

Survival rates were calculated as the percentage of survivors in each vial. The groups were compared on this trait using the *t*-test by the Benjamini-Hochberg method [24].

3. Results

3.1. Fertility of D. melanogaster Lines w153 and Bi90 Infected with the wMelPlus and wMel Wolbachia Strains, Correspondingly, and Control Uninfected Lines w153^T and Bi90^T

Earlier, we had shown that the wMelPlus strain of *Wolbachia* being transferred to Bi90^T line of *D. melanogaster* caused significant changes in host fertility level [14], so the first thing to study was the fertility of the "native" wMelPlus line w153 in comparison with the uninfected (tetracycline-treated) line w153^T (Figure 1).

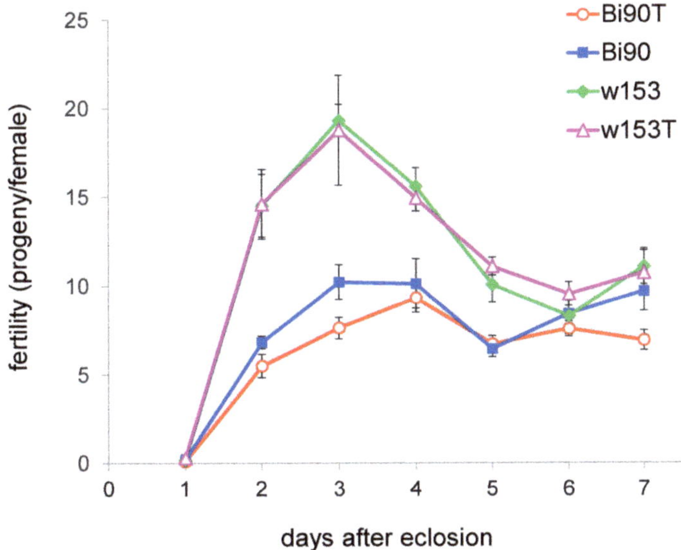

Figure 1. The fertility of *D. melanogaster* wild type lines Bi90 (infected with wMel *Wolbachia* strain), w153 (infected with wMelPlus *Wolbachia* strain), Bi90^T (uninfected), w153^T (uninfected). Each point represents an average value of 9–10 replicates (3 females per test) as means ± s.e.m.

The Bi90^T line, together with its precursor, the Bi90 line naturally infected with the wMel strain, were taken into analysis as well (Figure 1). Analysis of the fertility curve shows that, according to the level of fertility, this period can be divided into 2 sub-periods: (1) typical fertility peak reached between 2 and 4 days after eclosion [28] and (2) subsequent decrease in fertility level from 5 to 7 days after eclosion. In the first sub-period, lines with different genetic backgrounds (originating from the Bi90 line and originating from the w153 line) strongly differ from each other. In the second sub-period, the differences are smoothed out. We analyzed the differences between all four lines in these sub-periods using the principal component method (Figure 2).

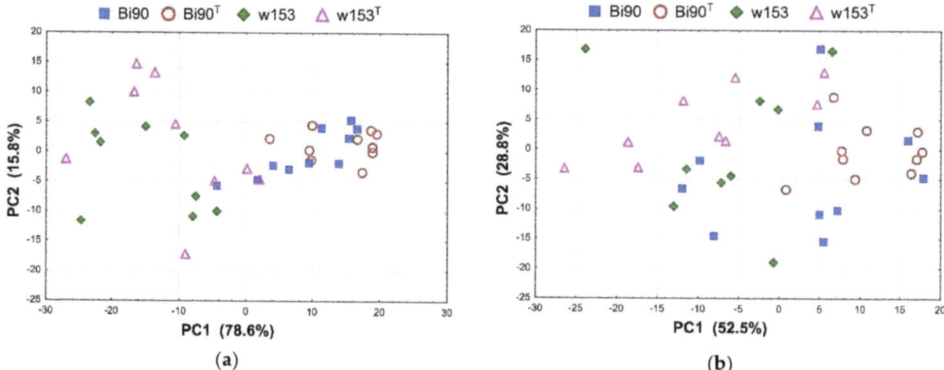

Figure 2. PCA plot showing variability of the fertility level per female per day in the Bi90 (infected with wMel *Wolbachia* strain), w153 (infected with wMelPlus *Wolbachia* strain), Bi90^T (uninfected), w153^T (uninfected) lines of *D. melanogaster*: (**a**) days 2–4 after eclosion; (**b**) days 5–7 after eclosion. Each data point represents one biological replicate (three females per replicate).

In the first sub-period, lines Bi90 and Bi90^T clearly differ from lines w153 and w153^T, implying differences in fertility caused by the genetic component of the host, i.e., by *D. melanogaster* genotype (Figure 2a). However, there are no differences observed between both the wMel-infected Bi90 line and the uninfected line Bi90^T, and between the wMelPlus-infected w153 line and the uninfected line w153^T. In other words, component analysis did not reveal any differences in fertility for this period resulting from the presence/absence of both *Wolbachia* strains under study.

Statistical assessment of the fertility level for days 2–4 is given in Tables S1 and S2, presenting a pairwise comparison of all lines under study using the *t*-test (below the diagonal) by the Benjamini–Hochberg method (above the diagonal). There is a high significance level for PC1 of the differences between the lines of Bi90 genotype compared with the lines of w153 genotype (Table S1); there are no significant differences for PC2 (Table S2).

The results of comparing fertility levels of the Bi90 (infected with wMel *Wolbachia* strain), w153 (infected with wMelPlus *Wolbachia* strain), Bi90^T (uninfected), and w153^T (uninfected) lines in the second sub-period (days 5–7) by the principal component method are presented in Figure 2b. Significant differences between lines can be seen, just as in the first sub-period. However, they are less prominent, which is evidenced by minimal (but significant) *t*-test by the Benjamini–Hochberg method for PC1 (Table S3). For PC2, differences are insignificant (Table S4).

3.2. Developmental Rate of D. melanogaster Lines w153 and Bi90 Infected with the wMelPlus and wMel Wolbachia Strains, Correspondingly, and Control Uninfected Lines w153^T and Bi90^T

Working with the Bi90 and w153 *D. melanogaster* lines infected with wMel and wMelPlus *Wolbachia* strains, correspondingly, we have noted that eclosion in the w153 line occurs one day later than in the Bi90 line. For this reason, we analyzed the developmental rates of these two lines as well as their uninfected versions, Bi90^T and w153^T. Figure 3 presents developmental rate curves expressed as a percentage of the number of eclosed flies: from the first to the last, counted at equal 12-h intervals. In the wMel-infected Bi90 line and the uninfected Bi90^T line, developmental rate curves almost match; in the wMelPlus-infected w153 line, the eclosion peak is delayed by 24 h compared to the Bi90 and Bi90^T lines, and the peak in the uninfected line w153^T is close to that of w153.

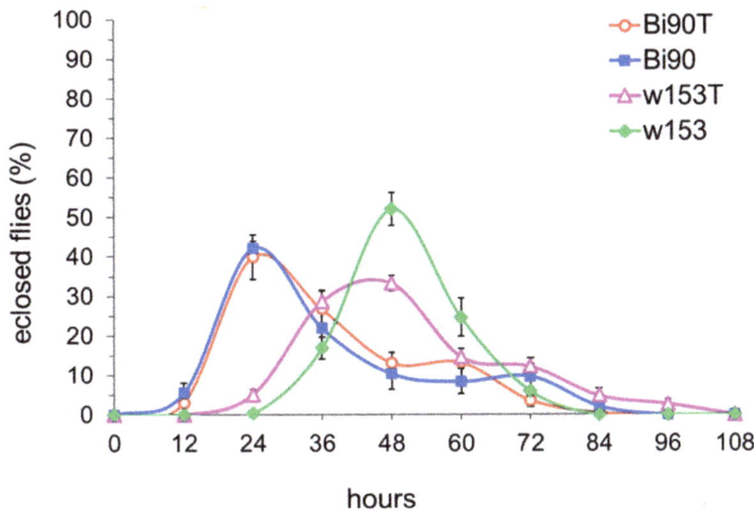

Figure 3. The developmental rate of *D. melanogaster* wild type lines Bi90 (infected with wMel *Wolbachia* strain), w153 (infected with wMelPlus *Wolbachia* strain), Bi90^T (uninfected), w153^T (uninfected). Each point represents the percentage of flies eclosed during 12 h (9–10 biological replicates per point; five flies in each replicate, means ± s.e.m.

Developmental rate analysis by the principal component method shows that while lines with the same genetic background (Bi90 and Bi90^T) do not differ from each other, they differ by PC1 from lines with a different genetic background (w153 and w153^T) with a high significance level (Figure 4).

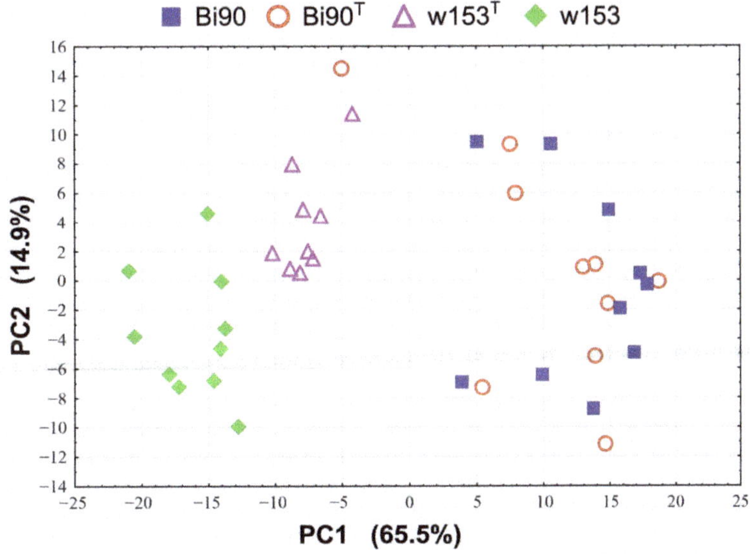

Figure 4. PCA plot showing the variability of developmental rate in the Bi90 (infected with wMel *Wolbachia* strain), w153 (infected with wMelPlus *Wolbachia* strain), Bi90^T (uninfected), w153^T (uninfected) lines of *D. melanogaster*. Each point (one replicate) represents the percentage of flies eclosed during 12 h (five flies in each replicate).

Tables S5 and S6 show *t*-test by the Benjamini–Hochberg method for PC1 and PC2, correspondingly. The significance is very high: for NpBH, it is exponential. Notably, developmental rates also significantly differ between lines w153 and w153^T both by PC1 (Table S5) and by PC2 (Table S6), which may imply that the *Wolbachia* strain wMelPlus influences the developmental rate of the w153 line.

3.3. Developmental Rate of the D. melanogaster Line Bi90 Infected with the wMelPlus Wolbachia Strain in Comparison with the Control Uninfected Line Bi90^T

In order to verify our assumption concerning the effect of the wMelPlus strain on host developmental rate, we studied this trait in the Bi90^{wMelPlus} line, which carries wMelPlus *Wolbachia* on Bi90^T nuclear background, in comparison with the uninfected line Bi90^T (Figure 5).

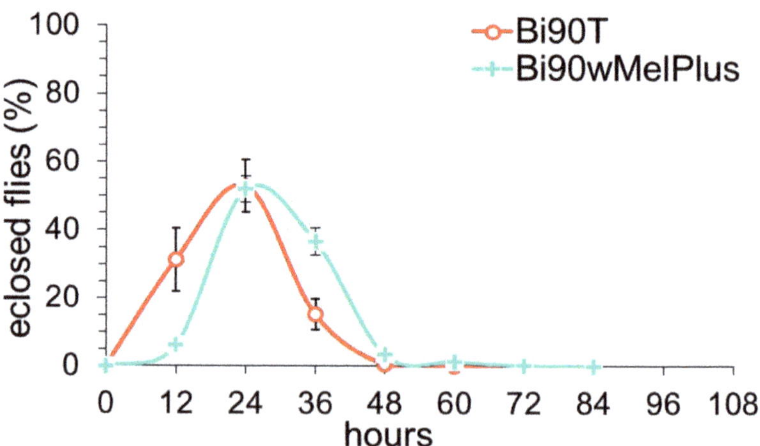

Figure 5. The developmental rate of *D. melanogaster* uninfected line Bi90^T and line Bi90^{wMelPlus}, carrying *Wolbachia* strain wMelPlus from line w153 on nuclear background of line Bi90^T. Each point represents the percentage of flies eclosed during 12 h (10 biological replicates per line; five flies in each replicate) as means ± s.e.m.

The PCA plot (Figure 6) allowed us to determine that the differences in developmental rate between the Bi90^{wMelPlus} and Bi90^T lines are significant in both PC1 and PC2 (Table S7); or significant at t = 3.73, NpBH = 0.00168 if we rotate the PCA plot by 30° (Figure S1).

3.4. Survival under Acute Heat Stress of D. melanogaster Lines w153 and Bi90 Infected with the wMelPlus and wMel Wolbachia Strains, Correspondingly, and the Bi90 Line Infected with the wMelPlus Strain in Comparrison with Control Uninfected Lines Bi90^T and w153^T

Earlier, we showed that the Bi90^{wMelPlus} line is characterized by increased viability under acute heat stress compared with the Bi90 lines infected with other *Wolbachia* strains [22], so it was of interest to find out if the wMelPlus strain causes the same effect in its "native" host, line w153. The analysis of survival under heat stress (38 °C, 4 h) carried out in the Bi90, Bi90^T, Bi90^{wMelPlus}, w153 and w153^T lines demonstrated that the line pairs Bi90–Bi90^T and w153–w153^T did not differ in stress resistance within the pairs (Figure 7, Table S8).

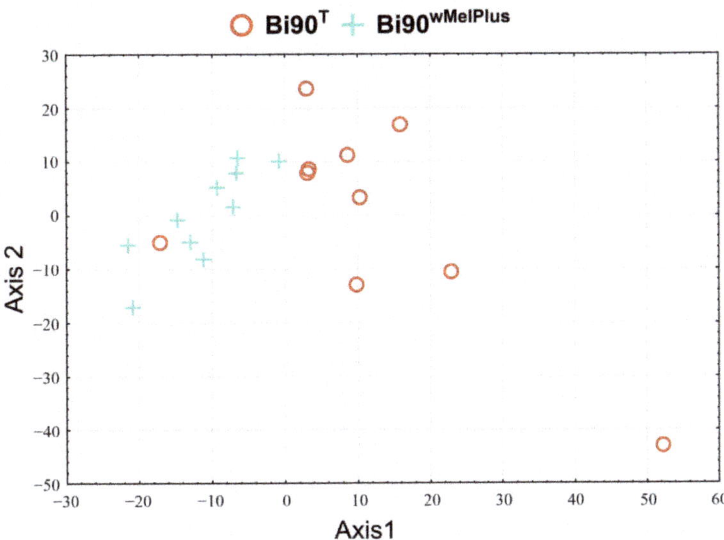

Figure 6. PCA plot showing the variability of developmental rate in *D. melanogaster* uninfected line Bi90^T and line Bi90^{wMelPlus}, carrying *Wolbachia* strain wMelPlus from line w153 on nuclear background of line Bi90^T. Each point represents the percentage of flies eclosed during 12 h (one biological replicates per point; five flies in each replicate).

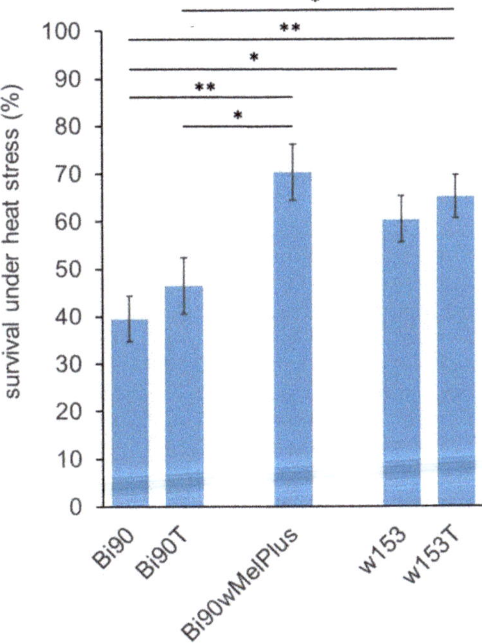

Figure 7. Survival under heat stress (38 °C, 4 h) of *D. melanogaster* females of lines Bi90 (infected with wMel *Wolbachia* strain), w153 (infected with wMelPlus *Wolbachia* strain), Bi90^T (uninfected), w153^T (uninfected). Each point represents an average value of 30–35 tests (N = 4 or 5 for each test) as means ± s.e.m. The asterisk indicates significant differences between the lines (**, $p < 0.01$; *, $p < 0.05$).

However, survival rate under heat stress in both lines with nuclear background of the wMelPlus-infected w153 line was significantly higher than that of the wMel-infected Bi90 line, and w153^T's survival rate was higher than that of Bi90^T (Figure 7, Table S8), which suggests a strong effect of the *D. melanogaster* genotype on the trait under study. On the other hand, the maximum survival rate was demonstrated by the Bi90^{wMelPlus} line differing significantly from the Bi90 and Bi90^T lines (Figure 7, Table S8), which could be evidence of *Wolbachia* genotype input to the trait value. At the same time, it should be noted that there was no significant difference in stress resistance between the Bi90^{wMelPlus} line and the w153 and w153^T lines (Figure 7, Table S8).

4. Discussion

Most studies performed on *D. melanogaster* have used only a few "wild-type" strains, representing very little genetic diversity. However, it is the genetic variation that is one of the main drivers of the evolution of life-history traits. Major life-history traits, which are subject to evolution by natural selection and are therefore vital to understanding adaptation, include developmental rate, size at eclosion, progeny number, life span, and various stress resistance traits [28]. Here, we present data on two wild-caught isofemale lines demonstrating significant differences in three life-history traits, namely, developmental rate, fertility (estimated as progeny per female), and survival under acute heat stress. We found out that lines Bi90^T and w153^T differ significantly in all traits under study. It was shown that fertility level could be highly variable among both laboratory lines and wild populations measured in the laboratory [29–31] but correlations between progeny number and developmental rate were usually positive [28]. However, females of the w153^T line demonstrated a higher fertility level and a lower developmental rate compared to the Bi90^T line (see Figures 1, 2a, 3 and 4), i.e., these two traits were negatively correlated, which was rather unexpected for us. Moreover, reduced fertility observed in the Bi90^T line (compared to the w153^T line) was shown to correlate to increased resistance to such types of stressors as desiccation and starvation [32,33], but in our experiments it went together with decreased resistance to heat stress (see Figure 7). This could mean the existence of a different mechanism of resistance to different types of stress, the specificity of lines under study, or even both. The fact that stands for the uniqueness of one of the studied lines, w153^T, is that it carries the rather unique *Wolbachia* strain wMelPlus, which not only alters its "native" host's developmental rate (see Figures 3 and 4), but also changes the fertility level, starvation, and heat stress resistance of the host when transferred to a new *D. melanogaster* line [14,22,34]. It should be noted that for other *Wolbachia* strains (except the well-known pathogenic strain wMelPop [35]) no such effects were found either in the "native" host or following transfer to a new one [13,14,22].

There is evidence that the interaction between *Wolbachia* and *Drosophila* has a complex nature. For example, it was found that a single *Wolbachia* strain wHa being transferred into three genetically distinct isofemale lines of *Drosophila simulans* with the use of microinjection methodology caused a dramatic fitness benefit in one of these lines and did not affect the fitness of two others [16]. The transfection of a single *Wolbachia* strain of wMel genotype into two different *Drosophila* species, *D. melanogaster* and *D. nigrosparsa*, resulted in completely different changes in the differential expression of genes [36].

On the other hand, different *Wolbachia* strains being transferred to the same *D. melanogaster* line Bi90^T caused different effects on fertility level, dopamine metabolism, and resistance to heat stress [13,14,22]. The results obtained in the present paper correspond with these data: the wMelPlus strain, which belongs to the wMelCS *Wolbachia* genotype and is shown to affect the fertility of females of line Bi90^{wMelCS} [14], does not influence this trait in its "native" host, the *D. melanogaster* line w153 (see Figures 1 and 2, Tables S1 and S2). Similarly, the wMelPlus strain increases resistance to acute heat stress when transferred to the Canton S or Bi90^T lines [22], but does not affect it in w153 (see Figure 7).

Several attempts to shed light on the molecular mechanisms of the effect of *Wolbachia* on host's physiology have been made. The transcriptome analyses of infected *D. melanogaster* females performed recently demonstrated changes in differential gene expression, which allowed to relate them to the Gene Ontology terms *Iron ion binding* and *Oxidation–reduction process* [36] or to create protein–protein interaction networks in STRING with the strongest interactions including *Metabolism, Ubiquitin, RNA binding and processing, Transcription and translation* and *Stress* [37]. The latter correlates with our findings concerning increased stress resistance of wMelPlus-infected females, and data on changes in metabolism correspond with results on increased lipid and glucose content found in both Bi90 and Bi90^{wMelPlus} females [34]. However, it is not obvious how the findings made in transcriptome analyses are connected with the effect of *Wolbachia* on fertility or developmental rate.

It is well-known that effects of many mutations found in one genetic background are often suppressed or enhanced in other backgrounds [38]. According to the data obtained here and presented in other investigations, it seems possible to suggest that an epistatic interaction of this kind could be discovered in genetic interaction between a host and a symbiotic bacterium, *Wolbachia* in particular.

However, not all effects of the wMelPlus strain depend on the host genotype: one can see that it slows down developmental rate and postpones eclosion in both w153 and Bi90^{wMelCS} lines (see Figures 3–6). It was shown in the end of the last century and the beginning of the present one that considerable variation in egg-to-adult development time could occur among wild-type strains of various *Drosophila* species [39,40]. As testing flies for the presence of *Wolbachia* was not common practice at the time, and *Drosophila* has been shown to have high rates of infection [2], one cannot be sure that at least part of the variables observed in these studies were not caused by *Wolbachia*. Another possible explanation for our findings could be the uniqueness of the wMelPlus *Wolbachia* strain, which is the only one to be found to increase resistance to acute heat stress [14,22] and to change the host's developmental rate (see Figures 3–6). The latter assumption is indirectly confirmed by the data of Strunov et al. [41], who found that the wMelCS type of infection and the wMel type did not influence any developmental life-history traits.

It should be noted that the results which demonstrate that wMelCS-infected flies were more fertile than wMel-infected flies, while the latter did not differ in fertility from uninfected flies [41], also agree with our data showing that the wMel *Wolbachia* strain, which infected the Bi90 line, does not cause any effects on the life-history traits under study. No changes in fertility level, developmental rate, and stress resistance in the Bi90 line compared to the uninfected Bi90^T line were observed, while the w153 line infected with *Wolbachia* of the wMelCS type has increased early life fertility level compared to the Bi90 and Bi90^T lines (Figures 1–7). Increased usefulness of the wMelCS type of *Wolbachia* compared to the wMel type in terms of enhancing the host's fertility was also demonstrated in the experiments with fertility rescue in flies with the *bag of marbles* (*bam*) hypomorphic mutation [42]. Moreover, wMelCS-like *Wolbachia* variants were shown to provide stronger protection against *Drosophila* Flock House and C viruses compared to wMel-like variants as well [43].

Thus, it could be concluded that *Wolbachia–Drosophila* interaction depends on the genotypes of both the host and the symbiont. However, taking into account that some of wMelPlus effects on life-history traits occur in the infected Bi90 and Canton S lines and not in the "native" line for this strain, w153, it could be hypothesized that at least some of the effects which occur in a *D. melanogaster* host infected with *Wolbachia* depend not on the genotype of the symbiont, but on the fact of its recent transfer. And one more supposition is possible as the w153^T line is characterized by increased early life fertility and stress resistance (see Figures 1, 2 and 7) even in the absence of the wMelPlus *Wolbachia* strain (see Figures 1, 2 and 7). We suppose that it could be an evidence of a successful co-evolution of the host line w153 and the symbiont *Wolbachia* strain wMelPlus. It can also be said that our data provides some insight into the prospects for the use of *Wolbachia* in pest control, indicating the need for thorough genetic studies of *Wolbachia* strains in pest species, such as *D. suzuki* or mosquitoes of the *Aedes* genus, aimed at finding the genetic variations of the bacterium most suitable for IIT and PBT.

Supplementary Materials: The following supporting information can be downloaded at: https://www.mdpi.com/article/10.3390/insects14120928/s1, Table S1. Significance level for PC1 of the fertility differences between the Bi90 (infected with wMel *Wolbachia* strain), w153 (infected with wMelPlus strain), Bi90^T (uninfected), w153^T (uninfected) lines for days 2–4 after eclosion using the *t*-test by the Benjamini–Hochberg method. Table S2. Significance level for PC2 of the fertility differences between the Bi90 (infected with wMel *Wolbachia* strain), w153 (infected with wMelPlus strain), Bi90^T (uninfected), w153^T (uninfected) lines for days 2–4 after eclosion using the *t*-test by the Benjamini–Hochberg method. Table S3. Significance level for PC1 of the fertility differences between the Bi90 (infected with wMel *Wolbachia* strain), w153 (infected with wMelPlus strain), Bi90^T (uninfected), w153^T (uninfected) lines for days 5–7 after eclosion using the *t*-test by the Benjamini–Hochberg method. Table S4. Significance level for PC2 of the fertility differences between Bi90 (infected with wMel *Wolbachia* strain), w153 (infected with wMelPlus strain), Bi90^T (uninfected), w153^T (uninfected) lines for days 5–7 after eclosion using the *t*-test by the Benjamini–Hochberg method. Table S5. Significance level for PC1 of the differences in developmental rate between the Bi90 (infected with wMel *Wolbachia* strain), w153 (infected with wMelPlus strain), Bi90^T (uninfected), w153^T (uninfected) lines with the *t*-test by the Benjamini–Hochberg method. Table S6. Significance level for PC2 of the differences in developmental rate between the Bi90 (infected with wMel *Wolbachia* strain), w153 (infected with wMelPlus strain), Bi90^T (uninfected), w153^T (uninfected) lines using the *t*-test by the Benjamini–Hochberg method. Table S7. Significance level for PC1 and PC2 of the differences in developmental rate between the Bi90^T and Bi90^{wMelPlus} lines using the *t*-test by the Benjamini–Hochberg method. Figure S1. PCA plot (after a 30° rotation) showing the variability of developmental rate in the Bi90^T and Bi90^{wMelPlus} lines of *D. melanogaster*. Each point represents the percentage of flies eclosed during 12 h (one biological replicates per point; five flies in each replicate). Table S8. Significance level of the differences in survival under acute heat stress between the Bi90 (infected with wMel *Wolbachia* strain), w153 (infected with wMelPlus *Wolbachia* strain), Bi90^T (uninfected), w153^T (uninfected) lines using the *t*-test by the Benjamini–Hochberg method.

Author Contributions: Conceptualization, N.V.A. and N.E.G.; methodology, N.V.A. and V.M.E.; software, V.M.E.; validation, V.M.E.; formal analysis, N.E.G. and V.M.E.; investigation, N.V.A.; writing—original draft preparation, N.V.A., V.M.E. and N.E.G.; writing—review and editing, N.E.G.; visualization, N.V.A. and V.M.E.; supervision, N.E.G.; project administration, N.E.G.; funding acquisition, N.E.G. All authors have read and agreed to the published version of the manuscript.

Funding: The study is supported by RSF grant No. 21-14-00090.

Data Availability Statement: Data are contained within the article.

Acknowledgments: The maintenance of experimental *D. melanogaster* lines was carried out in the *Drosophila* collection of the Institute of Cytology and Genetics SB RAS and was supported by BP #FWNR-2022-0019 of the Ministry of Science and Higher Education of the Russian Federation. We thank Darya Kochetova for language editing.

Conflicts of Interest: The authors declare no conflict of interest.

References

1. Werren, J.H.; Baldo, L.; Clarkm, M.E. *Wolbachia*: Master manipulators of invertebrate biology. *Nat. Rev. Microbiol.* **2008**, *6*, 741–751. [CrossRef] [PubMed]
2. Kriesner, P.; Conner, W.R.; Weeks, A.R.; Turelli, M.; Hoffmann, A.A. Persistence of a *Wolbachia* infection frequency cline in *Drosophila melanogaster* and the possible role of reproductive dormancy. *Evolution* **2016**, *70*, 979–997. [CrossRef] [PubMed]
3. Zug, R.; Hammerstein, P. Still a host of hosts for *Wolbachia*: Analysis of recent data suggests that 40% of terrestrial arthropod species are infected. *PLoS ONE* **2012**, *7*, e38544. [CrossRef] [PubMed]
4. Burdina, E.V.; Gruntenko, N.E. Physiological Aspects of *Wolbachia pipientis–Drosophila melanogaster* Relationship. *J. Evol. Biochem. Phys.* **2022**, *58*, 303–317. [CrossRef]
5. Shropshire, J.D.; Leigh, B.; Bordenstein, S.R. Symbiont-mediated cytoplasmic incompatibility: What have we learned in 50 years? *eLife* **2020**, *9*, e61989. [CrossRef]
6. Pimentel, A.C.; Cesar, C.S.; Martins, M.; Cogni, R. The antiviral effects of the symbiont bacteria *Wolbachia* in insects. *Front. Immunol.* **2020**, *11*, 626329. [CrossRef]
7. Caragata, E.P.; Dutra, H.L.C.; Sucupira, P.H.F.; Ferreira, A.G.A.; Moreira, L.A. *Wolbachia* as translational science: Controlling mosquitoborne pathogens. *Trends Parasitol.* **2021**, *37*, 1050–1067. [CrossRef] [PubMed]

8. Kaur, R.; Shropshire, J.D.; Cross, K.L.; Leigh, B.; Mansueto, A.J.; Stewart, V.; Bordenstein, S.R.; Bordenstein, S.R. Living in the endosymbiotic world of *Wolbachia*: A centennial review. *Cell Host. Microbe* **2021**, *29*, 879–893. [CrossRef]

9. Gong, J.T.; Li, T.P.; Wang, M.K.; Hong, X.Y. *Wolbachia*-based strategies for control of agricultural pests. *Curr. Opin. Insect. Sci.* **2023**, *57*, 101039. [CrossRef]

10. Nikolouli, K.; Colinet, H.; Renault, D.; Enriquez, T.; Mouton, L.; Gibert, P.; Sassu, F.; Cáceres, C.; Stauffer, C.; Pereira, R.; et al. Sterile insect technique and *Wolbachia* symbiosis as potential tools for the control of the invasive species *Drosophila suzukii*. *J. Pest Sci.* **2018**, *91*, 489–503. [CrossRef] [PubMed]

11. Nikolouli, K.; Sassù, F.; Mouton, L.; Stauffer, C.; Bourtzis, K. Combining sterile and incompatible insect techniques for the population suppression of *Drosophila suzukii*. *J. Pest Sci.* **2020**, *93*, 647–661. [CrossRef] [PubMed]

12. Martinez, J.; Tolosana, I.; Ok, S.; Smith, S.; Snoeck, K.; Day, J.P.; Jiggins, F.M. Symbiont strain is the main determinant of variation in *Wolbachia*-mediated protection against viruses across *Drosophila* species. *Mol. Ecol.* **2017**, *26*, 4072–4084. [CrossRef] [PubMed]

13. Gruntenko, N.E.; Ilinsky, Y.Y.; Adonyeva, N.V.; Burdina, E.V.; Bykov, R.A.; Menshanov, P.N.; Rauschenbach, I.Y. Various *Wolbachia* genotypes differently influence host *Drosophila* dopamine metabolism and survival under heat stress conditions. *BMC Evol. Biol.* **2017**, *28* (Suppl. S2), 252. [CrossRef] [PubMed]

14. Gruntenko, N.E.; Karpova, E.K.; Adonyeva, N.V.; Andreenkova, O.V.; Burdina, E.V.; Ilinsky, Y.Y.; Bykov, R.A.; Menshanov, P.N.; Rauschenbach, I.Y. *Drosophila* female fertility and juvenile hormone metabolism depends on the type of *Wolbachia* infection. *J. Exp. Biol.* **2019**, *22*, jeb195347. [CrossRef] [PubMed]

15. Fry, A.J.; Rand, D.M. *Wolbachia* interactions that determine *Drosophila melanogaster* survival. *Evolution* **2002**, *56*, 1976–1981. [PubMed]

16. Dean, M.D. A *Wolbachia*-associated fitness benefit depends on genetic background in *Drosophila simulans*. *Proc. R. Soc. B Biol. Sci.* **2006**, *273*, 1415–1420. [CrossRef]

17. Riegler, M.; Sidhu, M.; Miller, W.J.; O'Neill, S.L. Evidence for a global *Wolbachia* replacement in *Drosophila melanogaster*. *Curr. Biol.* **2005**, *15*, 1428–1433. [CrossRef]

18. Nunes, M.D.; Nolte, V.; Schlotterer, C. Nonrandom *Wolbachia* infection status of *Drosophila melanogaster* strains with different mtDNA haplotypes. *Mol. Biol. Evol.* **2008**, *25*, 2493–2498. [CrossRef] [PubMed]

19. Ilinsky, Y.Y. Coevolution of *Drosophila melanogaster* mtDNA and *Wolbachia* genotypes. *PLoS ONE* **2013**, *8*, e54373. [CrossRef] [PubMed]

20. Richardson, M.F.; Weinert, L.A.; Welch, J.J.; Linheiro, R.S.; Magwire, M.M.; Jiggins, F.M.; Bergman, C.M. Population genomics of the *Wolbachia* endosymbiont in *Drosophila melanogaster*. *PLoS Genet.* **2012**, *8*, e1003129. [CrossRef] [PubMed]

21. Korenskaia, A.E.; Shishkina, O.D.; Klimenko, A.I.; Andreenkova, O.V.; Bobrovskikh, M.A.; Shatskaya, N.V.; Vasiliev, G.V.; Gruntenko, N.E. New *Wolbachia pipientis* Genotype Increasing Heat Stress Resistance of *Drosophila melanogaster* Host Is Characterized by a Large Chromosomal Inversion. *Int. J. Mol. Sci.* **2022**, *23*, 16212. [CrossRef] [PubMed]

22. Burdina, E.V.; Bykov, R.A.; Menshanov, P.N.; Ilinsky, Y.Y.; Gruntenko, N.E. Unique *Wolbachia* strain wMelPlus increases heat stress resistance in *Drosophila melanogaster*. *Arch. Insect. Biochem. Physiol.* **2021**, *106*, e21776. [CrossRef] [PubMed]

23. Gower, J.C. Some distance properties of latent root and vector methods used in multivariate analysis. *Biometrika* **1966**, *53*, 238–325. [CrossRef]

24. Benjamini, Y.; Hochberg, Y. Controlling the false discovery rate: A practical and powerful approach to multiple testing. *J. R. Stat. Soc. Ser. B Methodol.* **1995**, *57*, 289–300. [CrossRef]

25. Narkevich, A.N.; Vinogradov, K.A.; Grjibovski, A.M. Multiple Comparisons in Biomedical Research: The Problem and its Solutions. *Ekol. Cheloveka* **2020**, *10*, 55–64. [CrossRef]

26. Kendall, M.G.; Stuart, A. *The Advanced Theory of Statistics: Inference and Relationship*, 2nd ed.; C. Griffin: London, UK, 1961; Volume 2, p. 312, paragraph 26.35.

27. Wasserstein, R.L.; Schirm, A.L.; Lazar, N.A. Moving to a world beyond "p < 0.05". *Am. Stat.* **2019**, *73* (Suppl. S1), 1–19.

28. Flatt, T. Life-History Evolution and the Genetics of Fitness Components in *Drosophila melanogaster*. *Genetics* **2020**, *214*, 3–48. [CrossRef]

29. Klepsatel, P.; Gáliková, M.; De Maio, N.; Ricci, S.; Schlötterer, C.; Flatt, T. Reproductive and post-reproductive life history of wild-caught *Drosophila melanogaster* under laboratory conditions. *J. Evol. Biol.* **2013**, *26*, 1508–1520. [CrossRef]

30. Durham, M.F.; Magwire, M.M.; Stone, E.A.; Leips, J. Genome-wide analysis in *Drosophila* reveals age-specific effects of SNPs on fitness traits. *Nat. Commun.* **2014**, *5*, 4338. [CrossRef] [PubMed]

31. Fabian, D.K.; Lack, J.B.; Mathur, V.; Schlötterer, C.; Schmidt, P.S.; Pool, J.E.; Flatt, T. Spatially varying selection shapes life history clines among populations of *Drosophila melanogaster* from sub-Saharan Africa. *J. Evol. Biol.* **2015**, *28*, 826–840. [CrossRef]

32. Hoffmann, A.A.; Harshman, L.G. Desiccation and starvation resistance in *Drosophila*: Patterns of variation at the species, population and intrapopulation levels. *Heredity* **1999**, *83* (Pt 6), 637–643. [CrossRef]

33. Rion, S.; Kawecki, T.J. Evolutionary biology of starvation resistance: What we have learned from *Drosophila*. *J. Evol. Biol.* **2007**, *20*, 1655–1664. [CrossRef] [PubMed]

34. Karpova, E.K.; Bobrovskikh, M.A.; Deryuzhenko, M.A.; Shishkina, O.D.; Gruntenko, N.E. *Wolbachia* Effect on *Drosophila melanogaster* Lipid and Carbohydrate Metabolism. *Insects* **2023**, *14*, 357. [CrossRef] [PubMed]

35. Min, K.T.; Benzer, S. *Wolbachia*, normally a symbiont of *Drosophila*, can be virulent, causing degeneration and early death. *Proc. Natl. Acad. Sci. USA* **1997**, *94*, 10792–10796. [CrossRef] [PubMed]

36. Detcharoen, M.; Schilling, M.P.; Arthofer, W.; Schlick-Steiner, B.C.; Steiner, F.M. Differential gene expression in *Drosophila melanogaster* and *D. nigrosparsa* infected with the same *Wolbachia* strain. *Sci. Rep.* **2021**, *11*, 11336. [CrossRef] [PubMed]
37. Lindsey, A.R.I.; Bhattacharya, T.; Hardy, R.W.; Newton, I.L.G. Wolbachia and Virus Alter the Host Transcriptome at the Interface of Nucleotide Metabolism Pathways. *mBio* **2021**, *12*, e03472-20. [CrossRef]
38. Mackay, T.F.C. Epistasis and quantitative traits: Using model organisms to study gene-gene interactions. *Nat. Rev. Genet.* **2014**, *15*, 22–33. [CrossRef]
39. Welbergen, P.; Sokolowski, M.B. Development time and pupation behavior in the *Drosophila melanogaster* subgroup (Diptera: Drosophilidae). *J. Insect. Behav.* **1994**, *7*, 263–277. [CrossRef]
40. Flatt, T.; Kawecki, T.J. Pleiotropic effects of *methoprene-tolerant* (*Met*), a gene involved in juvenile hormone metabolism, on life history traits in *Drosophila melanogaster*. *Genetica* **2004**, *122*, 141–160. [CrossRef]
41. Strunov, A.; Lerch, S.; Blanckenhorn, W.U.; Miller, W.J.; Kapun, M. Complex effects of environment and *Wolbachia* infections on the life history of *Drosophila melanogaster* hosts. *J. Evol. Biol.* **2022**, *35*, 788–802. [CrossRef]
42. Bubnell, J.E.; Fernandez-Begne, P.; Ulbing, C.K.S.; Aquadro, C.F. Diverse wMel variants of *Wolbachia pipientis* differentially rescue fertility and cytological defects of the bag of marbles partial loss of function mutation in *Drosophila melanogaster*. *G3* **2021**, *11*, jkab312. [CrossRef] [PubMed]
43. Chrostek, E.; Marialva, M.S.; Esteves, S.S.; Weinert, L.A.; Martinez, J.; Jiggins, F.M.; Teixeira, L. *Wolbachia* variants induce differential protection to viruses in *Drosophila melanogaster*: A phenotypic and phylogenomic analysis. *PLoS Genet.* **2013**, *9*, e1003896. [CrossRef] [PubMed]

Article

The Roles of Mating, Age, and Diet in Starvation Resistance in *Bactrocera oleae* (Olive Fruit Fly)

Evangelia I. Balampekou, Dimitrios S. Koveos, Apostolos Kapranas, Georgios C. Menexes and Nikos A. Kouloussis *

School of Agriculture, Aristotle University of Thessaloniki, 54124 Thessaloniki, Greece; evibal@agro.auth.gr (E.I.B.); koveos@agro.auth.gr (D.S.K.); akapranas@agro.auth.gr (A.K.); gmenexes@agro.auth.gr (G.C.M.)
* Correspondence: nikoul@agro.auth.gr

Simple Summary: The olive fruit fly (*Bactrocera oleae* (Rossi) (Diptera: Tephritidae)) is a pest of major economic importance that threatens the olive industry. Studying several factors affecting the survival ability of this insect during food deprivation, such as its mating status, age, and diet, may provide important insights into the biology of *B. oleae* that are useful for its effective control. The starvation resistance (hours of survival after the removal of food) of adult olive fruit flies was measured in four age classes in virgin and mated adults fed a full diet (water/sugar/yeast hydrolysate as protein in a 5:4:1 ratio) or a restricted, sugar-based diet lacking in protein, examining both males and females. The pattern of starvation resistance was the same for both genders under the same conditions (mating status, age, and diet) in the laboratory. Specifically, (a) mated adults showed much less resistance to starvation compared to virgin adults; (b) younger adults endured longer starvation periods compared to older adults; and (c) adults fed the restricted diet endured longer starvation periods than those fed the full diet. We conclude that mating, a full diet, and aging reduce starvation resistance.

Abstract: The olive fruit fly (*Bactrocera oleae* (Rossi) (Diptera: Tephritidae)), although a pest of major economic importance for the olive industry, has not been sufficiently studied with respect to the factors affecting its survival resistance to food deprivation. In the present study, we examined the effect of the interaction between mating status (virgin/mated), age class (11–20/21–30/31–40/41–50), and diet quality (protein plus sugar or only sugar) on starvation resistance in *B. oleae* under constant laboratory conditions. We conducted a total of 16 treatments ($2 \times 4 \times 2 = 16$) for each gender. Our results showed that starvation resistance in *B. oleae* did not differ significantly between females and males. The main conclusions of our study regarding mating status, age, and diet indicated that mated adults showed much less starvation resistance compared to virgins, younger adults endured longer, and the adults fed a restricted diet endured longer than those fed a full diet. A three-way interaction between mating status, diet, and age class was also identified and was the same for both genders. The interaction between mating status, age class, and diet also had a significant influence on starvation resistance in both sexes.

Keywords: stress; aging; food type; lifespan; pest management; sustainability; *Bactrocera oleae*; olive fruit fly

Citation: Balampekou, E.I.; Koveos, D.S.; Kapranas, A.; Menexes, G.C.; Kouloussis, N.A. The Roles of Mating, Age, and Diet in Starvation Resistance in *Bactrocera oleae* (Olive Fruit Fly). *Insects* **2023**, *14*, 841. https://doi.org/10.3390/insects14110841

Academic Editors: Sérgio M. Ovruski and Flávio Roberto Mello Garcia

Received: 8 September 2023
Revised: 12 October 2023
Accepted: 18 October 2023
Published: 29 October 2023

1. Introduction

Bactrocera oleae (Rossi) (Diptera: Tephritidae) (olive fruit fly) is a pest of major importance in the olive fruit production industry, as it causes up to 30% of the yield damage to olive crops worldwide (along with fungi and weeds) [1,2]. Female *B. oleae* lay their eggs inside the olive fruit [3], reducing the quality of both the olive fruit and the olive oil produced [4]. Controlling the olive fruit fly is difficult because the larvae feed inside the olive fruit, in a protected environment; therefore, pest control can be effective only before oviposition takes place. Given the potential losses of olive crops due to *B. oleae*, olive farmers in the past constantly used conventional chemical insecticides to reduce yield losses [5]. For many years, the use of insecticides was the only available approach to suppressing the population of insects [1,2]. However, several concerns were soon raised concerning an increase in the insects' resilience to chemical substances, as well as the health of humans or other mammals due to the presence of pesticide residues, which were very often detected in olive oil [6]. In addition, it has been reported that the use of insecticides has a negative effect on the natural enemies of *B. oleae*, such as *Chrisopids* (Neuroptera: Chrysopidae) [7] and *Psyttalia concolor* (Szépligeti) (Hymenoptera: Braconidae) [8]. To address these issues related to the use of chemicals, most olive-growing countries have adopted the concept of integrated pest management (IPM) as a sustainable strategy for olive crop protection to reduce the amounts of chemicals used to control pests in olive groves, and they constantly seek other methods consistent with economic, ecological, and toxicological requirements to maintain pests below the economic threshold while giving priority to natural limiting factors [9].

Lately, there has been an increased interest in exploring more sustainable control methods, such as the sterile insect technique (SIT) [10,11]. The efficacy of this method relies on producing mass-reared male insects, which, when released into the wild in large numbers, are more competitive than wild males and are also capable of greater endurance under stressful environmental conditions [10,11], such as food deprivation. Starvation resistance (SR) is considered an important trait in pest management [12–16]; starvation resistance and thermal stress [12] have been identified as two of the most common environmental conditions that insects may face in their lifetimes [17]. Key factors that affect the starvation resistance in insects are their mating status, age, and diet [15,16,18–20]. For males, mating or even only courtship can lead to significant energy expenditure and consequently shorten the insect's lifespan [21–24]. Females mainly face energy losses due to egg maturation [18]. An insect's mating status is also influenced by environmental conditions and strain [25–27]. So far, the study of aging in insects has shown that, regarding their starvation resistance, their survival ability decreases with age [14,16,28]. Starvation resistance in adult insects fed different diets affects their fitness [29]. Survival and sexual signaling were shown to be crucially influenced by diet quality early in a male's life [24]. The evaluation of insect survival under varying conditions of food availability, diet, and quality provides insights into the factors that drive the evolution of different feeding strategies and helps us better understand the biology and ecology of insects [14,29]. The effects of food deprivation have been studied very little in model organisms, such as *Drosophila melanogaster* (Meigen) (Diptera: Drosophilidae) [30,31], and insects of agricultural importance, like *Ceratitis capitata* (Wiedemann) (Diptera: Tephritidae) [16]. A recent study on *D. melanogaster* showed that feeding adults a protein-based diet for twenty generations led to lower body weights and wing reductions in male adults [32], and the trade-off between reproduction and lifespan is still an area under investigation [33]. There are several studies on topics related to *B. oleae* regarding the physiology of the insect (e.g., the determination of volatile substances in olives and their effect on reproduction [34], mating competition between wild and artificially reared adults [35] and altered activity and rest patterns [36]), but, to our knowledge, starvation resistance in *B. oleae* has not yet been studied. Additionally, the only existing work in the literature addressing starvation resistance in Tephritidae that examines how aging and diet affect starvation resistance was conducted on virgin *C. capitata* [16], which is a polyphagous, cosmopolitan insect. There are some studies on the topic of starvation resistance in Tephritidae, evaluating the effect of diet on several physiological aspects of the adult olive fruit fly, such as lipid reserves, the onset of oviposition, lifetime egg production and the longevity of adults [37], field survival [38], male sexual performance in relation to insect vulnerability to starvation [38,39], or starvation resistance in different time intervals [40], but none of them focus on the interaction between different parameters and how they influence adults' resistance to starvation. Furthermore, useful insights on starvation resistance in insects and the importance of factors affecting their adaptation to periods of food scarcity have already been identified in studies on *D. melanogaster* [41].

The olive fruit fly is an oligophagous insect species relying on olives for its survival and reproduction [3]. Thus, the survival of this species depends solely on the olive fruit. Additionally, the olive tree commonly produces a much greater-than-average crop in one year and a much lower-than-average crop in the following year in olive cultivation [42]. Therefore, these features, in conjunction with starvation resistance in *B. oleae*, as a result, shape the population dynamics of the olive fly [43]. During years characterized by low olive yields, it becomes crucial for the olive fruit fly to live for a longer period and, therefore, to maximize its reproductive potential by producing a greater number of offspring [42,43]. In the above context, the effects of mating status, age, and diet on starvation resistance in *B. oleae* were studied for both males and females. The aims of this study were to identify which mating status (virgin, mated), age class in days (11–20, 21–30, 31–40, 41–50), and diet (full, restricted) make *B. oleae* more vulnerable, namely, more susceptible to several stresses, and to study the effect of the interaction between these three factors on starvation resistance in this pest, namely, the combination of values of the three factors that resulted in increased insect vulnerability due to starvation. We hope that studying the key factors affecting *B. oleae* under food deprivation will provide insights that will help improve existing pest control management or even formulate new, more sustainable and effective strategies.

2. Materials and Methods

The experiment was conducted in three main stages (Figure 1): (1) collecting non-infested olives for oviposition and infested olives to obtain adults for the experiments and maintain the colony; (2) preparing experimental insects that would be used in the experiments; (3) recording the deaths every 4 h daily from the 11th day up to the 50th day after the experimental insects were subjected to starvation (80 insects daily, with 3200 insects in total).

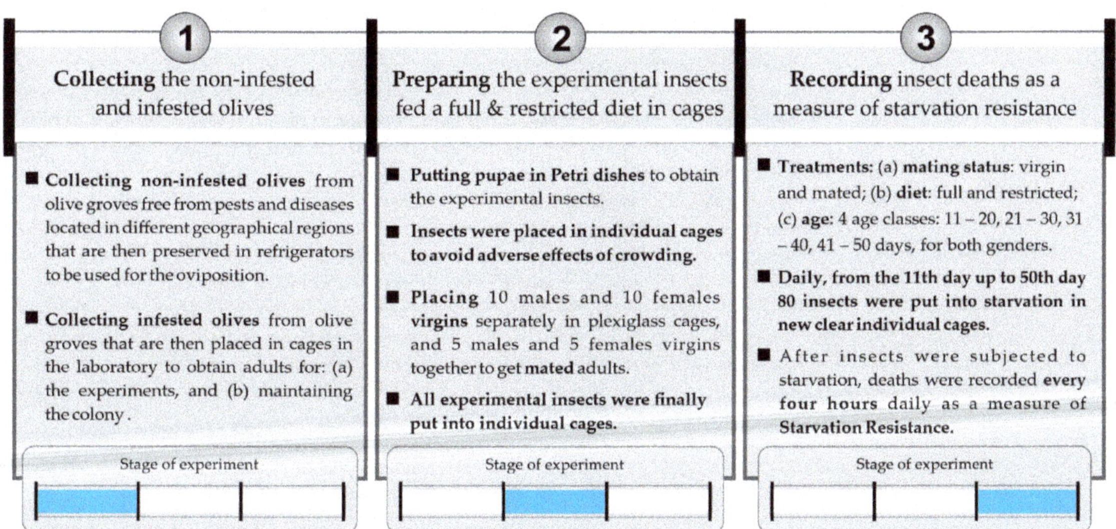

Figure 1. Stages of the experiment: (1) collecting non-infested and infested olives; (2) preparing experimental insects fed full (protein) and restricted (sugar) diets; (3) recording insect deaths as a measure of starvation resistance.

2.1. Origin and Handling of Experimental Insects and Olive Fruit

Adult survival following food deprivation, which is an index of starvation resistance, was assessed in stable laboratory conditions (temperature: 25 ± 2 °C; relative humidity: 65 ± 5%; and photoperiod: L14:D10). Larvae were reared in (non-infested) olive fruit collected in the wild from olive trees free of pests and diseases. The original total number of insects involved was approximately 20,000 adults, while the final number of insects participating in the study (due to the insects' mortality) was 3200 flies. In all cases, adults were carefully transferred using an aspirator, with particular attention paid to not disturbing the insects. The aspirator was used as a transportation means, allowing insects to enter and exit (from the same point), almost walking, without any force being applied to them. In the case that any insect suffered any disturbance or injury during transportation, it was removed and replaced.

The stages for acquiring and handling the olive fruit were the following:

(a) *Insect cage types used in the experiments:* The wild adult olive fly lines were housed in custom-made insect cages based on the model of BugDorm cages (Model DP1000B) usually used in entomological experiments [44]. The types of insect cages used in the laboratory for the needs of the experiments were (A) BugDorm-type cages, 30 × 30 × 30 cm^3 (colony cages for rearing of the olive fruit fly) (Figure 2a); (B) transparent plexiglass cages with dimensions 20 × 20 × 20 cm^3 (Figure 2b), where (i) pupae gathered from infested olives in basins were placed into Petri dishes and then were transferred into the plexiglass cages for the adults to emerge (see experimental procedures and protocol for further information), and (ii) flies on the 10th day of their lives were transferred to mate or to be together with other flies of the same sex; (C) individual plastic cages (Figure 2c), where (i) flies were transferred individually upon their emergence, with either full or restricted diet and water, and (ii) flies were transferred at the appropriate age, each time to a new individual plastic cage that was thoroughly clean of any trace of food to measure the hours until death (starvation resistance).

(b) *Diet food types for adult fruit flies: Two different diets were used:* (A) a full diet consisting of a mixture of hydrolyzed yeast (protein) in a ratio of 5:4:1 (water/sugar/yeast hydrolysate as protein) or (B) a restricted diet containing only sugar but deprived of protein. Water was supplied to all cohorts through a wetted cotton wick.

(c) *Harvesting (non-infested) olive fruit for the rearing of olive fruit flies:* Olive fruits used in the experiments were collected from olive groves located in the region of Chalkidiki and Northern Greece. The olives were selected one-by-one by hand in the above regions from trees that were as free from pests and diseases as possible. To maintain the number of insects needed for the experiment and the genetic diversity of the experimental insect population close to that of natural populations, infested olives were constantly collected from olive groves for a period of approximately three months, and wild insects were constantly introduced to the colony. The total amount of olives needed for the experiment was roughly 200 kg (Figure 3).

(d) *Maintaining the (non-infested) olive fruit flies:* Immediately after harvesting, the olive fruits were placed in glass jars in the refrigerator at 6 ± 1 °C (Figure 3).

(e) *Collecting infested olive fruits:* Mc Cain traps with an appropriate food attractant were used in the aforementioned regions to identify the period of the first adult flights and the onset of infestation in the field. Olive fruits that had been infested by the olive flies were collected from the trees and transported to the laboratory (Figure 3).

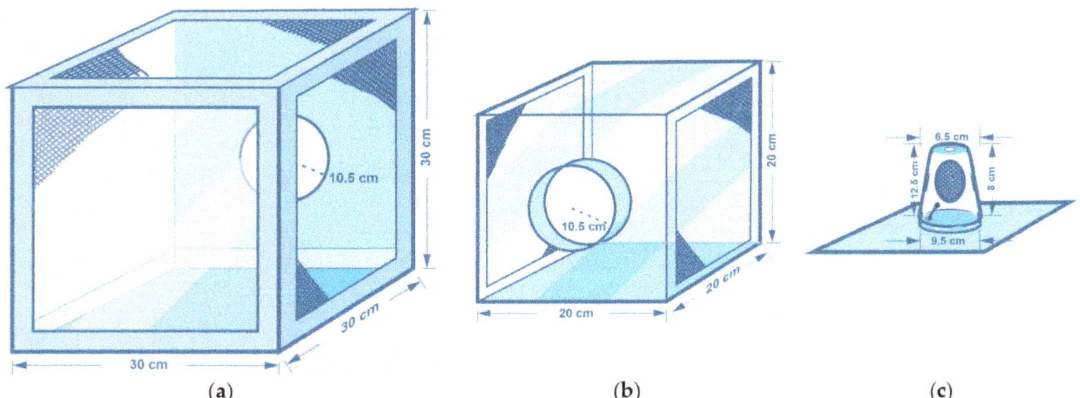

Figure 2. Blueprints of the experimental cages used for fly rearing under constant laboratory conditions: (**a**) BugDorm-type (custom-made) cages for rearing; (**b**) transparent plexiglass cages to which flies were transferred (on the 10th day) to mate or to be together with others of the same sex; (**c**) plastic cages (also custom-made from plastic cups) as individual cages.

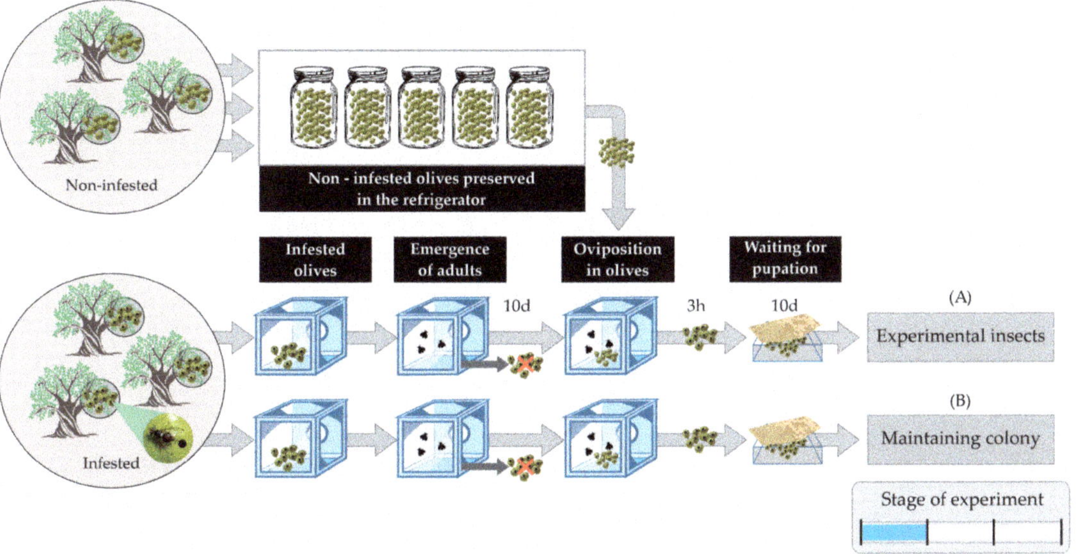

Figure 3. Collecting healthy and infested olives to rear (A) the experimental insects and (B) flies for maintaining the colony.

2.2. Experimental Design

The response variable measured in the experiments was the duration of time for which the insects survived after food deprivation (starvation resistance, measured in hours). Insect deaths were recorded every 4 h daily by the same human observer. The starvation resistance was then calculated based on the date and time of death. The experimental units that were studied individually for starvation resistance were 3200 adult olive fruit flies (10 adults ×40 days ×2 diets ×2 mating status ×2 gender = 3200) that originated from the larvae in the infested olives. The initial number of flies was at least sixfold that of the flies used in the experiments. This was due to the mortality of the flies observed before they were included in the experiment. All adults that took part in each treatment were derived from larvae reared in olive fruit. The experiments on starvation resistance were carried out from the eleventh to the fiftieth day of the insect's life. The treatments examined were the following: (a) mating status factor with two levels (virgin and mated); (b) age factor with four levels (age classes: 11–20, 21–30, 31–40, and 41–50 days); and (c) diet factor with two levels (full and restricted diets). There were 16 combinations of treatments in total ($2 \times 4 \times 2 = 16$) for each gender. There were also 10 replicates, namely, 10 insects of the same gender and same mating status and at the same age (day of life), also fed the same diet, that were randomly selected for each treatment, as described in detail in the experimental procedures and protocol. The wild *B. oleae* is mature for mating after the 7th day [35]; this is why the 10th day was selected for mating. The period from the 11th to the 50th day was selected, because after an age of 50 days, it is difficult to maintain the number of insects needed for each treatment.

The aim of the experimental design was to measure starvation resistance in olive fruit flies under different conditions (mating status, food/diet, and age) to identify the status at which insects are more vulnerable or more durable.

2.3. Experimental Procedures and Protocol

The experimental processes followed can be grouped into the following stages:

(a) *Rearing the parents of the experimental insects:* Adults from infested olive fruits that had been collected from trees hatched inside wooden cages with plenty of water and protein food. After the completion of their hatching, the olives of their origin were removed, and fresh olives were added (ones that we had collected and maintained in the refrigerator). After mating, the females laid their eggs in the olive fruit. These infested olives were removed from the cages, laid into basins, and covered with a suitable cloth to ensure the appropriate humidity and temperature conditions (Figure 3). After pupation and before the emergence of the adults, pupae were transferred to plexiglass cages with dimensions 20×20 cm, awaiting the appearance of the adults (Figure 2b). In these plexiglass cages, there was either a full or restricted diet and water (Figure 4).

(b) *Handling the experimental insects before the experiment:* Upon emergence, adults were placed in individual plastic cages (Figure 2c) with water and food (either the full or restricted diet). At the age of 10 days, groups of 10 adults of either only females and males (both virgin) or 5 virgin males and 5 virgin females (mated) were allowed to be together in larger cages (20×20 cm) for one day (Figure 2b) before being placed back into individual cages (Figure 2c). After this period, the flies were placed back into individual cages (Figure 2c) to eliminate crowding and social interactions (Figure 4). Flies that had been kept with conspecifics of the opposite sex were monitored by a human observer to verify mating. We observed the flies for mating from 16:00 to 21:00 because, in this species, mate searching and courtship take place during the late evening [34,36]. Individuals that had not mated were removed from the experiment and were replaced with others that had mated.

(c) *Preparing the flies to undergo starvation (food deprivation):* The steps followed (Figure 4): (1) initially, experimental insects (pupae) were placed in Petri dishes in plexiglass cages; (2) upon adult emergence, they were transferred individually to plastic cages with water, half with the full diet and half with the restricted diet; (3) on the 10th day, all insects were transferred to 8 plexiglass cages: (a) 4 cages with the full diet (40 adults in total: 1 cage with 10 males, 1 with 10 females, and 2 cages with 5 males and 5 females in each cage) and (b) 4 cages with the restricted diet (40 adults in total: 1 cage with 10 males, 1 with 10 females, and 2 cages with 5 males and 5 females in each cage); (4) at the end of the 10th day, insects were transferred back to their individual cages with water and with the same diet that they were fed in the plexiglass cage.

(d) *Recording of deaths—Calculating starvation resistance:* Upon reaching the eleventh day of their adult life, ten individual adults from each treatment at a specific age (11th, 12th, up to 50th day of life) were each transferred to a new individual plastic cage (Figure 2c) thoroughly clean of any trace of food [16]. The insects' deaths were recorded every four hours due to food deprivation during the light period (four times per day: 08:00, 12:00, 16:00, 20:00). In Figure 5, a schematic representation showing the feeding and starvation stages is given. From the 11th day up to the 50th day of their lifespan, 80 insects (3200 in total) fed the full or restricted diet (40 adults fed the full diet and 40 adults the restricted diet, in each case: 10 virgin males, 10 virgin females, 10 mated males, and 10 mated females) were subjected to starvation in new clear individual cages. Within the period from the 11th to the 50th day, every four hours, the deaths were recorded as a measure of starvation resistance. In case there was difficulty in identifying an insect's death, a fine paintbrush was used to gently move the insect and confirm its death. Rotation of the plastic cages was performed daily to reduce potential experimental errors. Starvation resistance was finally calculated as the difference between the date and time of death and the date and time of the moment the insects were subjected to food deprivation.

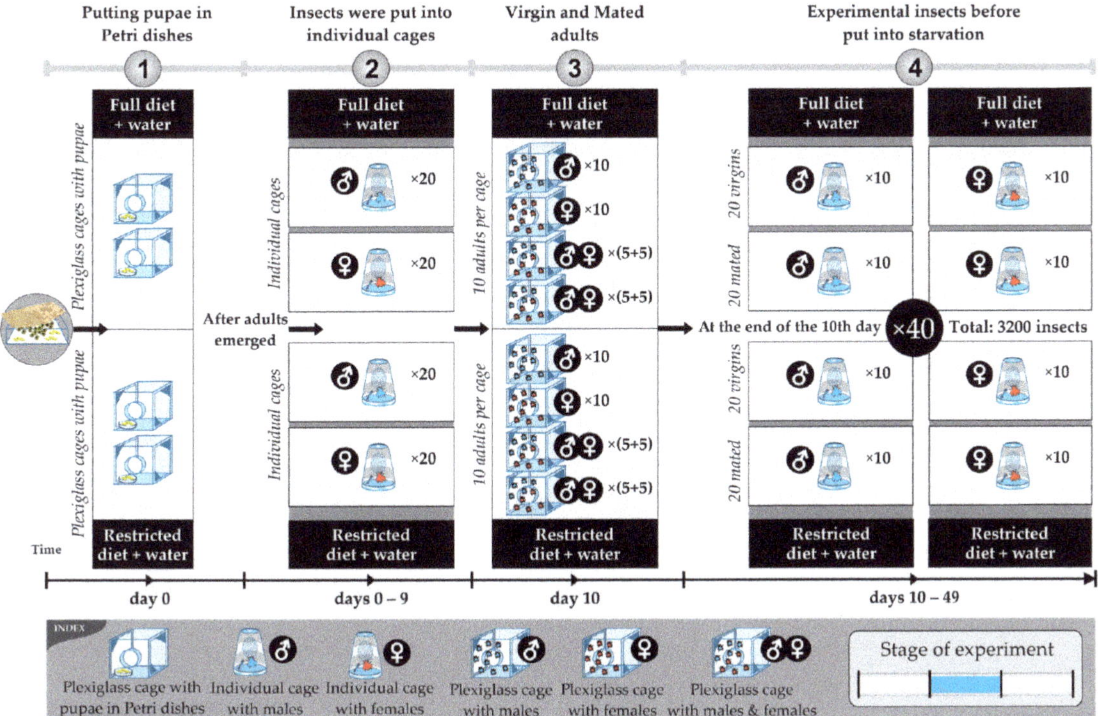

Figure 4. Schematic representation of the processes for the preparation of the experimental insects before they underwent starvation: (1) pupae were put in Petri dishes in plexiglass cages; (2) after adults emerged, insects were put into individual plastic cages to avoid crowding; (3) on the 10th day, all insects were transferred to 8 plexiglass cages: 4 cages with full diet (40 adults in total: 1 cage with 10 males, 1 with 10 females, and 2 cages with 5 males and 5 females in each cage); (4) at the end of the 10th day, insects were transferred back to their individual cages.

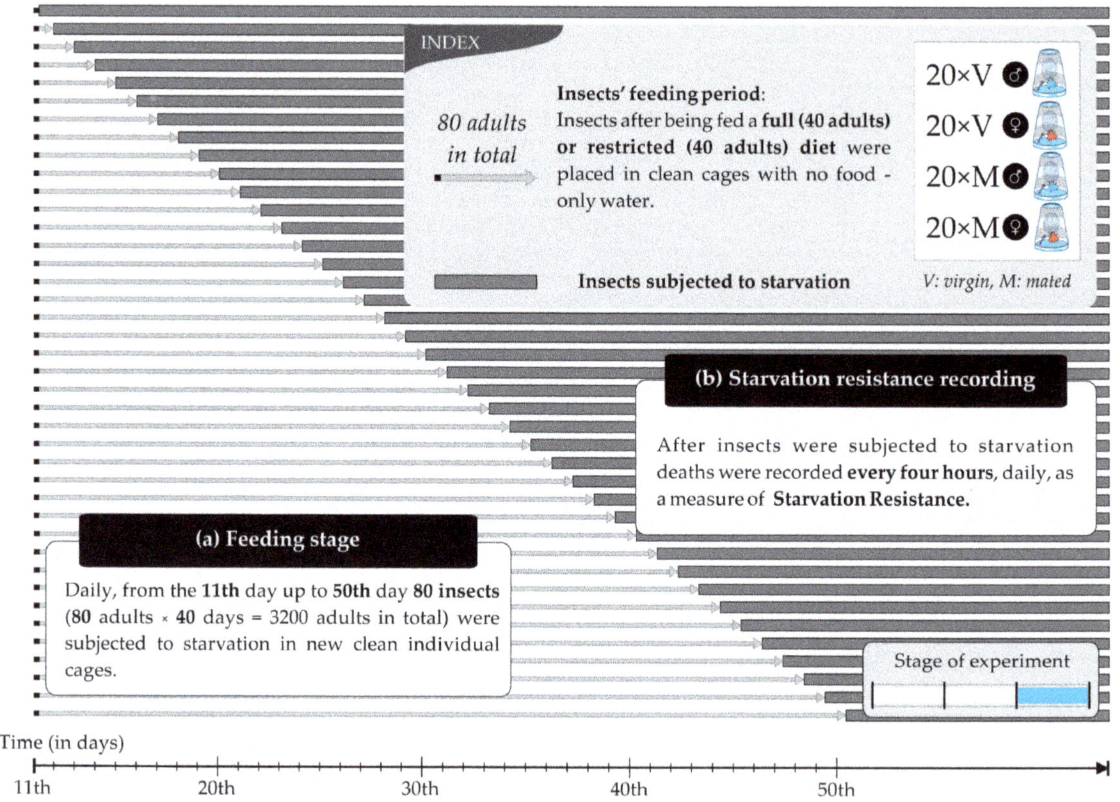

Figure 5. Schematic representation of the experimental design: (a) feeding stage: from the 11th day up to the 50th day of their lifespan, 80 insects (3200 in total) fed with the full or restricted diet were subjected to starvation in new clear individual cages; (b) starvation resistance recording: within the period of the 11th to 50th day, deaths were recorded every 4 h.

We adopted an age clustering of the results (starvation resistance—time in hours to death) on a ten-day basis, as was presented in similar work on Tephritidae [16], to simplify the results' interpretation. The experiments conducted were analyzed for each gender individually, because the inherent differences in the effects of mating and fecundity on aging and longevity are usually studied separately for males and females due to their physiology [45]. Additionally, the same pattern of starvation resistance has been identified for both genders.

2.4. Statistical Analyses

The survival resistance of adult insects for each gender was analyzed with the ANOVA method within the methodological framework of General Linear Models. The ANOVA model included the effects (three main effects, three two-way interactions, and one three-way interaction) of three factors: the mating factor with two levels (virgin and mated), the diet factor with two levels (protein-rich food and sugar-only food without protein), and the age factor with four levels (age classes: 11–20, 21–30, 31–40, and 41–50 days) [16]. There were 16 combinations in total of the three factors' levels ($2 \times 4 \times 2 = 16$). The ANOVA method was mainly used to estimate the correct standard errors of the differences between the mean values of the factor levels' combinations. Tukey's multiple-comparisons procedure [46] was used to test the significance of the differences between the compared mean values. Linear models' residuals were tested for normality and homoscedasticity. The residuals' normality assumption was examined by visually inspecting the corresponding histogram and boxplot, comparing the residuals' median values with the value of 0 (zero), assessing the corresponding skewness and kurtosis indices, and analyzing the results of the Kolmogorov–Smirnov test for normality. The homoscedasticity assumption was examined by visually inspecting the residuals' scatter plot against the model's predicted values and assessing the magnitude of Spearman's rho rank correlation coefficients between the residuals' absolute values and the model's predicted values. No serious violations of these two assumptions were detected. Data are presented as mean $\pm$ standard error (SE). Additional descriptive statistical indices are presented in Supplementary Tables S1–S6. Since only the terminal survival time of the insects was recorded, there was no specific need to examine the data using a survival analysis model. All statistical analyses were performed with the IBM SPSS Statistics ver. 26.0 Software (IBM Corp., Armonk, NY, USA). The significance level in all statistical hypothesis testing procedures was preset at $a = 0.05$ ($p \leq 0.05$).

Finally, we calculated the differences (in percent) between the values (days for which insects endured starvation) for each of the 16 treatments (mating status: virgin, mated) $\times$ (age class: 11–20, 21–30, 31–40, 41–50) $\times$ (diet: full, restricted) and the grand mean of all flies in each gender regardless of their mating status, age class, and diet to better represent the influence of each of the three factors on insect starvation resistance based on the following formula (Equation (1)):

$$\left(\overline{X}_T - \overline{X}\right) * 100 / \overline{X} \tag{1}$$

where $\overline{X}_T$ is the mean starvation resistance (in hours) for each treatment, and $\overline{X}$ is the total mean for each gender (in hours).

3. Results

3.1. Study of Starvation Resistance in Males

Tukey post hoc was applied to examine whether statistically significant differences exist between the treatments (Supplementary Table S1). Also, descriptive statistics and the ANOVA results are also available (Supplementary Tables S3 and S5, respectively). There is a significant three-way interaction in males between the mating status, the diet, and the age class, as indicated by ANOVA ($F(3, 15) = 6.5$, $p < 0.001$); see also Supplementary Table S5 for a summary of the full model ANOVA results. Consequently, the focus is mainly on examining this interaction using the simple–simple effects analysis approach. More specifically, the two-way interaction between the four age classes and the two mated statuses was examined within each diet [47]. The comparison between age classes is based on the mean starvation resistance of the insects measured in hours to death (values in parentheses).

In virgin males fed the full diet (protein), resistance to starvation decreased with age (Figure 6A, Supplementary Tables S1 and S3). In the 11–20 age class, virgin males fed the full diet lived longer compared to the 31–40 age class (in the 11–20 class: 72.0 ± 3.6 h; in the 31–40 class: 48.5 ± 3.2 h; $p = 0.003$) and also compared to the 41–50 age class (in the 41–50 class: 39.6 ± 1.4 h, $p < 0.001$). Insects in the 21–30 age class lived longer than those in the 41–50 age class (in the 21–30 class: 63.3 ± 5.7 h; in the 41–50 class: 39.6 ± 1.4 h; $p = 0.002$). Therefore, in virgin males fed the full diet (protein), the ability to resist stress (starvation resistance) tended to decline as they aged. On the contrary, in mated males fed the full diet, there was no decline in starvation resistance as they aged. In mated males fed the restricted diet (sugar), resistance to starvation decreased with age, following a similar pattern to the one identified for virgin males. Insects in the 11–20 age class lived longer than those in the 31–40 age class (in the 11–20 class: 65.8 ± 4.6 h; in the 31–40 class: 33.8 ± 3.1 h; $p < 0.001$). In virgin males fed the restricted diet (sugar), no decline in resistance was observed as they aged (Figure 6B).

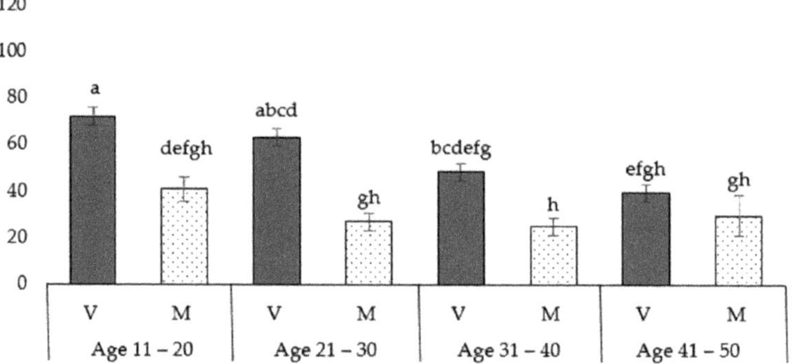

(**A**) Mean starvation resistance for males fed the full diet.

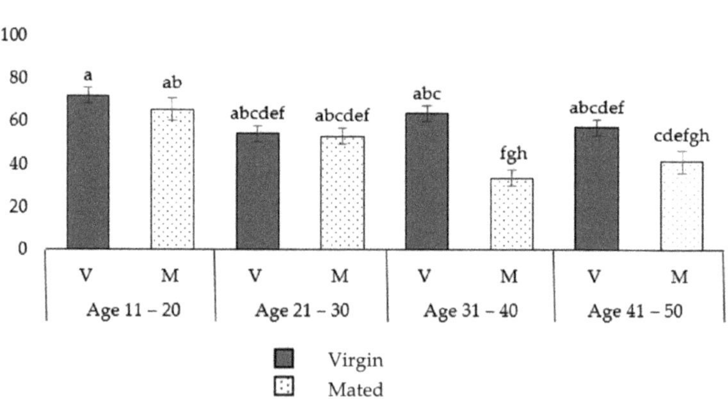

(**B**) Mean starvation resistance for males fed the restricted diet.

Figure 6. Starvation resistance (in hours to death) for males: (**A**) fed the full diet (protein); (**B**) fed the restricted diet (sugar). For both figures (**A**,**B**), bars with different lower-case letters above them correspond to differences in mean values that are statistically significant, at a significance level of $a = 0.05$, according to the results of Tukey's test. Error bars correspond to standard errors of the mean. All 16 mean values (16 treatments: 2 diets $\times$ 4 age classes $\times$ 2 mated statuses) are comparable across the two figure sections.

In the following age classes, virgin males fed the full diet were more resistant than mated ones. Specifically, in the 11–20 age class, virgin males fed the full diet were more resistant to starvation than mated ones (in the 11–20 class, virgins: 72.0 ± 3.6 h; mated: 41.0 ± 5.5 h; $p = 0.001$), and in the 21–30 age class, virgin males fed the full diet were more resistant to starvation than mated ones (in 21–30 class, virgins: 63.3 ± 5.7 h; mated: 27.2 ± 1.8 h; $p < 0.001$). In the 31–40 age class, virgin males fed the full diet were more resistant to starvation than mated ones (in the 31–40 class, virgin males: 48.5 ± 3.2 h; mated males: 25.1 ± 2.5 h; $p = 0.003$) (Figure 6A). In the 31–40 age class, virgin males fed the restricted diet were more resistant than mated ones (virgins: 64.0 ± 6.7 h; mated: 33.8 ± 3.1 h; $p < 0.001$). No differences were observed between virgin and mated males fed the restricted diet in the other three age classes (Figure 6B).

In the following age classes, mated males that were fed the full diet were less resistant than those fed the restricted diet. Specifically, in the 11–20 age class, mated males that were fed the full diet were less resistant to starvation than mated adults fed the restricted diet (males on full diet: 41.0 ± 5.5 h; males fed the restricted diet: 65.8 ± 4.6 h; $p = 0.05$), and in the 21–30 age class, mated males fed the full diet were more resistant to starvation than mated ones that were fed the restricted diet (mated males fed the full diet: 27.2 ± 1.8 h; mated males fed the restricted diet: 53.3 ± 4.8 h; $p < 0.001$). On the contrary, no significant differences in starvation resistance were observed between virgin males that received a full diet and those that followed a restricted diet in any age group (Figure 6A,B).

3.2. Study of Starvation Resistance in Females

Tukey post hoc was applied to examine whether statistically significant differences exist between the treatments (Supplementary Table S2). Also, descriptive statistics and the ANOVA results are also available (Supplementary Tables S4 and S6, respectively). There is a significant three-way interaction in females between mating status, diet, and age class, as determined by ANOVA ($F(3, 15) = 3.6$, $p = 0.016$); see also Supplementary Table S6 for a summary of the full model ANOVA results. Consequently, the focus is mainly on examining this interaction using the simple–simple main effects analysis approach. More specifically, the two-way interaction between the four age classes and the two mating statuses was examined within each diet [47]. The comparison between age classes is based on the mean starvation resistance of the insects, measured in hours to death (values in parentheses).

In both virgin and mated females fed the full diet, the resistance to starvation decreased with age (Figure 7A, Supplementary Tables S2 and S4). In both cases, females in the 11–20 age class lived longer compared to the 41–50 age class. Specifically, in the 11–20 age class, virgin females fed the full diet lived longer compared to full-diet-fed virgins in the 41–50 age class (in the 11–20 class: 74.4 ± 2.4 h; in the 41–50 class: 46.5 ± 1.5; $p = 0.001$). In the 11–20 age class, mated females fed the full diet (mean starvation resistance in hours to death: 57.4 ± 5.0 h) lived longer compared to their counterparts in the 41–50 age class (mean starvation resistance in hours to death: 32.0 ± 4.6 h, $p = 0.05$). Therefore, in both virgin and mated females that were fed the full diet (protein), their ability to resist stress (starvation resistance) tended to decline with age.

Furthermore, in both virgin and mated females fed the restricted diet, the resistance to starvation decreased with age. Specifically, virgin females in the 11–20 age class fed the restricted diet exhibited significantly higher resistance to stress compared to those in the 31–40 age class (virgin females in the 11–20 class: 107.8 ± 3.7 h; in the 31–40 class: 69.0 ± 4.8 h; $p < 0.001$). In the 11–20 age class, mated females fed the restricted diet exhibited significantly higher resistance to stress compared to those in the 31–40 age class (mated females in the 11–20 class: 78.9 ± 7.8 h; in the 31–40 class: 42.3 ± 2.7 h; $p < 0.001$) and also to those in the 41–50 age class (in the 41–50 class: 33.4 ± 3.6 h, $p < 0.001$). Similarly, in the 21–30 age class, mated females fed the restricted diet lived longer compared to those in the 31–40 age class (mated females in the 21–30 class: 70.9 ± 6.9 h; in the 31–40 age class: 42.3 ± 2.7 h; $p = 0.001$) and also to those in the 41–50 age class (in the 41–50 class: 33.4 ± 3.6 h, $p < 0.001$) (Figure 7B, Supplementary Tables S2 and S4).

In all four age classes, no significant differences in starvation resistance were observed between virgin and mated females that were fed the full diet (Figure 7A). In some age classes, virgin females fed the restricted diet were more resistant than mated ones. Specifically, in the 11–20 age class, virgin females fed the restricted diet were more resistant than mated females (virgins: 107.8 ± 3.7 h; mated: 78.9 ± 7.8 h; $p = 0.012$). In the 31–40 age class, virgin females fed the restricted diet were more resistant to starvation than mated females (virgins: 69.0 ± 4.8 h; mated: 42.3 ± 2.7 h; $p = 0.002$), and in the 41–50 age class, virgin females fed the restricted diet were also more resistant to starvation than mated ones (virgins: 88.3 ± 7.1 h; mated: 33.4 ± 3.6 h; $p < 0.001$) (Figure 7B).

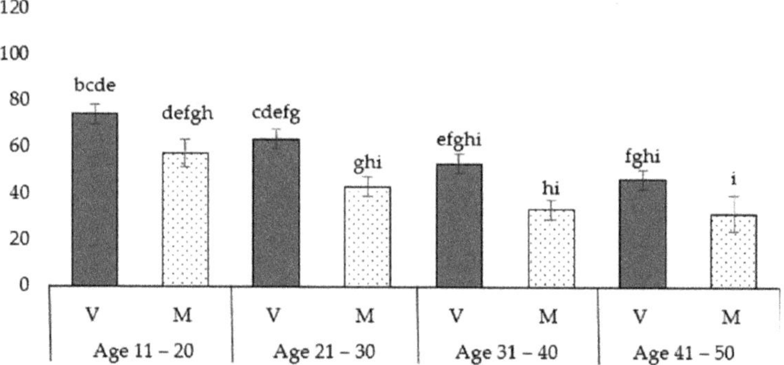

(**A**) Mean starvation resistance for females fed the full diet.

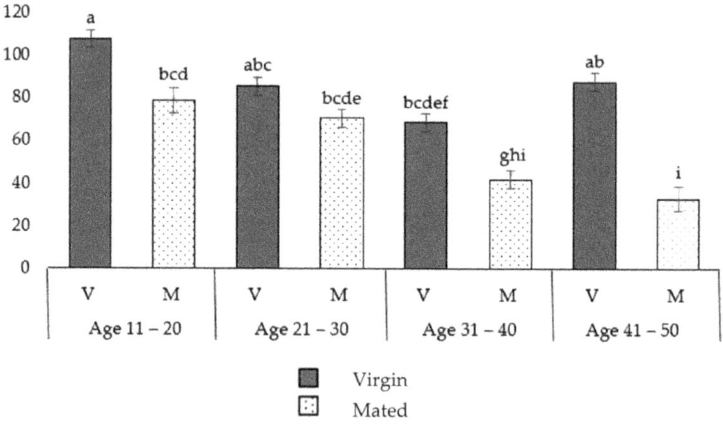

(**B**) Mean starvation resistance for females fed the restricted diet.

Figure 7. Starvation resistance (in hours to death) for females: (**A**) fed the full diet (protein); (**B**) fed the restricted diet (sugar). For both figures (**A**,**B**), bars with different lower-case letters above them correspond to differences in mean values that are statistically significant, at a significance level of $a = 0.05$, according to the results of Tukey's test. Error bars correspond to the standard errors of the mean values. All 16 mean values (16 treatments: 2 diets × 4 age classes × 2 mated statuses) are comparable across the two figure sections.

In the 11–20 age class, virgin females fed the full diet had reduced resistance to starvation compared to those fed sugar (virgins fed the full diet: 74.4 ± 2.4 h; virgins fed the restricted diet—sugar: 107.8 ± 3.7 h; $p < 0.001$). Similarly, in the 41–50 age class, virgin females fed the full diet exhibited reduced resistance to starvation compared to those fed sugar (virgins fed the full diet: 46.5 ± 1.5 h; virgins fed the restricted diet: 88.3 ± 7.1 h; $p = 0.05$). Additionally, in the 21–30 age class, mated females fed the full diet had decreased resistance to starvation compared to their sugar-fed counterparts (mated and fed the full diet: 43.4 ± 2.2 h; mated and fed the restricted diet: 70.9 ± 6.9 h, $p = 0.001$) (Figure 7A,B).

3.3. Percentage Starvation Resistance Differences from the Corresponding Gender Mean

In Figure 8, the starvation resistance of males and females was also calculated as the percent difference from the corresponding mean of each gender (Equation (1)). Based on these results, the trend identified for both genders regarding starvation resistance is almost the same, meaning that gender is not a factor that affects the longevity of the insects under stress conditions. In both genders and for all age classes, the virgin adults treated with a restricted diet (sugar) had a higher starvation resistance (for males: from 10.3 up to 46.5%; for females: 12.6 up to 75.9%) compared to the corresponding (male, female) mean. Also, in younger age classes (11–20 and 21–30), in both genders, virgin adults treated with the full diet (protein) (for males: from 28.4 up to 46.0%; for females: 3.9 up to 21.4%) and mated adults treated with the restricted diet (sugar) (for males: from 8.1 up to 33.5%; for females: 15.7 up to 28.7%) also had higher starvation resistance compared to the corresponding (male, female) mean. Conversely, lower starvation resistance was similar for both genders when compared to their (male, female) mean. In both genders and for all age classes, the mated adults treated with the full diet (protein) showed lower starvation resistance (for males: from −16.8 up to −49.1%; for females: −6.4 up to −47.8%) compared to the corresponding mean of their gender. Also, in older age classes (31–40 and 41–50), in both genders, virgin adults treated with the full diet (protein) (for males: from −1.6 up to −19.7%; for females: −13.1 up to −24.1%) and mated adults treated with the restricted diet (sugar) (for males: from −16.6 up to −31.4%; for females: −31.0 up to −45.5%) also had lower starvation resistance compared to their (male, female) mean.

| | | | Starvation resistance differences (%) from each mean | |
Mating status	Diet	Age Class	Males	Females
Virgin (V)	Full (F)	11 – 20	46.0	21.4
		21 – 30	28.4	3.9
		31 – 40	-1.6	-13.1
		41 – 50	-19.7	-24.1
	Restricted (R)	11 – 20	46.5	75.9
		21 – 30	10.3	40.3
		31 – 40	29.8	12.6
		41 – 50	16.6	44.0
Mated (M)	Full (F)	11 – 20	-16.8	-6.4
		21 – 30	-44.8	-29.2
		31 – 40	-49.1	-45.2
		41 – 50	-39.1	-47.8
	Restricted (R)	11 – 20	33.5	28.7
		21 – 30	8.1	15.7
		31 – 40	-31.4	-31.0
		41 – 50	-16.6	-45.5

Figure 8. Percentage starvation resistance differences for males and females from the corresponding gender mean.

In sum, the results show that (a) resistance to food deprivation decreases with age, (b) in each age class, virgins are more resistant than the corresponding mated flies, (c) in each age class, adults on the restricted diet are more resistant than those fed a full diet. Mean values also indicate that females withstand food deprivation more than males in almost all treatments.

4. Discussion

The results of our study revealed crucial insights into the conditions of age, diet, and mating status, under which *B. oleae* is more susceptible to food deprivation. These findings can lead to the implementation of more effective and environmentally sound pest management strategies that focus on the conditions under which *B. oleae* is more vulnerable. These findings can not only provide critical information for more efficient pest control strategies but also improve the understanding of the physiology of the insect to develop integrated pest management strategies that focus on the weaknesses of *B. oleae*. In this way, new, effective, and sustainable strategies can be developed that are not harmful to beneficial insects, human health, or the environment. This new sustainable approach would be of great interest to olive-growing countries that are constantly seeking the formulation of more effective pest control strategies.

To our knowledge, our work is the first and unique study in the starvation scientific literature that investigates how starvation resistance in the oligophagous insect *B. oleae* is influenced by aging (across each day and for the age range from the 11th to the 50th day). Due to the lack of other similar studies, there is difficulty in the assessment of our findings. The only study that is close to the topic of our work but concerning a different insect from the same family (Tephritidae) is one that was carried out and published by our laboratory on the cosmopolitan and polyphagous species *C. capitata*, in which it was examined how resistance to food deprivation changes across each day and throughout the entire adult lifespan of the insect [16]. Overall, our findings are like those on *C. capitata*; they specifically indicate that starvation resistance declines with age in both genders, and younger adults endure longer. In addition, our results on the effect of diet on starvation resistance in *B. oleae* are also in line with the results from the study for *C. capitata*; adults fed the restricted diet show a higher starvation resistance than those fed the full diet. In our work, olive fruit flies fed the restricted diet show a higher starvation resistance in the first age class (11–20 days), followed by an abrupt decline in the second age class (21–30 days) and, finally, a small increase in the other two older age classes (31–40 and 41–50 days), which, as a result, is in line with the results of the study on *C. capitata* [16]. A comparison with other studies is not considered feasible due to the heterogeneity of the studies (i.e., with studies on *D. melanogaster* or other insects that belong to the same order but to different families). However, due to the lack of similar studies, a comparison will be attempted in the same order-based context to achieve some sort of assessment. Moreover, virgin adults of *D. melanogaster* fed a full diet had—in relative terms—a lower resistance to starvation [48]. Furthermore, *D. melanogaster* adults fed a restricted diet showed the reverse pattern of starvation resistance, with age positively affecting resistance in females and without any effects on males [46].

Experiments with wild *B. oleae* are inherently difficult to conduct, as this insect requires fresh olive fruit to complete its life cycle. As a result, olives must be uninfested and should be harvested in time to effectively rear olive fruit flies. These olives can be preserved in scientific refrigerators but for no longer than 4 months, setting a barrier to the completion of the experiments. Also, it is difficult to obtain large numbers of insects simultaneously for a given period. To maintain the number of insects needed for the experiment, infested olives were constantly collected from olive groves for a period of approximately three months. In addition, specific constant laboratory conditions are required for rearing and for conducting experiments with *B. oleae* adults, such as temperature ($25 \pm 2\,^\circ$C), relative humidity ($65 \pm 5\%$), and photoperiod (L14:D10).

Moreover, it must be stressed that the insects cannot be easily synchronized; this means that to obtain the number of insects for each treatment, at least 2–3 days were needed for younger insects, and at least 20–25 were required for the older ones. In addition, not all insects underwent starvation simultaneously. For each day from the 11th to the 50th day of their lifespan, 80 insects per treatment were subjected to starvation, increasing the complexity of the experiment and the difficulty in the management of the measurements. The pattern of subjecting insects to starvation daily is based on a similar approach adopted from a previous study on Tephritidae [16]. Therefore, conducting experiments with wild *B. oleae* is a challenging process, and this might be the reason why no similar experiments have been conducted on starvation resistance for this species so far.

During the stage of collecting non-infested and infested olives, an effort was made to obtain olives and flies located in the region of Chalkidiki and Northern Greece to have more representative samples. To reduce the subjective bias errors in the measurements, the same person from our scientific team was responsible for recording insect deaths due to starvation. In addition, the rotation of the individual cages (change in their position in the laboratory on a circular basis) was performed daily to reduce potential experimental errors. As a result, all of these efforts to reduce biased errors further increased the complexity of the experiment and the time needed for its completion.

Mating-specific effects on starvation resistance. In our study, virgin adults exhibited greater starvation resistance compared to mated adults in all age classes. The results of our study show that mating can be energetically costly for both males and females, and as a result, virgin adults exhibit higher starvation resistance compared to mated ones. These results are in line with previous studies on insect mating status [24,49,50]. Male *C. capitata* that courted but did not mate have a similar lifespan to those that courted and then mated, meaning that courtship alone is metabolically costly [50]. It was also found that mating decreased the locomotor activity levels of males; these findings provide evidence that both courting and mating are metabolically costly [36]. Males may need to compete for access to mates or invest in courtship displays. Wing vibration associated with courtship [51,52] and the spreading of pheromones [53] result in a shorter lifespan for males [50]. Moreover, energetically expensive traits such as wings with special morphological characteristics that signal male quality and attractiveness to potential mates are costly to produce and maintain [51,52]. In *Bactrocera tryoni* (Froggatt) (Diptera: Tephritidae), mated females seem to be less resistant to starvation than virgins, and that may happen because of mating attempts and due to the mating procedure [54]. These mating procedures cost more energy in mated females than virgins, the latter being more resistant to starvation. A recent study showed that in two different strains of *Anastrepha ludens* (Loew) (Diptera: Tephritidae), sexual maturation reduced survival in both genders [55].

Age-specific effects on starvation resistance. The results of the present study show that younger individuals exhibit higher starvation resistance than those in older age classes. This can be explained by the fact that younger adults may have greater energy reserves from the larval stages than older ones or more efficient metabolic processes that allow them to better endure periods of food deprivation. Gerofotis et al., following a similar methodology to that used in the present study, found that in *C. capitata*, starvation resistance declines with age; age and adult diet were the most significant determinants of starvation resistance, followed by gender [16]. Belyi et al. observed a strong negative correlation between age and starvation resistance in *D. melanogaster* [56]. Experiments on *B. tryoni* showed that resistance to starvation and desiccation in both males and females decreases with age, although there were statistically significant differences in the pattern and extent of this decline [57]. Throughout aging, other biological functions also decrease. In *B. tryoni*, the effect of age on the olfactory response and exploratory activity was found to be important; specifically, the probability of an olfactory response in both genders to test odors declined with age [28].

Diet-specific effects on starvation resistance. Our experiments revealed a clear relationship between diet and starvation resistance across all age classes, and specifically, adults fed a protein-rich diet displayed a lower resistance to starvation compared to adults fed a sugar-rich diet. Dietary restriction has already been studied in many species. Experiments on *A. ludens* showed that a restricted diet not only extended their longevity but also reduced their reproductive output [58]. Similarly, in females of *Anastrepha fraterculus* (Wiedemann) (Diptera: Tephritidae), protein restriction expanded longevity and decreased egg production [59]. Also, in *B. tryoni*, lifespan and egg production were often closely linked to diet and the consumption of micronutrients [13]. In addition, protein consumption accelerates sexual maturation, leading to further energy losses due to mating and, as a result, to a shorter lifespan [21–24]. In the case of sugar consumption, the female produces a smaller number of eggs and, therefore, expends less energy [16,60]. These patterns are in line with our findings and explain the higher starvation resistance of virgin adults when fed a restricted diet. In females, the consumption of a protein diet plays an important role in the maturation of their eggs. Specifically, in *B. tryoni* this laboratory adaptation has been found to significantly enhance fecundity, resulting in a notable 4- to 5-fold rise in the rate at which dietary protein is converted into eggs; mated females seem to be less resistant to starvation than virgin ones on different diets, and that may result from mating attempts and due to the mating procedure [54]. In the case of *C. capitata*, it has been observed that mating enhances egg production in protein-fed females, but this is counterbalanced by reduced survival [61]. In *Anastrepha serpentina* (Wiedemann) (Diptera: Tephritidae), egg production was highest in flies maintained on a protein-rich diet; in flies kept on a restricted diet, egg production was lower [37]. It has been found that in *B. oleae*, mating negatively affects female longevity [62].

In Tephritid flies, protein digestion can be more energetically expensive than sugar digestion, leaving fewer resources for other essential functions, such as maintaining energy reserves [10,11,63–65]. Specifically, in *C. capitata*, nutritional status is a dominant factor known to affect the male's signaling performance and determine the female's decision to accept a male as a sexual partner [63]. Mating and a full diet (protein) shorten longevity in both genders. In *C. capitata*, wild males fed protein had a mating advantage over protein-deprived males [64]. These activities can require significant energy expenditure, leaving fewer resources available for essential functions, such as maintaining energy reserves, thus leading to shorter lifespans [10,11,65]. Research on *D. melanogaster* indicated that a seminal fluid protein in stored sperm, the molecule Acp26Aa, is responsible for an initial elevation in egg laying; females mated to mutant males that lack the molecule Acp26Aa lay fewer eggs than those mated with wild males [66].

A recent study on *Bactrocera dorsalis* (Hendel) (Diptera: Tephritidae) showed that a restricted diet leads to changes in phenotype, antioxidant response, and gene expression and a prolongation of lifespan in this species [67]. It would be interesting to study whether *B. oleae* may experience similar negative effects when fed a restricted diet. *Anastrepha ludens* and *Anastrepha obliqua* (Macquart) (Diptera: Tephritidae) fruit flies exposed to a combination of sugar and fresh mango fruit pulp showed greater longevity and field survival and better mating performance [38]. Experiments on *B. oleae* with a diet based on fresh fruits, their derivatives, or fruit pulps may lead to similarly useful insights. According to our findings, insects fed a full diet exhibited reduced starvation resistance in laboratory conditions. Therefore, in future experiments on *B. oleae*, it is suggested that fruit-based diets be preferred to protein to examine field survival and mating performance while limiting the potential negative effects of protein on starvation resistance that have been identified under laboratory conditions. These results could be applied to sustainable and environmentally friendly control methods for *B. oleae*, such as the sterile insect technique or other biotechnological methods.

Three-factor interactions in starvation resistance. In a study on *C. capitata*, it was found that as the cohort aged, there was a noticeable trend of smaller consecutive lipid crests, and as the cohort reached advanced ages and approached the maximum age, the lipid contents experienced a significant decline [68]. These findings are in line with our results: between the ages of 11 and 20 days, mated males that had consumed a full diet exhibited a lower resistance to starvation. As mated adults reached the age of 21 to 30 days, differences in starvation resistance were observed between those that had consumed full or restricted food. This variation arises from the ongoing maturation of their eggs, as well as the influence of courtship, competition, and mating in both genders. However, based on our results, after reaching 30 days of age, the diet consumed does not appear to affect starvation resistance in mated adults. Conversely, in the age group of 41 to 50 days, there is a discrepancy in starvation resistance between virgin males and females, but no significant difference is observed among mated individuals. This suggests that a restricted diet enhances the resistance of adults in this age range when it comes to starvation resistance.

Further research should focus on the isolation and comprehensive investigation of genes associated with fertility, development, and ovarian function. For instance, a study conducted on *Zeugodacus cucurbitae* (Coquillett) (Diptera: Tephritidae) provided genetic insights into the intricate processes underlying ovarian development and reproductive outcomes affected by nutritional factors [69]. The tangible next steps in *B. oleae* research will first be studies in natural olive grove conditions, where the influence of factors affecting insect starvation resistance may vary. Secondly, studying lipid contents in *B. oleae* could provide additional useful insights into the proximal physiological processes underlying starvation resistance.

Our results have strong practical implications for pest control. The study of adult stress resistance in insects could ultimately lead to (early-life) rearing protocols that would enhance the physiological traits of sterile mass-reared males (e.g., starvation and desiccation resistance). Based on the results of our study, we recommend further research on the use of sugar with fruit derivatives as a diet for mass-reared males (which is also more cost-effective compared to yeast hydrolysate diets) because it could boost starvation resistance and conceivably other fitness traits, ultimately benefiting the efficacy of the sterile insect technique. The same applies to other "genetic" methods where the release of mass-reared insects is required. The findings of the current study can therefore be used to formulate more effective strategies based on better knowledge of the insect's biology. Following up on our findings, additional studies should seek to investigate the physiological mechanisms behind starvation resistance, thus providing essential benefits for the management of species of insects of agricultural and medical importance.

5. Conclusions

The results of the current study reveal insights regarding the key factors affecting starvation resistance in *B. oleae*: mating status, age, and diet. This information is important for improving existing or formulating new, more effective, and environmentally sound pest control strategies.

The main conclusions of the current study regarding the factors affecting starvation resistance are the following:

Mating status: virgin adults exhibit higher starvation resistance compared to mated adults in all age classes.

Age: younger adults exhibit higher starvation resistance in almost all treatments.

Diet: adults that are fed a full diet containing protein show notably lower starvation resistance compared to sugar-fed ones.

Gender: the same pattern of starvation resistance has been identified for both genders.

Regarding the interactions between the above factors, it can be deduced that in both genders, younger virgin adults fed the restricted diet show higher resistance in conditions of food deprivation.

We expect that the findings from our study on the critical factors of age, diet, and mating status that affect starvation resistance in *B. oleae* will provide valuable information on the vulnerability of this insect to food deprivation. Field studies and further research must be conducted to confirm the results of our study on a larger scale.

Supplementary Materials: The following supporting information can be downloaded at https://www.mdpi.com/article/10.3390/insects14110841/s1: Table S1: "Tukey (HSD) results for males. Mean values followed by different numbers are statistically significantly different at significance level $a = 0.05$ ($p \leq 0.05$), according to the results of the Tukey's multiple comparisons procedure"; Table S2: "Tukey (HSD) results for females. Mean values followed by different numbers are statistically significantly different at significance level $a = 0.05$ ($p \leq 0.05$), according to the results of the Tukey's multiple comparisons procedure"; Table S3: "Additional descriptive statistical indices for each treatment combination for starvation resistance for males"; Table S4: "Additional descriptive statistical indices for each treatment combination for starvation resistance for females"; Table S5: "Tests of between-subject effects for males (dependent variable: starvation resistance in hours)";

Table S6: "Tests of between-subject effects for females (dependent variable: starvation resistance in hours)".

Author Contributions: Conceptualization, N.A.K. and E.I.B.; methodology, N.A.K., D.S.K. and E.I.B.; software, E.I.B.; validation, N.A.K., D.S.K., A.K., G.C.M. and E.I.B.; formal analysis, E.I.B. and G.C.M.; investigation, E.I.B. and N.A.K.; data curation, E.I.B., G.C.M. and N.A.K.; writing—original draft preparation, E.I.B.; writing—review and editing, N.A.K., E.I.B., D.S.K., G.C.M. and A.K.; visualization, N.A.K., E.I.B., D.S.K., G.C.M. and A.K.; supervision, N.A.K. All authors have read and agreed to the published version of the manuscript.

Funding: This research received no external funding.

Data Availability Statement: Data will be available upon reasonable request.

Acknowledgments: We would like to thank Eleni Koutsogeorgiou and Eleni Giakoumi, who reviewed the English of the manuscript.

Conflicts of Interest: The authors declare no conflict of interest. The funders had no role in the design of the study; in the collection, analyses, or interpretation of data; in the writing of the manuscript; or in the decision to publish the results.

References

1. Bengochea, P.; Christiaens, O.; Amor, F.; Viñuela, E.; Rougé, P.; Medina, P.; Smagghe, G. Insect Growth Regulators as Potential Insecticides to Control Olive Fruit Fly (*Bactrocera oleae* Rossi): Insect Toxicity Bioassays and Molecular Docking Approach. *Pest. Manag. Sci.* **2013**, *69*, 27–34. [CrossRef]
2. Varikou, K.; Garantonakis, N.; Marketaki, M.; Charalampous, A.; Anagnostopoulos, C.; Bempelou, E. Residual Degradation and Toxicity of Insecticides against *Bactrocera oleae*. *Environ. Sci. Pollut. Res.* **2018**, *25*, 479–489. [CrossRef]
3. Nardi, F.; Carapelli, A.; Dallai, R.; Roderick, G.K.; Frati, F. Population Structure and Colonization History of the Olive Fly, *Bactrocera oleae* (Diptera, Tephritidae). *Mol. Ecol.* **2005**, *14*, 2729–2738. [CrossRef]
4. Malheiro, R.; Casal, S.; Baptista, P.; Pereira, J.A. A Review of *Bactrocera oleae* (Rossi) Impact in Olive Products: From the Tree to the Table. *Trends Food Sci. Technol.* **2015**, *44*, 226–242. [CrossRef]
5. Berg, H.; Maneas, G.; Engström, A.S. A Comparison between Organic and Conventional Olive Farming in Messenia, Greece. *Horticulturae* **2018**, *4*, 15. [CrossRef]
6. Sánchez, A.G.; Martos, N.R.; Ballesteros, E. Multiresidue Analysis of Pesticides in Olive Oil by Gel Permeation Chromatography Followed by Gas Chromatography-Tandem Mass-Spectrometric Determination. *Anal. Chim. Acta* **2006**, *558*, 53–61. [CrossRef]
7. Raina, B. *Olives*; Elsevier Science Ltd.: Jammu, India, 2003.
8. Pinheiro, L.A.; Dáder, B.; Wanumen, A.C.; Pereira, J.A.; Santos, S.A.P.; Medina, P. Side Effects of Pesticides on the Olive Fruit Fly Parasitoid Psyttalia Concolor (Szépligeti): A Review. *Agronomy* **2020**, *10*, 1755. [CrossRef]
9. Bueno, A.M.; Jones, O. Alternative Methods for Controlling the Olive Fly, *Bactrocera oleae*, Involving Semiochemicals. In *Use of Pheromones and Other Semiochemicals in Integrated Production*; IOBC Wprs Bulletin: Andalusia, Spain, 2002; Volume 25.
10. Ant, T.; Koukidou, M.; Rempoulakis, P.; Gong, H.F.; Economopoulos, A.; Vontas, J.; Alphey, L. Control of the Olive Fruit Fly Using Genetics-Enhanced Sterile Insect Technique. *BMC Biol.* **2012**, *10*, 51. [CrossRef]
11. Ahmad, S.; Haq, I.U.; Cáceres, C.; Tomas, U.S.; Dammalage, T.; Gembinsky, K.; Paulus, H.; Vreysen, M.J.B.; Rempoulakis, P. One for All: Mating Compatibility among Various Populations of Olive Fruit Fly (Diptera: Tephritidae) for Application of the Sterile Insect Technique. *PLoS ONE* **2018**, *13*, e0206739. [CrossRef]
12. Choi, K.S.; Samayoa, A.C.; Hwang, S.Y.; Huang, Y.B.; Ahn, J.J. Thermal Effect on the Fecundity and Longevity of *Bactrocera dorsalis* Adults and Their Improved Oviposition Model. *PLoS ONE* **2020**, *15*, e0235910. [CrossRef]
13. Fanson, B.G.; Taylor, P.W. Additive and Interactive Effects of Nutrient Classes on Longevity, Reproduction, and Diet Consumption in the Queensland Fruit Fly (*Bactrocera tryoni*). *J. Insect Physiol.* **2012**, *58*, 327–334. [CrossRef] [PubMed]
14. Clarke, A.R.; Powell, K.S.; Weldon, C.W.; Taylor, P.W. The Ecology of *Bactrocera tryoni* (Diptera: Tephritidae): What Do We Know to Assist Pest Management? *Ann. Appl. Biol.* **2011**, *158*, 26–54. [CrossRef]
15. Aita, R.C.; Kees, A.M.; Aukema, B.H.; Hutchison, W.D.; Koch, R.L. Effects of Starvation, Age, and Mating Status on Flight Capacity of Laboratory-Reared Brown Marmorated Stink Bug (Hemiptera: Pentatomidae). *Environ. Entomol.* **2021**, *50*, 532–540. [CrossRef]
16. Gerofotis, C.D.; Kouloussis, N.A.; Koukougiannidou, C.; Papadopoulos, N.T.; Damos, P.; Koveos, D.S.; Carey, J.R. Age, Sex, Adult and Larval Diet Shape Starvation Resistance in the Mediterranean Fruit Fly: An Ecological and Gerontological Perspective. *Sci. Rep.* **2019**, *9*, 10704. [CrossRef] [PubMed]
17. Rosen, J.Y.A.; Hoy, M.A. Genetic Improvement of a Parasitoid Biological Control Agent: Artificial Selection for Insecticide Resistance in *Aphytis Melinus* (Hymenoptera: Aphelinidae). *J. Econ. Entomol.* **1988**, *81*, 1539–1550.
18. Harwood, J.F.; Chen, K.; Müller, H.G.; Wang, J.L.; Vargas, R.I.; Carey, J.R. Effects of Diet and Host Access on Fecundity and Lifespan in Two Fruit Fly Species with Different Life-History Patterns. *Physiol. Entomol.* **2013**, *38*, 81–88. [CrossRef] [PubMed]

19. Perez-Mendoza, J.; Campbell, J.F.; Throne, J.E. Influence of Age, Mating Status, Sex, Quantity of Food, and Long-Term Food Deprivation on Red Flour Beetle (Coleoptera: Tenebrionidae) Flight Initiation. *J. Econ. Entomol.* **2011**, *104*, 2078–2086. [CrossRef]
20. Zhang, D.W.; Xiao, Z.J.; Zeng, B.P.; Li, K.; Tang, Y.L. Insect Behavior and Physiological Adaptation Mechanisms under Starvation Stress. *Front. Physiol.* **2019**, *10*, 163. [CrossRef]
21. Benelli, G.; Romano, D.; Messing, R.H.; Canale, A. Population-Level Lateralized Aggressive and Courtship Displays Make Better Fighters Not Lovers: Evidence from a Fly. *Behav. Process.* **2015**, *115*, 163–168. [CrossRef]
22. Kouloussis, N.A.; Damos, P.T.; Ioannou, C.S.; Tsitsoulas, C.; Papadopoulos, N.T.; Nestel, D.; Koveos, D.S. Age Related Assessment of Sugar and Protein Intake of *Ceratitis capitata* in Ad Libitum Conditions and Modeling Its Relation to Reproduction. *Front. Physiol.* **2017**, *8*, 271. [CrossRef]
23. Adnan, S.M.; Mendez, V.; Morelli, R.; Akter, H.; Farhana, I.; Taylor, P.W. Dietary Methoprene Supplement Promotes Early Sexual Maturation of Male Queensland Fruit Fly *Bactrocera tryoni*. *J. Pest. Sci.* **2018**, *91*, 1441–1454. [CrossRef]
24. Papanastasiou, S.A.; Carey, J.R.; Papadopoulos, N.T. Effects of Early-Life Protein Starvation on Longevity and Sexual Performance of Male Medfly. *PLoS ONE* **2019**, *14*, e0219518. [CrossRef]
25. Goenaga, J.; Mensch, J.; Fanara, J.J.; Hasson, E. The Effect of Mating on Starvation Resistance in Natural Populations of *Drosophila melanogaster*. *Evol. Ecol.* **2012**, *26*, 813–823. [CrossRef]
26. Service, P.M. The Effect of Mating Status on Lifespan, Egg Laying, and Starvation Resistance in *Drosophila melanogaster* in Relation to Selection on Longevity. *J. Insect Physiol.* **1989**, *35*, 447–452. [CrossRef]
27. Hicks, S.K.; Hagenbuch, K.L.; Meffert, L.M. Variable Costs of Mating, Longevity, and Starvation Resistance in *Musca Domestica* (Diptera: Muscidae). *Environ. Entomol.* **2004**, *33*, 779–786. [CrossRef]
28. Tasnin, M.S.; Merkel, K.; Clarke, A.R. Effects of Advanced Age on Olfactory Response of Male and Female Queensland Fruit Fly, *Bactrocera tryoni* (Froggatt) (Diptera: Tephritidae). *J. Insect Physiol.* **2020**, *122*, 104024. [CrossRef]
29. Llandres, A.L.; Marques, G.M.; Maino, J.L.; Kooijman, S.A.L.M.; Kearney, M.R.; Casas, J. A Dynamic Energy Budget for the Whole Life-Cycle of Holometabolous Insects. *Ecol. Monogr.* **2015**, *85*, 353–371. [CrossRef]
30. Edmunds, D.; Wigby, S.; Perry, J.C. 'Hangry' *Drosophila*: Food Deprivation Increases Male Aggression. *Anim. Behav.* **2021**, *177*, 183–190. [CrossRef] [PubMed]
31. Knoppien, P.; Van Der Pers, J.N.C.; Van Delden, W. Quantification of Locomotion and the Effect of Food Deprivation on Locomotor Activity in Drosophila. *J. Insect Behav.* **2000**, *13*, 27–43. [CrossRef]
32. Krittika, S.; Yadav, P. Trans-Generational Effect of Protein Restricted Diet on Adult Body and Wing Size of *Drosophila melanogaster*. *R. Soc. Open Sci.* **2022**, *9*, 211325. [CrossRef]
33. Flatt, T. Survival Costs of Reproduction in *Drosophila*. *Exp. Gerontol.* **2011**, *46*, 369–375. [CrossRef]
34. Kokkari, A.I.; Milonas, P.G.; Anastasaki, E.; Floros, G.D.; Kouloussis, N.A.; Koveos, D.S. Determination of Volatile Substances in Olives and Their Effect on Reproduction of the Olive Fruit Fly. *J. Appl. Entomol.* **2021**, *145*, 841–855. [CrossRef]
35. Terzidou, A.; Koveos, D.; Kouloussis, N. Mating Competition between Wild and Artificially Reared Olive Fruit Flies. *Crops* **2022**, *2*, 247–257. [CrossRef]
36. Terzidou, A.M.; Koveos, D.S.; Papadopoulos, N.T.; Carey, J.R.; Kouloussis, N.A. Artificial Diet Alters Activity and Rest Patterns in the Olive Fruit Fly. *PLoS ONE* **2023**, *18*, e0274586. [CrossRef]
37. Jacome, I.; Aluja, M.; Liedo, P.; Nestel, D. The Influence of Adult Diet and Age on Lipid Reserves in the Tropical Fruit Fly *Anastrepha serpentina* (Diptera: Tephritidae). *J. Insect Physiol.* **1995**, *41*, 1079–1086. [CrossRef]
38. Utgés, M.E.; Vilardi, J.C.; Oropeza, A.; Toledo, J.; Liedo, P. Pre-Release Diet Effect on Field Survival and Dispersal of *Anastrepha ludens* and *Anastrepha obliqua* (Diptera: Tephritidae). *J. Appl. Entomol.* **2013**, *137*, 163–177. [CrossRef]
39. Yuval, B.; Maor, M.; Levy, K.; Kaspi, R.; Taylor, P.; Shelly, T. Breakfast of Champions or Kiss of Death? Survival and Sexual Performance of Protein-Fed, Sterile Mediterranean Fruit Flies (Diptera: Tephritidae). *Fla. Entomol.* **2007**, *90*, 115–122. [CrossRef]
40. Shelly, T.E.; Kennelly, S.S. Starvation and the Mating Success of Wild Male Mediterranean Fruit Flies (Diptera: Tephritidae). *J. Insect Behav.* **2003**, *16*, 171–179. [CrossRef]
41. McCue, M.D. Starvation Physiology: Reviewing the Different Strategies Animals Use to Survive a Common Challenge. *Comp. Biochem. Physiol. A Mol. Integr. Physiol.* **2010**, *156*, 1–18. [PubMed]
42. Jose Gonzalez Minero, F.; Candau, P.; Morales, J.; Tomas, C. Forecasting Olive Crop Production Based on Ten Consecutive Years of Monitoring Airborne Pollen in Andalusia (Southern Spain). *Agric. Ecosyst. Environ.* **1998**, *69*, 201–215. [CrossRef]
43. Rossini, L.; Bruzzone, O.A.; Contarini, M.; Bufacchi, L.; Speranza, S. A Physiologically Based ODE Model for an Old Pest: Modeling Life Cycle and Population Dynamics of *Bactrocera oleae* (Rossi). *Agronomy* **2022**, *12*, 2298. [CrossRef]
44. Marubbi, T.; Cassidy, C.; Miller, E.; Koukidou, M.; Martin-Rendon, E.; Warner, S.; Loni, A.; Beech, C. Exposure to Genetically Engineered Olive Fly (*Bactrocera oleae*) Has No Negative Impact on Three Non-Target Organisms. *Sci. Rep.* **2017**, *7*, 11478. [CrossRef]
45. Austad, S.N.; Fischer, K.E. Sex Differences in Lifespan. *Cell Metab.* **2016**, *23*, 1022–1033. [PubMed]
46. Toothaker, L.E. *Multiple Comparison Procedures*; Sage Publications, Inc.: Thousand Oaks, CA, USA, 1993.
47. Field, A. *Discovering Statistics Using IBM SPSS Statistics: And Sex and Drugs and Rock "N" Roll*, 4th ed.; Sage: Los Angeles, CA, USA; London, UK; New Delhi, India, 2013.
48. Rush, B.; Sandver, S.; Bruer, J.; Roche, R.; Wells, M.; Giebultowicz, J. Mating Increases Starvation Resistance and Decreases Oxidative Stress Resistance in *Drosophila melanogaster* Females. *Aging Cell* **2007**, *6*, 723–726. [CrossRef] [PubMed]

49. Ekanayake, E.W.M.T.D.; Clarke, A.R.; Schutze, M.K. Effect of Body Size, Age, and Premating Experience on Male Mating Success in *Bactrocera tryoni* (Diptera: Tephritidae). *J. Econ. Entomol.* **2017**, *110*, 2278–2281. [CrossRef] [PubMed]

50. Papadopoulos, N.T.; Liedo, P.; Müller, H.G.; Wang, J.L.; Molleman, F.; Carey, J.R. Cost of Reproduction in Male Medflies: The Primacy of Sexual Courting in Extreme Longevity Reduction. *J. Insect Physiol.* **2010**, *56*, 283–287. [CrossRef] [PubMed]

51. Terzidou, A.; Kouloussis, N.; Papanikolaou, G.; Koveos, D. Acoustic Characteristics of Sound Produced by Males *of Bactrocera oleae* Change in the Presence of Conspecifics. *Sci. Rep.* **2022**, *12*, 13086. [CrossRef]

52. Benelli, G.; Canale, A.; Bonsignori, G.; Ragni, G.; Stefanini, C.; Raspi, A. Male Wing Vibration in the Mating Behavior of the Olive Fruit Fly *Bactrocera oleae* (Rossi) (Diptera: Tephritidae). *J. Insect Behav.* **2012**, *25*, 590–603. [CrossRef]

53. Benelli, G.; Daane, K.M.; Canale, A.; Niu, C.Y.; Messing, R.H.; Vargas, R.I. Sexual Communication and Related Behaviours in Tephritidae: Current Knowledge and Potential Applications for Integrated Pest Management. *J. Pest. Sci.* **2014**, *87*, 385–405.

54. Meats, A.; Holmes, H.M.; Kelly, G.L. Laboratory Adaptation of *Bactrocera tryoni* (Diptera: Tephritidae) Decreases Mating Age and Increases Protein Consumption and Number of Eggs Produced per Milligram of Protein. *Bull. Entomol. Res.* **2004**, *94*, 517–524. [CrossRef]

55. Aceves-Aparicio, E.; Pérez-Staples, D.; Arredondo, J.; Corona-Morales, A.; Morales-Mávil, J.; Díaz-Fleischer, F. Combined Effects of Methoprene and Metformin on Reproduction, Longevity, and Stress Resistance in *Anastrepha ludens* (Diptera: Tephritidae): Implications for the Sterile Insect Technique. *J. Econ. Entomol.* **2021**, *114*, 142–151. [CrossRef]

56. Belyi, A.A.; Alekseev, A.A.; Fedintsev, A.Y.; Balybin, S.N.; Proshkina, E.N.; Shaposhnikov, M.V.; Moskalev, A.A. The Resistance of *Drosophila melanogaster* to Oxidative, Genotoxic, Proteotoxic, Osmotic Stress, Infection, and Starvation Depends on Age According to the Stress Factor. *Antioxidants* **2020**, *9*, 1239. [CrossRef] [PubMed]

57. Weldon, C.W.; Taylor, P.W. Desiccation Resistance of Adult Queensland Fruit Flies *Bactrocera tryoni* Decreases with Age. *Physiol. Entomol.* **2010**, *35*, 385–390. [CrossRef]

58. Carey, J.R.; Harshman, L.G.; Liedo, P.; Müller, H.G.; Wang, J.L.; Zhang, Z. Longevity-Fertility Trade-Offs in the Tephritid Fruit Fly, *Anastrepha ludens*, across Dietary-Restriction Gradients. *Aging Cell* **2008**, *7*, 470–477. [CrossRef]

59. Oviedo, A.; Nestel, D.; Papadopoulos, N.T.; Ruiz, M.J.; Prieto, S.C.; Willink, E.; Vera, M.T. Management of Protein Intake in the Fruit Fly *Anastrepha fraterculus*. *J. Insect Physiol.* **2011**, *57*, 1622–1630. [CrossRef] [PubMed]

60. Jaleel, W.; Tao, X.; Wang, D.; Lu, L.; He, Y. Using Two-Sex Life Table Traits to Assess the Fruit Preference and Fitness *of Bactrocera dorsalis* (Diptera: Tephritidae). *J. Econ. Entomol.* **2018**, *111*, 2936–2945. [CrossRef]

61. Papanastasiou, S.A.; Nakas, C.T.; Carey, J.R.; Papadopoulos, N.T. Condition-Dependent Effects of Mating on Longevity and Fecundity of Female Medflies: The Interplay between Nutrition and Age of Mating. *PLoS ONE* **2013**, *8*, e70181. [CrossRef]

62. Gerofotis, C.D.; Yuval, B.; Ioannou, C.S.; Nakas, C.T.; Papadopoulos, N.T. Polygyny in the Olive Fly—Effects on Male and Female Fitness. *Behav. Ecol. Sociobiol.* **2015**, *69*, 1323–1332. [CrossRef]

63. Kyritsis, G.A.; Koskinioti, P.; Bourtzis, K.; Papadopoulos, N.T. Effect of Wolbachia Infection and Adult Food on the Sexual Signaling of Males of the Mediterranean Fruit Fly *Ceratitis capitata*. *Insects* **2022**, *13*, 737. [CrossRef] [PubMed]

64. Shelly, T.E.; Kennelly, S. Influence of male diet on male mating success and longevity and female remating in the mediterranean fruit fly (diptera: Tephritidae) under laboratory conditions. *Fla. Entomol.* **2002**, *85*, 572–579. [CrossRef]

65. Gregoriou, M.E.; Reczko, M.; Kakani, E.G.; Tsoumani, K.T.; Mathiopoulos, K.D. Decoding the Reproductive System of the Olive Fruit Fly, *Bactrocera oleae*. *Genes* **2021**, *12*, 355. [CrossRef] [PubMed]

66. Herndon, L.A.; Wolfnert, M.F. A Drosophila Seminal Fluid Protein, Acp26Aa, Stimulates Egg Laying in Females for 1 Day after Mating. *Proc. Natl. Acad. Sci. USA* **1995**, *92*, 10114–10118. [CrossRef]

67. Chen, E.H.; Hou, Q.L.; Wei, D.D.; Jiang, H.B.; Wang, J.J. Phenotypes, Antioxidant Responses, and Gene Expression Changes Accompanying a Sugar-Only Diet in *Bactrocera dorsalis* (Hendel) (Diptera: Tephritidae). *BMC Evol. Biol.* **2017**, *17*, 194. [CrossRef] [PubMed]

68. Nestel, D.; Papadopoulos, N.T.; Liedo, P.; Gonzales-Ceron, L.; Carey, J.R. Trends in Lipid and Protein Contents during Medfly Aging: An Harmonic Path to Death. *Arch. Insect Biochem. Physiol.* **2005**, *60*, 130–139. [CrossRef]

69. Chen, D.; Han, H.L.; Li, W.J.; Wang, J.J.; Wei, D. Expression and Role of Vitellogenin Genes in Ovarian Development of *Zeugodacus cucurbitae*. *Insects* **2022**, *13*, 452. [CrossRef] [PubMed]

Article

Novel Lactone-Based Insecticides and *Drosophila suzukii* Management: Synthesis, Potential Action Mechanisms and Selectivity for Non-Target Parasitoids

Javier G. Mantilla Afanador [1], Sabrina H. C. Araujo [2,*], Milena G. Teixeira [3], Dayane T. Lopes [3], Cristiane I. Cerceau [3], Felipe Andreazza [2], Daiana C. Oliveira [4], Daniel Bernardi [4], Wellington S. Moura [5], Raimundo W. S. Aguiar [6], Ana C. S. S. Oliveira [6], Gil R. Santos [6], Elson S. Alvarenga [3] and Eugenio E. Oliveira [2,*]

[1] Research Institute in Microbiology and Agroindustrial Biotechnology, Universidad Católica de Manizales, Carrera 23, 60–63, Manizales 170002, Colombia; jmantilla@ucm.edu.co

[2] Departamento de Entomologia, Universidade Federal de Viçosa (UFV), Viçosa 36570-900, MG, Brazil; andreazzafelipe@yahoo.com.br

[3] Departamento de Química, Universidade Federal de Viçosa (UFV), Viçosa 36570-900, MG, Brazil; elson@ufv.br (E.S.A.)

[4] Department of Plant Protection, Federal University of Pelotas, Pelotas, Mailbox 354, Capão-do-Leão 96010-900, RS, Brazil; dbernardi2004@yahoo.com.br (D.B.)

[5] Programa de Pós-Graduação em Biodiversidade e Biotecnologia—Rede Bionorte, Universidade Federal do Tocantins (UFT), Gurupi 77402-970, TO, Brazil

[6] Programa de Pós-Graduação em Biotecnologia, Universidade Federal do Tocantins (UFT), Gurupi 77402-970, TO, Brazil; rwsa@uft.edu.br (R.W.S.A.); gilrsan@uft.edu.br (G.R.S.)

* Correspondence: shcaraujo@gmail.com (S.H.C.A.); eugenio@ufv.br (E.E.O.); Tel.: +55-31-3612-5282 (S.H.C.A.); +55-313-612-5280 (E.E.O.)

Citation: Mantilla Afanador, J.G.; Araujo, S.H.C.; Teixeira, M.G.; Lopes, D.T.; Cerceau, C.I.; Andreazza, F.; Oliveira, D.C.; Bernardi, D.; Moura, W.S.; Aguiar, R.W.S.; et al. Novel Lactone-Based Insecticides and *Drosophila suzukii* Management: Synthesis, Potential Action Mechanisms and Selectivity for Non-Target Parasitoids. *Insects* **2023**, *14*, 697. https://doi.org/10.3390/insects14080697

Academic Editor: Denis J. Wright

Received: 14 June 2023
Revised: 28 July 2023
Accepted: 6 August 2023
Published: 9 August 2023

Simple Summary: *Drosophila suzukii* is an insect of global economic importance, including in the Neotropical region. Due to the difficulty in controlling this insect pest with conventional insecticidal molecules, it is necessary to search for novel alternatives. Here, we present the potential of synthetic lactone-based insecticides to control *D. suzukii*. Additionally, we demonstrate molecular predictions regarding the actions of these molecules on the nervous system of the target pest and on the nervous system of its parasitoid, *Trichopria anastrephae*. By using in silico approaches, we demonstrate that the lactone derivatives (*rac*)-8 and compound **4** predominantly affect the TRP channels of *D. suzukii* (TRPM) and exhibit less stable interactions with the TRP channels expressed in *T. anastrephae*.

Abstract: *Drosophila suzukii*, an invasive insect pest, poses a significant threat to various fruit crops. The use of broad-spectrum insecticides to control this pest can reduce the effectiveness of biological control agents, such as the parasitoid *Trichopria anastrephae*. Here, we evaluated the toxicity of newly synthesized lactone derivatives on *D. suzukii* and their selectivity towards *T. anastrephae*. We used in silico approaches to identify potential targets from the most promising molecules in the *D. suzukii* nervous system and to understand potential differences in susceptibilities between *D. suzukii* and its parasitoid. Of the nine molecules tested, (*rac*)-8 and compound **4** demonstrated efficacy against the fly. Exposure to the estimated LC_{90} of (*rac*)-8 and compound **4** resulted in a mortality rate of less than 20% for *T. anastrephae* without impairing the parasitoid's functional parasitism. The in silico predictions suggest that (*rac*)-8 and compound **4** target gamma amino butyric acid (GABA) receptors and transient receptor potential (TRP) channels of *D. suzukii*. However, only the reduced interaction with TRP channels in *T. anastrephae* demonstrated a potential reason for the selectivity of these compounds on the parasitoid. Our findings suggest the potential for integrating (*rac*)-8 and compound **4** into *D. suzukii* management practices.

Keywords: spotted wing *Drosophila*; *Trichopria anastrephae*; in silico approaches; pesticide mode of action

1. Introduction

The spotted wing drosophila, *Drosophila suzukii*, is a significant insect species that reduces flesh fruit productivity in the Neotropical region [1,2]. Originally from Asia, *D. suzukii* has now spread worldwide [3], and despite its recent invasion into orchards in the Neotropical region, 28 plant species have been identified as hosts for *D. suzukii* [2]. For instance, *D. suzukii* infestation has led to estimated productivity losses of around 30% for Neotropical strawberry production [4].

The control of *D. suzukii* in the Neotropical region, as already described for Europe and the USA [5,6], is heavily dependent on the use of a few molecules (e.g., organophosphates, pyrethroids and the spinosyns) with very-well-characterized undesired effects on non-target organisms, including those that can provide naturally occurring biological control [5–7]. A possible alternative to foliar spraying is the use of toxic baits or low-volume, reduced-risk sprays in conjunction with feeding attractants [8,9]. However, although the use of these devices can substantially reduce the amount of insecticide applied, the efficiency can be strongly influenced by factors such as the high density of insects, unharvested fruits, and other alternative host fruits in the field, in addition to the physiological aspects (e.g., reproductive maturity, age, mating status) of insects [10,11].

In support of sustainable control options and compatible production methods of small fleshy fruits, the use of parasitoids has been widely investigated [12,13]. The pupal idiobiont parasitoid *Trichopria anastrephae* has been proposed as an effective biological control agent for *D. suzukii* [2]. This parasitoid is naturally distributed in Brazilian regions with occurrence on blackberry and strawberry fruits attacked by *D. suzukii* [14,15]. It is able to achieve a parasitism rate of over 90% at some sites in Southern France [16]. Thus, generating alternative pesticides compatible with the conservation of beneficial insects can be a robust factor for the control of *D. suzukii* populations.

Macrocyclic lactones, such as avermectins and milbemycins, have been widely used as insecticides to control a variety of insect pests with reported low risk to non-target insects [17–20]. These lactones are derived from naturally occurring compounds produced by soil-dwelling bacteria belonging to the genera Streptomyces (for avermectins) and Streptomyces and Streptomyces avermitilis (for milbemycins) [18]. However, there is a knowledge gap regarding the potential modes of action of synthetic lactone derivatives in target and non-target organisms. For instance, previous investigations have demonstrated the actions of some macrocyclic lactones on ligand-activated receptors (e.g., GABA receptors) and transient potential receptor (TRP) channels expressed in invertebrate nervous systems [21–23]. Indeed, considering the fact that the differential actions of novel insecticides on GABA receptors and TRP channels have been demonstrated for insect pests and their natural enemies [24–27], it would be reasonable to expect that such differential activities might be related to the lactone derivatives.

Here, we synthesized novel lactone derivatives and evaluated the toxicity of lactone derivatives on *D. suzukii* and its parasitoid, *T. anastrephae*. We further conducted in silico approaches to identify potential physiological targets in the *D. suzukii* nervous system for the actions of the most promising lactone derivatives. Such molecular prediction approaches helped to assess the action targets with higher selectivity potential for *T. anastrephae*.

2. Materials and Methods

2.1. Chemicals and Synthesis Process

We synthesized nine lactone derivatives. The identification of the compounds, as well as their molecular structures, is described in Table S1. The progress of reactions to obtain all the molecules used in this study was monitored by thin-layer chromatography (TLC) plates, and purification was performed by column chromatography on silica gel 70–230 mesh. When necessary, solvents and reagents were purified according to the literature [19]. Complete and detailed synthesis of the molecules is described in Teixeira et al. [28] and Näsman [20].

2.2. Chemical Solutions Preparation

The solid crystals of each molecule were weighed in 25 mL scintillation glass vials at masses that would allow the desired concentration (i.e., 1000 mg L^{-1}) to be reached after the addition of the solvents, i.e., dimethyl sulfoxide (DMSO) and sugar water solution at 20% *m/v*. To dilute the molecules, first, an amount of DMSO that represented 2% of the final volume of the solution, according to the exact molecular mass present in each sample, was added to the vial. The DMSO + molecule mix was then gently hand-stirred preventing the unnecessary spread of the solids on the vial walls. After allowing these first mixes to rest for at least 5 min, or enough for all the crystals solubilize in the DMSO, the remaining volume needed to reach the final solution volume was completed using a prediluted 20% *m/v* sugar water solution. The addition of the sugar water into the DMSO + molecule mix must be performed very gently and slowly by releasing the sugar water at the walls of the vial, preventing turmoil or strong disturbance in the solution, followed by gentle stir using a metal spatula. Failure in this step results in the molecule reprecipitating at the bottom or surface of the solution, preventing even exposure to the chemical later on. The control treatment consisted of a solution of 20% *m/v* sugar water containing 2% DMSO.

2.3. Toxicities on D. suzukii

The toxicity ratios between the compounds were estimated following the methodology proposed by Andreazza et al. [29]. Briefly, the initial assessment of the toxicities of the lactone derivatives in *D. suzukii* adults was conducted by exposing adult flies to a discriminatory concentration of 3 g L^{-1} for a 24 h period. For those lactone derivatives that killed more than 80% at the initial test, we formed concentration–mortality curves. For both the initial discriminatory test and the subsequent concentration–mortality curve assays, the exposure was completely randomized. Our experimental unit consisted of 25 unsexed 3–4-day-old flies placed into a 250 mL glass vial. To prepare each exposure unit, a dental cotton wick was placed inside a 250 mL glass vial, and 1.8 mL of the testing solution was applied to the cotton wicks. Subsequently, the vial was closed at the top with a foam plug. The fly release occurred by inserting a plastic tube between the plug and the vial's wall and puffing the flies into the vial. The insects could then feed on the solution ad libitum. At the end of 24 h period, the mortality was checked, and a fly was considered dead if it was not able to move upon being touched with a fine brush.

2.4. Toxicities on the Parasitoid T. anastrephae

Adult parasitoid *T. anastrephae*, up to 24 h old, were submitted to an ingestion bioassay for a 24 h period. For this, the insects were deprived of food for 8 h prior to the installation of the bioassays and placed inside plastic cages (100 mL) (10 pairs per cage), as described by Bernardi et al. [30]. The treatments were composed of compound **4** and (*rac*)-**8**, prepared as described in the "Chemical solutions preparation" section of this article. After 24 h of exposure, the insecticide-contaminated diets were removed, and the insects were provided with pure honey as a food source until the end of the bioassay. Insect mortality was evaluated for up to 120 h following the beginning of exposure, and the data were submitted to a survival analysis on Sigma Plot 12.5 (Systat software Inc., San Jose, CA, USA). The experimental design was completely randomized with seven replicates per treatment, with each replicate being composed of 10 pairs of *T. anastrephae* (n = 140).

To evaluate the sublethal effects of the treatments on the wasps' functional parasitism abilities, ten *D. suzukii* pupae (24 h old pupae) were offered per day for seven days (beginning at 120 h) to each surviving *T. anastrephae* female from the ingestion bioassay. The pupae were exposed to the wasps on a wet hydrophilic cotton layer on an acrylic petri dish. Daily, the pupae were removed and placed in plastic cups (100 mL) sealed on top with voile until the fly or wasp emerged. During the evaluation period, the wasps were fed with 80% (*w/v*) honey/water. The number of parasitoid offspring that emerged was recorded, and the percentage of parasitism was estimated for each treatment during the 7 days of pupae exposure.

The percentage of parasitism data used for the function of treatment and days of pupae exposure was submitted to a covariance analysis using Proc Mixed in SAS software v 12.0 (SAS Inc. 2013, Cary, NC, USA) with three levels for the first covariable (i.e., control, compound **4** and (*rac*)-**8**) and seven levels for the second covariable (i.e., first through seventh day). The covariant structure used was compound symmetry based on the smallest AICC (corrected Akaike's Information Criterion) obtained for this structure among several other covariant structures tested.

2.5. In Silico Evaluation of the Potential Target Receptors of Lactone Derivates on D. suzukii and T. anastrephae

2.5.1. Prediction of Putative Targets of Lactone Derivates

The selective lactone molecules in favor of parasitoid insects were drawn using Marvin Sketch 18.12.0 (ChemAxon, Budapest, Hungary) and saved in 3D mol2 format. Target receptor predictions of lactone derivates was carried out with the Similarity Ensemble Approach (SEA) and SwissTargetPrediction databases [31,32]. The genes of the predicted target receptors were downloaded from the NCBI and Uniprot databases and the better interactions against the selective lactone molecules determined from AutoDock Vina software (CCSB, Center for Computational Structural Biology, La Jolla, CA, USA) were used for the further analysis of molecular docking in both spotted wing drosophila and parasitoids.

2.5.2. Data Resources for the Selected Target Receptors of *D. suzukii* and *T. anastrephae*

The amino acid sequences of transient receptor potential (TRP) channels and gamma aminobutyric acid GABA receptors of *D. suzukii* were retrieved from the National Center for Biotechnology Information (NCBI) database. On the other hand, *T. anastrepha* has no sequenced data resource available. Therefore, the proteins of a closely related species, *Trichopria drosophilae*, were selected. The *T. drosophilae* proteins were obtained from the transcriptome data found in the original SRA RNA-seq reads available from the National Center for Biotechnology Information (NCBI). Sequence quality was assessed for each dataset through visualization in FastQC (released 0.11.5). Adapters were removed and low-quality regions were discarded using Trimmomatic (version 0.36). Low-quality readings (mean score of less than 20) and those with less than 50 nucleotides were excluded [33]. After processing the raw readings, we proceeded with their reconstitution through Trinity (version 2.5.1) with the default settings, resulting in contigs of the transcription sequences [34]. Then, we performed the prediction of coding sequences with more than 100 amino acids using TransDecoder [35]. We used Blast2GO to perform a functional annotation with default parameters and an InterProScan analysis of the TransDecoder to predict coding transcripts [36]. After obtaining the GO annotation for every coding transcript, the GABA receptor and TRP channel were identified. Protein domains for both the GABA receptor and TRP channel were identified using HMMER (release 3.0) with the PFAM database.

2.5.3. Generation and Validation of 3D Structures of Target Receptors

Homology modeling was used to construct the 3D structures of both the GABA receptor and TRP channel using The Swiss Model Workspace (https://swissmodel.expasy.org/ accessed on 17 January 2023). The templates were selected using the BLASTp tool, and the crystallographic structures were obtained from the Protein Data Bank (https://www.rcsb. org/ accessed on 17 January 2023). For the choice of the best structures, the experimental method used and the quality parameters (i.e., resolution) considered were the R-value and its complexing with a ligand. Clashes in crystallographic structures and amino acid positioning in the active site were checked using the Swiss model [37]. The validation of the stability of the generated models was performed by analyzing the Ramachandran plot [34,38], in which it was possible to analyze the distribution of the torsion angles of the backbone, Φ and ψ, which are responsible for the stereochemical quality of the protein studies, and the QMEAN factor was also analyzed [39].

2.5.4. Molecular Docking of Lactone Derivates against Target Receptors

Both selective lactone molecules designed by Marvin Sketch 18.12.0 (ChemAxon) and the target receptors modeled were specified to the pdbqt format and were prepared for the molecular docking process using Autodock ps 1.5.7 [40,41]. First, we added hydrogen atoms to the ligands in order to compute the protonation states as well as all possible bond torsions. The coordinates used for docking were generated by positioning the grid box inside the receptor's active pocket, and the crystallographic structures were used to design the grid boxes. Posteriorly, the docking calculations were performed using AutoDock Vina 4 [42], and nine docking positions for each ligand interacting with all receptors' active sites were generated. Affinity energies (kcal/Mol) for each interaction were also provided. The results were analyzed using PyMOL 2.0 [43] and Discovery Studio 4.5 [44], and the best interaction positions were selected. The following parameters were used to determine the best positions: ligand interactions with the amino acids from the active site, receptor–ligand affinity energies, the root-mean-square deviation (RMSD) between the initial and subsequent ligand structures and the nature of interactions considering the hydrogen bonds and non-covalent interactions for each complex according to 2D interaction maps.

2.5.5. Phylogenetic Analysis of TRP Channels

The analysis of the evolution of the *D. Suzukii* and *T. anastrephae* TRP channels was conducted using TRP channel genes of seven other species, i.e., *Drosophila melanogaster* (Dm), *Bombyx mori* (Bm), *Tribolium castaneum* (Tc), *Apis mellifera* (Am), *Nasonia vitripennis* (Nv), and *Pediculus humanus* (Ph) [45]. For this, the sequences were aligned using Muscle software, and the maximum likelihood method was used to calculate the tree based on the WAG amino acid substitution model and with 100 bootstrapped datasets using MEGA6 (Molecular Evolutionary Genetics Analysis) software [46]. The results were visualized and represented using FigTree software (http://tree.bio.ed.ac.uk/software/figtree/ accessed on 17 January 2023). The analysis involved 79 amino acid sequences. The amino acid subfamilies of TRPA (XP_016934147.1–XP_036671523.19) and TRPC (XP_016945947.1–XP_036675292.1) of *Drosophila suzukii* were obtained from the NCBI (National Center and Biotechnological Information).

3. Results

3.1. Insecticide Activity of Lactone Derivatives

Lactone derivatives exhibited varying toxicities ($F_{14,75} = 48.2$, $p < 0.001$) against adult *D. suzukii* (Figure 1A). Among the tested compounds, five molecules, (*rac*)-2, (*rac*)-3, compound **4**, (*rac*)-5 and (*rac*)-8, demonstrated the ability to kill over 40% of *D. suzukii* adults. Compound **4** and (*rac*)-8 displayed the highest potencies with mortality rates exceeding 75% at a concentration of 3 g/L (Figure 1A). However, compound **4** ($LC_{50} = 1.04$ (1.01–1.08) g/L) and (*rac*)-8 ($LC_{50} = 1.13$ (1.07–1.18) g/L showed statistically non-significant differences in terms of toxicity (Figure 1B).

3.2. Functional Selectivity of Compound 4 and (rac)-8 Lactone Derivates to T. anastrephae Adults

The survival analysis of parasitoid males and females indicated that individuals exposed to the estimated LC_{90} for compound **4** (1.46 g/L) and (*rac*)-8 (1.91 g/L) had significantly (*log-rank test*, $\chi^2 = 27.5$, $p < 0.001$) lower survival abilities that those individuals that were not exposed to the lactone derivatives (Figure 2A). However, at the end of the experiment (i.e., 120 h) the survival rate for all exposed insects was greater than 80%. Additionally, exposure to the LC_{90} of lactone derivatives did not affect the ability of *T. anastrephae* to parasitize *D. suzukii* pupae (Figure 2B).

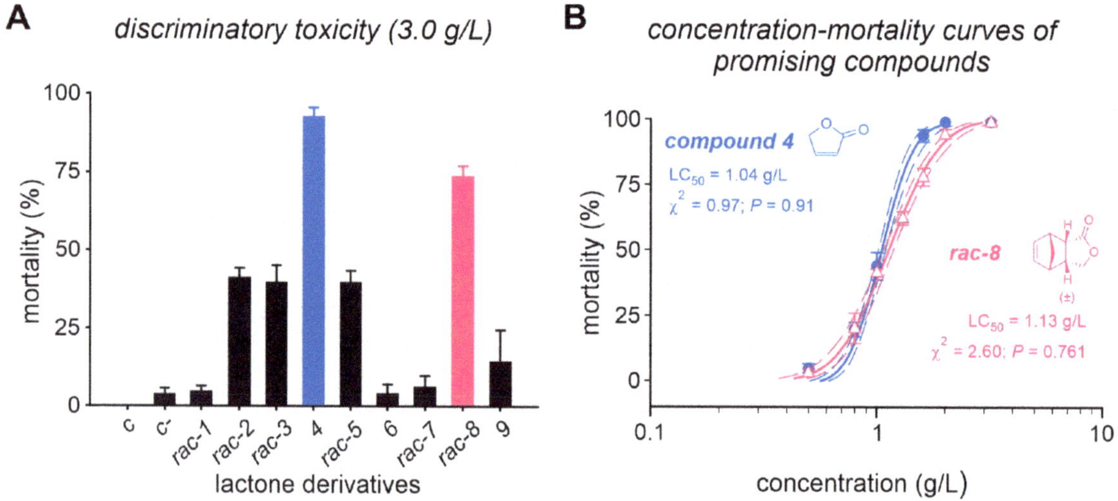

Figure 1. Toxicity screening bioassay of lactone derivatives on *Drosophila suzukii*. (**A**) Mortality of *D. suzukii* adults caused by nine lactone derivatives at a concentration of 3 g/L. (**B**) Concentration–mortality curves for the two most promising molecules (i.e., compound (**4**) and (*rac*)-8). Adult flies were exposed through the ingestion pathway, and the exposure period was 24 h. Control C represents insects treated with sugar solution. Control C represents insects treated with sugar solution containing 2% dimethyl sulfoxide (DMSO).

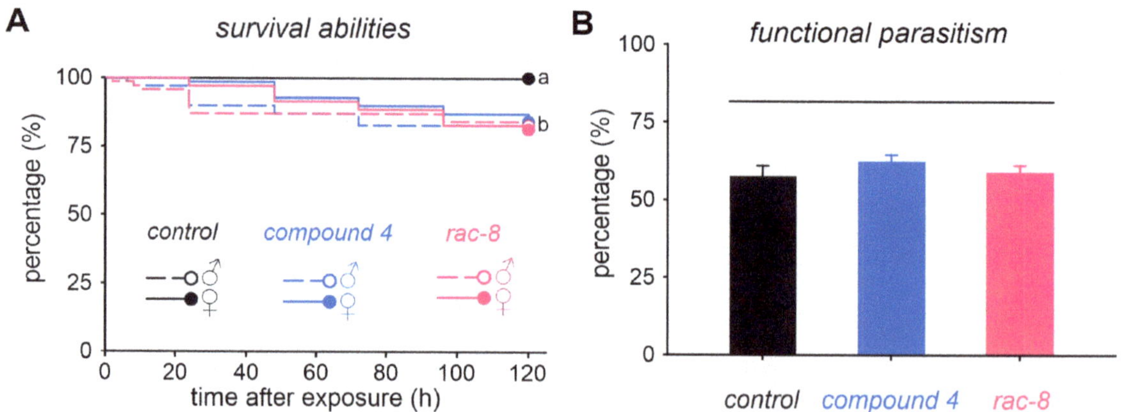

Figure 2. Selectivity of two lactone derivatives (i.e., compound (**4**) and (*rac*)-8) on males and females of the parasitoid *Trichopria anastrephae*. (**A**) Survival of *T. anastrephae* adults exposed to the LC_{90} of compound **4** (1.46 g/L) and (*rac*)-8 (1.91 g/L) estimated for *D. suzukii*. Survival curves followed by the same letter do not differ from each other (log rank test, $p > 0.05$). (**B**) Functional parasitism of *T. anastrephae* females after being exposed to compound **4** (1.46 g/L) and (*rac*)-8 (1.91 g/L). Columns represent the combined daily parasitism rate over a seven-day period after 24 h of exposure to the compounds. Columns under the same horizontal line do not differ from each other (Holm–Sidak test, $p > 0.05$). The control represents insects treated with sugar solution containing 2% dimethyl sulfoxide (DMSO).

3.3. Molecular Docking Analysis of the TRP Channels with Lactone Derivatives

The phylogenetic analysis revealed the evolution of TRP channels of *Drosophila suzukii* and the *Trichopria drosophilae* species, which is closely related to *Trichopria anastrephae* (Figure 3).

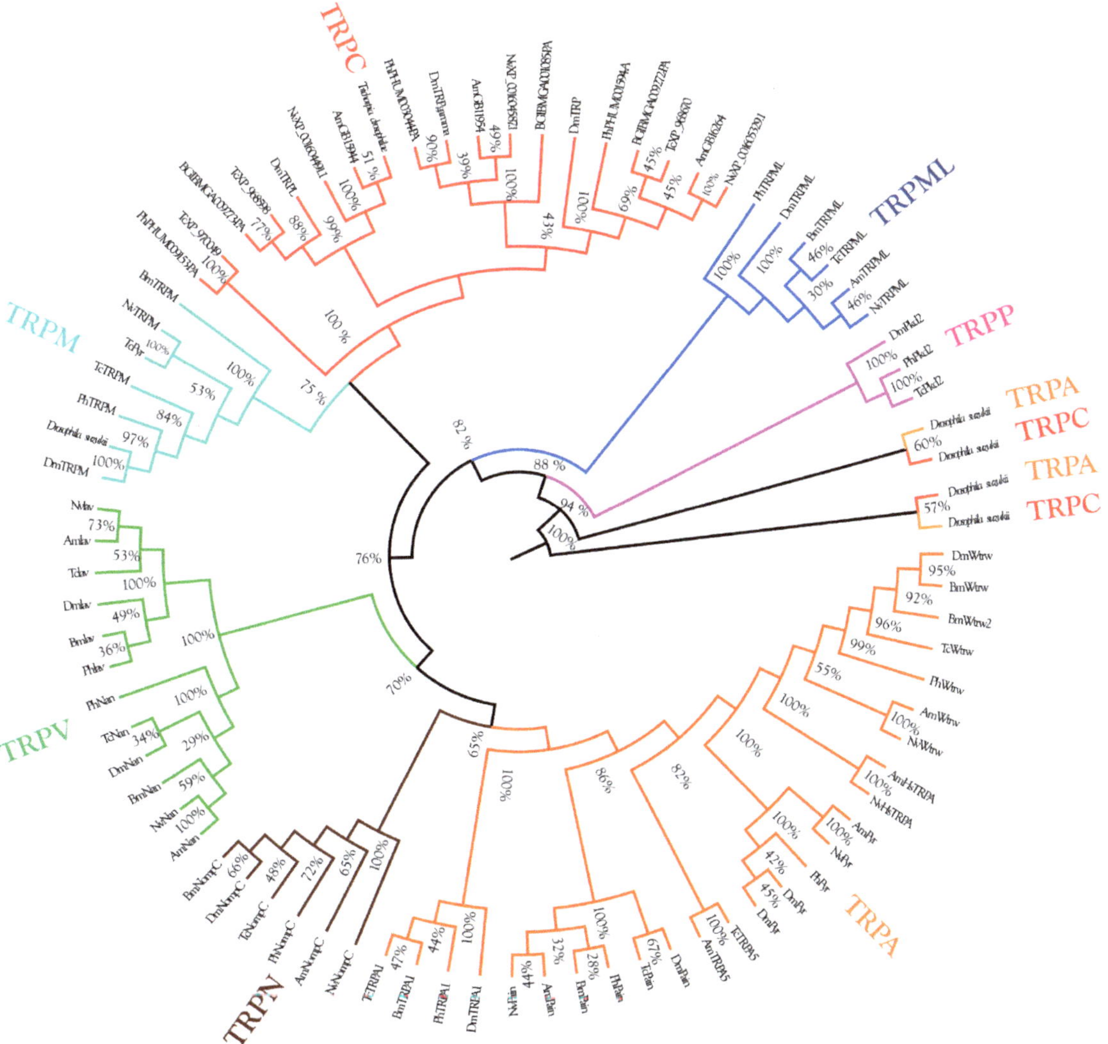

Figure 3. Analysis of the evolution of TRP channels of *Drosophila suzukii* and the *Trichopria drosophilae* species, which is closely related to *Trichopria anastrephae*. The identities of the TRP channels were determined using TRP channel genes of the *Drosophila melanogaster*, *Bombyx mori*, *Tribolium castaneum*, *Apis mellifera*, *Nasonia vitripennis*, and *Pediculus humanus* genomes.

Our in silico analysis indicated that the TRPM channels are potential targets for the actions of (*rac*)-8 and compound **4** in *D. suzukii* (Figure 4), but transcriptome analyses did not result in TRPM sequence availability in the parasitoid flies. *T. anastrephae* individuals are equipped with TRPC channels (Figure 4). The TRPM channels exhibited a Ramachandran value of 92.62% and a QMEAN factor of −2.95 (Figure 4A), while the TRPC channels showed similar results for the Ramachandran (91.8%) and QMEAN factor (−4.01) values (Figure 4A).

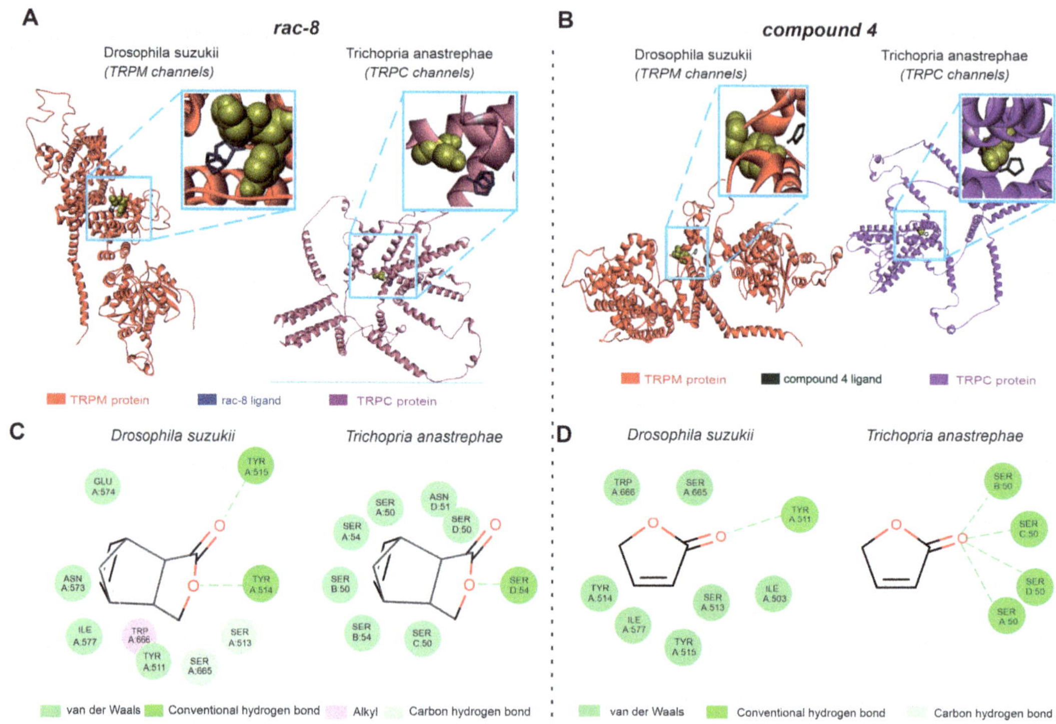

Figure 4. Predictions of (rac-8) and compound **4** lactone derivative binding to TRP channels related to *Drosophila suzukii* and the species closely related to the *Trichopria anastrephae* parasitoid, *Trichopria drosophilae*. (**A**) Structures of active sites of the *Drosophila suzukii* TRP channel (TRPM, left panel) and *Trichopria drosophilae* TRP channels (TRPC, right panel) interacting with (*rac*)-8 (blue). (**B**) Structures of active sites of the *D. suzukii* TRP channel (TRPM, left panel) and *Trichopria drosophilae* TRP channels (TRPC, right panel) interacting with compound **4** (green). (**C**) Two-dimensional interaction map representations of *D. suzukii* and *T. drosophilae* TRP channels with (*rac*)-8. (**D**) Two-dimensional interaction map representations of *D. suzukii* and *T. drosophilae* TRP channels with compound **4**. All detailed amino acids belonging to the lipid environment binding site are also represented. (For the interpretation of the references used to color in this figure legend, the reader is referred to the web version of this article).

The complex formed by (*rac*)-8 and the TRPM channels showed hydrogen bond interactions with TYR514 and TYR515, van der Waals interactions with TYR511, ILE577, ASN573 and GLU574, carbon hydrogen bond interactions with SER665 and SER513 and alkyl interactions with TRP666 (Figure 4B). The predicted binding interactions of (*rac*)-8 with the TRPC revealed higher instability as the predominant forces were van der Waals forces (Figure 4B). The (*rac*)-8 complex with the TRPC channel showed hydrogen bond interactions with SER54 and van der Waals interactions with SER C:50, SER B:54, SER B:50, SER A:54, SER A:50, ASN D:51 and SER D:50 (Figure 4B). Compound 4 exhibited a lower interaction energy (AutoDockVina affinity energy kcal mol^{-1}) between the TRPM channel (-3.9) compared to TRPC (-3.2) (Figure 4B). While the SWD-related compound **4** TRPM complex showed hydrogen bond interactions with TYR A: 511 and van de Waals interactions with TRP A: 666, SER A: 665, TYR A: 514, ILE A: 577, TYR A: 515, SER A: 513 and ILE A: 503, the complex formed by compound **4** and TRPC channels showed hydrogen bonds with SER A: 50, SER B: 50, SER C: 50 and SER D: 50 and carbon hydrogen bonds with SER A: 50 (Figure 4B).

3.4. Molecular Docking Analysis of the GABA Receptors with Lactone Derivatives

Our in silico analysis indicated that the GABA receptor is a potential target for the actions of only (*rac*)-8 in *D. suzukii* (Figure 5). The *D. suzukii* GABA receptors exhibited a Ramachandran value of 94.81% and a QMEAN factor of −3.8 (Figure 5A), while GABA receptors of the parasitoids showed similar Ramachandran (92.0%) and QMEAN factor (−3.9) values (Figure 5A). The molecular docking results predicted no significant differences in the interaction energy (AutoDockVina affinity energy kcal mol^{-1}) between the GABA receptors of *D. suzukii* (−6.1) and its parasitoid (−5.9) with (*rac*)-8 (Figure 5A). While the (*rac*)-8 complex with the *D. suzukii* GABA receptor showed hydrogen bond interactions with SER692 and LEU355, van de Waals interactions with GLY356, GLY354, SER591 and ASN590 and alkyl interactions with PHE596, MET428 and ILE596, the complex formed by (*rac*)-8 and the parasitoid-related GABA receptors revealed hydrogen bond interactions with ILE55, van der Waals interactions with THR348, VAL344, PHE46, LEU263, SER49, VAL51, LEU53 and ASN345, and alkyl interactions with ALA54. Similarly, the compound **4** complex with *D. suzukii* GABA receptors (−3.9) did not exhibit significant differences in interaction energy (AutoDockVina affinity energy kcal mol^{-1}) when compared to the interaction affinity recorded for the GABA receptors of the parasitoids (−4.0). While the compound **4** complex with the *D. suzuki* GABA receptors showed hydrogen bond interactions with SER813, and van der Waals interactions with TRP810, MET759, VAL756, ILE582, LEU814 and TYR817, the compound **4** complex with the parasitoid-related GABA receptors showed carbon hydrogen bond interactions with ALA123 and GLY105 and van der Waals interactions with ALA122, GLY119, GLY319, LEU318, ALA425, GLY421, PHE422 and SER126.

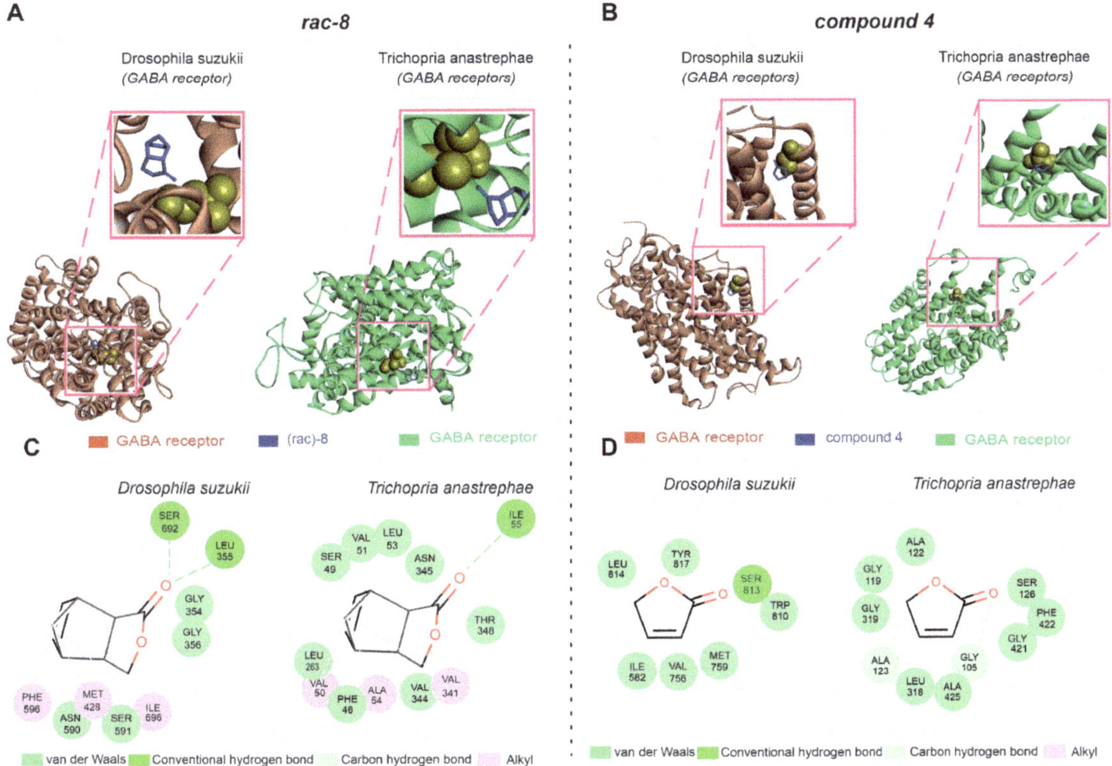

Figure 5. Predictions of (rac-8) and compound **4** lactone derivative binding to the GABA transporter of *Drosophila suzukii* and the species closely related to the *Trichopria anastrephae* parasitoid, *Trichopria*

drosophilae. (**A**) Structures of *Drosophila suzukii* (left panel) and *Trichopria drosophilae* (right panel) GABA receptor active sites interacting with (*rac*)-8 (blue). (**B**) Structures of *Drosophila suzukii* (left panel) and *Trichopria drosophilae* (right panel) GABA receptor active sites interacting with compound **4** (blue). (**C**) Two-dimensional interaction map representations of *D. suzukii* and *T. drosophilae* GABA receptors with (*rac*)-8. (**D**) Two-dimensional interaction map representations of *D. suzukii* and *T. drosophilae* GABA receptors with compound **4**. All detailed amino acids belonging to the lipid environment binding site are also represented. (For the interpretation of the references used to color in this figure legend, the reader is referred to the web version of this article).

4. Discussion

Here, we presented a description of novel lactone-derived molecules that exhibit potential for integration into *D. suzukii* management strategies. We demonstrated that two of these molecules, (*rac*)-8 and compound **4**, possess similar levels of efficacy for the killing of *D. suzukii* adults while leaving the parasitism functionalities of *T. anastrephae* unaffected. Additionally, through molecular docking analysis, we identified the mechanisms by which these molecules interact with the GABA receptors and TRP channels of *D. suzukii* and its parasitoids. This analysis shows that the effectiveness of these molecules against *D. suzukii*, as opposed to *T. anastrephae*, may be attributed to their distinct actions on the TRP channel subtypes present in these insect species.

It is already known that lactone-based compounds have toxic and antifeeding effects on pest insects [47–49]. For instance, Szczepanik et al. [48] demonstrated that lactone ring compounds cause feeding inhibition and behavioral deterrence during the larval growth of the lesser mealworm *Alphitobius diaperinus*. These compounds also showed strong antifeedant properties against adult *A. diaperinus*. Similar results have been described for natural lactones against the variegated cutworms *Peridroma saucia* [47] and *Spodoptera frugiperda* [49]. Our efforts reinforce such insecticide activities of lactone derivatives, demonstrating their potential to kill *D. suzukii* adults with lower toxicity and a complete absence of detrimental effects by *T. anastrephae* parasitism to its hosts. It is notable that *T. anastrephae* is one of the most promising biological agents for *D. suzukii* in Neotropical fruit orchards [30].

The potential integration of the (*rac*)-8 and compound **4** lactone derivatives into the management of *D. suzukii* would span a number of effective practices used for controlling *D. suzukii* in the Neotropical region. The reduced number of effective control practices for *D. suzukii* has been a serious problem for cherry and berry production in the Neotropical region [1,2,50–53]. The reliance on a reduced number of molecules can either be worsened by the fact that some of these molecules can also have detrimental effects on non-target organisms [54,55]. For instance, *T. anastrephae* has been shown to be susceptible to the conventional insecticides used in the management of *D. suzukii* [7,56,57].

Recent investigations that combined in vivo and in silico toxicological approaches have shown that both GABA receptors and TRP channels play relevant roles in the distinct actions of novel insecticides in insect pests and their natural enemies [24,25,27]. Here, our sequence phylogenetic analysis indicated that the *D. suzukii* and *T. anastrephae* might be equipped with different types of TRP channels. While *D. suzukii* expresses the TRPM channel type, which is involved in the removal functions of Mg^{2+} from hemolymph [58], repellent activities [59] and temperature avoidance [60], such channels are not present in *T. anastrephae*. The parasitoid expresses the TRPC channels, which were shown to have less stable molecular interactions in their lipid-binding environment with both (*rac*)-8 and compound **4** in our in silico predictions. This may explain the lower susceptibility of the parasitoid to both lactone derivatives. Interestingly, the stable interactions of both lactone derivatives and GABA receptors of *D. suzukii*, which were also recorded with *T. anastrephae* GABA receptors, did not allow the identification of the GABA receptors as a potential reason for the selectivity of (*rac*)-8 and compound **4** towards the parasitoid.

5. Conclusions

Despite further investigation aiming to evaluate further steps necessary to develop a pesticide product (e.g., formulation type, application method and evaluations of efficacy at field conditions), our findings represent a relevant and promising step that could lead to the development of novel tools for controlling *D. suzukii*. Our investigations demonstrate that lactone-derived molecules, (*rac*)-8 and compound **4** can effectively kill *D. suzukii* by targeting TRP channels and GABA receptors. Notably, these lactone derivatives exhibit reduced toxicity towards *T. anastrephae* with no adverse effects on functional parasitism. This selective efficacy against *D. suzukii* can be attributed to the expression of a specific TRP channel type (TRPM) in the fly, which facilitates more stable molecular interactions compared to the TRP channels expressed in the parasitoid (TRPC). Furthermore, the lactone derivatives' actions on GABA receptors were comparable in both insect species and thus do not contribute to the explanation of the lactone derivative's selectivity. Our findings demonstrate that both (*rac*)-8 and compound **4** exhibit the potential to be integrated into *D. suzukii* management.

Supplementary Materials: The following supporting information can be downloaded at: https://www.mdpi.com/article/10.3390/insects14080697/s1, Table S1: Lactone derivative compounds.

Author Contributions: Conceptualization, E.E.O., F.A., S.H.C.A., E.S.A., R.W.S.A. and G.R.S.; methodology, F.A., J.G.M.A., C.I.C., D.C.O., M.G.T., D.T.L., W.S.M., A.C.S.S.O. and E.E.O.; software, W.S.M., D.B., E.S.A., M.G.T., A.C.S.S.O., D.T.L., E.E.O. and C.I.C.; investigation, validation and formal analysis, E.E.O., F.A., J.G.M.A., C.I.C., D.C.O. and W.S.M.; resources, funds acquisition and supervision E.E.O., R.W.S.A., D.B. and G.R.S.; writing—original draft preparation, J.G.M.A., F.A., A.C.S.S.O. and S.H.C.A.; writing—review and editing, J.G.M.A., S.H.C.A., G.R.S., R.W.S.A. and E.E.O. All authors have read and agreed to the published version of the manuscript.

Funding: This research was funded by the National Council of Scientific and Technological Development (CNPq—process numbers: 313455/2019-8, 427304/2018-0, 309890/2022-5, 152366/2022-9), CAPES foundation (Finance Code 001), the Minas Gerais State Foundation for Research Aid (APQ 03771-18), the Tocantins State Foundation for Research Aid (FAPT-SESAU/TO-DECIT/SCTIE/MS_CNPQ/No. 01/2017) and the Federal University of Tocantins (PROPESQ)—EDITAL No. 19/2023 PROPESQ, and PPGBIOTEC/UFT/GURUPI—Chamada pública para auxílio de tradução e/ou publicação de artigos científicos.

Data Availability Statement: The data presented in this study are available on request from the corresponding authors.

Conflicts of Interest: The authors declare no conflict of interest. The funders had no role in the design of the study; in the collection, analyses, or interpretation of data; in the writing of the manuscript, or in the decision to publish the results.

References

1. Andreazza, F.; Bernardi, D.; Dos Santos, R.; Garcia, F.; Oliveira, E.E.; Botton, M.; Nava, D.E. Drosophila Suzukii in Southern Neotropical Region: Current Status and Future Perspectives. *Neotrop. Entomol.* **2017**, *16*, 591–605. [CrossRef]
2. Garcia, F.R.M.; Lasa, R.; Funes, C.F.; Buzzetti, K. Drosophila Suzukii Management in Latin America: Current Status and Perspectives. *J. Econ. Entomol.* **2022**, *115*, 1008–1023. [CrossRef] [PubMed]
3. Bolda, M.P.; Goodhue, R.E.; Zalom, F.G. Spotted Wing Drosophila: Potential Economic Impact of a Newly Estab-lished Pest. *Agric. Resour. Econ. Update* **2010**, *13*, 5–8.
4. Wollmann, J.; Schlesener, D.C.H.; Mendes, S.R.; Krüger, A.P.; Martins, L.N.; Bernardi, D.; Garcia, M.S.; Garcia, F.R.M. Infestation Index of Drosophila Suzukii (Diptera: Drosophilidae) in Small Fruit in Southern Brazil. *Arq. Inst. Biol.* **2020**, *87*, e0432018. [CrossRef]
5. Diepenbrock, L.M.; Hardin, J.A.; Burrack, H.J. Season-Long Programs for Control of Drosophila Suzukii in South-eastern U.S. Blackberries. *Crop Prot.* **2017**, *98*, 149–156. [CrossRef]
6. Shaw, B.; Hemer, S.; Cannon, M.F.L.; Rogai, F.; Fountain, M.T. Insecticide Control of Drosophila Suzukii in Com-mercial Sweet Cherry Crops under Cladding. *Insects* **2019**, *10*, 196. [CrossRef] [PubMed]
7. Schlesener, D.C.H.; Wollmann, J.; Pazini, J.D.B.; Padilha, A.C.; Grützmacher, A.D.; Garcia, F.R.M. Insecticide Toxicity to Drosophila Suzukii (Diptera: Drosophilidae) Parasitoids: Trichopria Anastrephae (Hymenoptera: Diapriidae) and Pachycrepoideus Vindemmiae (Hymenoptera: Pteromalidae). *J. Econ. Entomol.* **2019**, *112*, 1197–1206. [CrossRef]

8.   Andreazza, F.; Bernardi, D.; Baronio, C.A.; Pasinato, J.; Nava, D.E.; Botton, M. Toxicities and Effects of Insecticidal Toxic Baits to Control Drosophila Suzukii and Zaprionus Indianus (Diptera: Drosophilidae). *Pest. Manag. Sci.* **2017**, *73*, 146–152. [CrossRef]

9.   Noble, R.; Shaw, B.; Walker, A.; Whitfield, E.C.; Deakin, G.; Harris, A.; Dobrovin-Pennington, A.; Fountain, M.T. Control of Spotted Wing Drosophila (Drosophila Suzukii) in Sweet Cherry and Raspberry Using Bait Sprays. *J. Pest Sci.* **2023**, *96*, 623–633. [CrossRef]

10. Babu, A.; Rodriguez-Saona, C.; Sial, A.A. Factors Influencing the Efficacy of Novel Attract-and-Kill (ACTTRA SWD) Formulations Against Drosophila Suzukii. *J. Econ. Entomol.* **2022**, *115*, 981–989. [CrossRef]

11. Urbaneja-Bernat, P.; Holdcraft, R.; Hernández-Cumplido, J.; Rhodes, E.M.; Liburd, O.E.; Sial, A.A.; Mafra-Neto, A.; Rodriguez-Saona, C. Field, Semi-Field and Greenhouse Testing of HOOK SWD, a SPLAT-Based Attract-and-Kill Formulation to Manage Spotted-Wing Drosophila. *J. Appl. Entomol.* **2022**, *146*, 1230–1242. [CrossRef]

12. Hogg, B.N.; Lee, J.C.; Rogers, M.A.; Worth, L.; Nieto, D.J.; Stahl, J.M.; Daane, K.M. Releases of the Parasitoid Pachycrepoideus Vindemmiae for Augmentative Biological Control of Spotted Wing Drosophila, Drosophila Suzukii. *Biol. Control* **2022**, *168*, 104865. [CrossRef]

13. Seehausen, M.L.; Valenti, R.; Fontes, J.; Meier, M.; Marazzi, C.; Mazzi, D.; Kenis, M. Large-Arena Field Cage Releases of a Candidate Classical Biological Control Agent for Spotted Wing Drosophila Suggest Low Risk to Non-Target Species. *J. Pest Sci.* **2022**, *95*, 1057–1065. [CrossRef]

14. Cruz, P.P.; Neutzling, A.S.; Garcia, F.R.M. Primeiro Registro de Trichopria Anastrephae, Parasitoide de Moscas-Das-Frutas, No Rio Grande Do Sul. *Ciência Rural.* **2011**, *41*, 1297–1299. [CrossRef]

15. Garcia, F.R.M.; Ricalde, M.P. Augmentative Biological Control Using Parasitoids for Fruit Fly Management in Brazil. *Insects* **2013**, *4*, 55–70. [CrossRef]

16. Chabert, S.; Allemand, R.; Poyet, M.; Eslin, P.; Gibert, P. Ability of European Parasitoids (Hymenoptera) to Control a New Invasive Asiatic Pest, Drosophila Suzukii. *Biol. Control* **2012**, *63*, 40–47. [CrossRef]

17. Lumaret, J.-P.; Errouissi, F.; Floate, K.; Rombke, J.; Wardhaugh, K. A Review on the Toxicity and Non-Target Effects of Macrocyclic Lactones in Terrestrial and Aquatic Environments. *Curr. Pharm. Biotechnol.* **2012**, *13*, 1004–1060. [CrossRef]

18. Batiha, G.E.S.; Alqahtani, A.; Ilesanmi, O.B.; Saati, A.A.; El-Mleeh, A.; Hetta, H.F.; Beshbishy, A.M. Avermectin Derivatives, Pharmacokinetics, Therapeutic and Toxic Dosages, Mechanism of Action, and Their Biological Effects. *Pharmaceuticals* **2020**, *13*, 196. [CrossRef]

19. Perrin, D.D.W.L.F. Armarego Purification of Laboratory Chemicals. In *Recueil des Travaux Chimiques des Pays-Bas*; John Wiley & Sons, Ltd.: Oxford, UK, 1988; Volume 107, p. 685.

20. Näsman, J.H. 3-Methyl-2(5H)-Furanone. *Org. Synth.* **1990**, *68*, 162. [CrossRef]

21. Nakao, T.; Banba, S.; Hirase, K. Comparison between the Modes of Action of Novel Meta-Diamide and Macrocyclic Lactone Insecticides on the RDL GABA Receptor. *Pestic. Biochem. Physiol.* **2015**, *120*, 101–108. [CrossRef]

22. Hadiatullah, H.; Zhang, Y.; Samurkas, A.; Xie, Y.; Sundarraj, R.; Zuilhof, H.; Qiao, J.; Yuchi, Z. Recent Progress in the Structural Study of Ion Channels as Insecticide Targets. *Insect Sci.* **2022**, *29*, 1522–1551. [CrossRef]

23. Chalivendra, S. Microbial Toxins in Insect and Nematode Pest Biocontrol. *Int. J. Mol. Sci.* **2021**, *22*, 7657. [CrossRef] [PubMed]

24. Toledo, P.F.S.; Ferreira, T.P.; Bastos, I.M.A.S.; Rezende, S.M.; Viteri Jumbo, L.O.; Didonet, J.; Andrade, B.S.; Melo, T.S.; Smagghe, G.; Oliveira, E.E.; et al. Essential Oil from Negramina (Siparuna Guianensis) Plants Controls Aphids without Impairing Survival and Predatory Abilities of Non-Target Ladybeetles. *Environ. Pollut.* **2019**, *255*, 113153. [CrossRef] [PubMed]

25. Toledo, P.F.S.; Viteri Jumbo, L.O.; Rezende, S.M.; Haddi, K.; Silva, B.A.; Mello, T.S.; Della Lucia, T.M.C.; Aguiar, R.W.S.; Smagghe, G.; Oliveira, E.E. Disentangling the Ecotoxicological Selectivity of Clove Essential Oil against Aphids and Non-Target Ladybeetles. *Sci. Total Environ.* **2020**, *718*, 137328. [CrossRef] [PubMed]

26. Nesterov, A.; Spalthoff, C.; Kandasamy, R.; Katana, R.; Rankl, N.B.; Andrés, M.; Jähde, P.; Dorsch, J.A.; Stam, L.F.; Braun, F.J.; et al. TRP Channels in Insect Stretch Receptors as Insecticide Targets. *Neuron* **2015**, *86*, 665–671. [CrossRef]

27. Salgado, V.L. Insect TRP Channels as Targets for Insecticides and Repellents. *J. Pestic. Sci.* **2017**, *42*, 1–6. [CrossRef]

28. Teixeira, M.G.; Alvarenga, E.S.; Pimentel, M.F.; Picanço, M.C. Synthesis and Insecticidal Activity of Lactones Derived from Furan-2(5H)-One. *J. Braz. Chem. Soc.* **2015**, *26*, 2279–2289. [CrossRef]

29. Andreazza, F.; Vacacela Ajila, H.E.; Haddi, K.; Colares, F.; Pallini, A.; Oliveira, E.E. Toxicity to and Egg-Laying Avoidance of Drosophila Suzukii (Diptera: Drosophilidae) Caused by an Old Alternative Inorganic Insecticide Preparation. *Pest. Manag. Sci.* **2018**, *74*, 861–867. [CrossRef]

30. Bernardi, D.; Ribeiro, L.; Andreazza, F.; Neitzke, C.; Oliveira, E.E.; Botton, M.; Nava, D.E.; Vendramim, J.D. Potential Use of Annona by Products to Control Drosophila Suzukii and Toxicity to Its Parasitoid Trichopria Anastrephae. *Ind. Crop. Prod.* **2017**, *110*, 30–35. [CrossRef]

31. Keiser, M.J.; Roth, B.L.; Armbruster, B.N.; Ernsberger, P.; Irwin, J.J.; Shoichet, B.K. Relating Protein Pharmacology by Ligand Chemistry. *Nat. Biotechnol.* **2007**, *25*, 197–206. [CrossRef]

32. Daina, A.; Michielin, O.; Zoete, V. SwissTargetPrediction: Updated Data and New Features for Efficient Prediction of Protein Targets of Small Molecules. *Nucleic Acids Res.* **2019**, *47*, W357–W3664. [CrossRef] [PubMed]

33. Bolger, A.M.; Lohse, M.; Usadel, B. Trimmomatic: A Flexible Trimmer for Illumina Sequence Data. *Bioinformatics* **2014**, *30*, 2114–2120. [CrossRef] [PubMed]

34. Grabherr, M.G.; Haas, B.J.; Yassour, M.; Levin, J.Z.; Thompson, D.A.; Amit, I.; Adiconis, X.; Fan, L.; Raychowdhury, R.; Zeng, Q.; et al. Full-Length Transcriptome Assembly from RNA-Seq Data without a Reference Genome. *Nat. Biotechnol.* **2011**, *29*, 644–652. [CrossRef]

35. Haas, B.J.; Papanicolaou, A.; Yassour, M.; Grabherr, M.; Blood, P.D.; Bowden, J.; Couger, M.B.; Eccles, D.; Li, B.; Lieber, M.; et al. De Novo Transcript Sequence Reconstruction from RNA-Seq Using the Trinity Platform for Ref-erence Generation and Analysis. *Nat. Protoc.* **2013**, *8*, 1494–1512. [CrossRef]
36. Conesa, A.; Götz, S. Blast2GO: A Comprehensive Suite for Functional Analysis in Plant Genomics. *Int. J. Plant Genomics* **2008**, *2008*, 619832. [CrossRef] [PubMed]
37. Waterhouse, A.; Bertoni, M.; Bienert, S.; Studer, G.; Tauriello, G.; Gumienny, R.; Heer, F.T.; De Beer, T.A.P.; Rempfer, C.; Bordoli, L.; et al. SWISS-MODEL: Homology Modelling of Protein Structures and Complexes. *Nucleic Acids Res.* **2018**, *46*, W296–W303. [CrossRef]
38. Ramachandran, G.N.; Sasisekharan, V. Conformation of Polypeptides and Proteins. *Adv. Protein Chem.* **1968**, *23*, 283–437. [CrossRef]
39. Benkert, P.; Biasini, M.; Schwede, T. Toward the Estimation of the Absolute Quality of Individual Protein Structure Models. *Bioinformatics* **2011**, *27*, 343–350. [CrossRef]
40. Sanner, M.F. Python: A Programming Language for Software Integration and Development. *J. Mol. Graph. Model.* **1999**, *17*, 57–61.
41. Forli, S.; Huey, R.; Pique, M.E.; Sanner, M.F.; Goodsell, D.S.; Olson, A.J. Computational Protein-Ligand Docking and Virtual Drug Screening with the AutoDock Suite. *Nat. Protoc.* **2016**, *11*, 905–919. [CrossRef]
42. Trott, O.; Olson, A.J. AutoDock Vina: Improving the Speed and Accuracy of Docking with a New Scoring Function, Efficient Optimization and Multithreading. *J. Comput. Chem.* **2010**, *31*, 461. [CrossRef] [PubMed]
43. Schrödinger, L.; DeLano, W. PyMOL | Pymol.Org. Available online: https://pymol.org/2/ (accessed on 26 April 2023).
44. Biovia, D. Biovia Discovery Studio® 2016 Comprehensive Modeling and Simulations. *Biovia Discov. Studio* **2016**, *1*, 4.
45. Matsuura, H.; Sokabe, T.; Kohno, K.; Tominaga, M.; Kadowaki, T. Evolutionary Conservation and Changes in Insect TRP Channels. *BMC Evol. Biol.* **2009**, *9*, 228. [CrossRef]
46. Tamura, K.; Stecher, G.; Peterson, D.; Filipski, A.; Kumar, S. MEGA6: Molecular Evolutionary Genetics Analysis Version 6.0. *Mol. Biol. Evol.* **2013**, *30*, 2725–2729. [CrossRef]
47. Isman, M.B.; Rodriguez, E. Larval Growth Inhibitors from Species of Prthenium (Asteraceae). *Phytochemistry* **1983**, *22*, 2709–2713. [CrossRef]
48. Szczepanik, M.; Gliszczyńska, A.; Hnatejko, M.; Zawitowska, B. Effects of Halolactones with Strong Feed-ing-Deterrent Activity on the Growth and Development of Larvae of the Lesser Mealworm, Alphitobius Diaperinus (Coleoptera: Tenebrionidae). *Appl. Entomol. Zool.* **2016**, *51*, 393–401. [CrossRef]
49. Sosa, A.; Diaz, M.; Salvatore, A.; Bardon, A.; Borkosky, S.; Vera, N. Insecticidal Effects of Vernonanthura Nebularum against Two Economically Important Pest Insects. *Saudi J. Biol. Sci.* **2019**, *26*, 881–889. [CrossRef]
50. Andreazza, F.; Haddi, K.; Oliveira, E.E.; Ferreira, J.A.M. Drosophila Suzukii (Diptera: Drosophilidae) Arrives at Minas Gerais State, a Main Strawberry Production Region in Brazil. *Fla. Entomol.* **2016**, *99*, 796–798. [CrossRef]
51. Mendonca, L.D.P.; Oliveira, E.E.; Andreazza, F.; Rezende, S.M.; Faroni, L.R.D.A.; Guedes, R.N.C.; Haddi, K. Host Potential and Adaptive Responses of Drosophila Suzukii (Diptera: Drosophilidae) to Barbados Cherries. *J. Econ. Entomol.* **2019**, *112*, 3002–3006. [CrossRef]
52. Ferronato, P.; Woch, A.L.; Soares, P.L.; Bernardi, D.; Botton, M.; Andreazza, F.; Oliveira, E.E.; Corrêa, A.S. A Phy-logeographic Approach to the Drosophila Suzukii (Diptera: Drosophilidae) Invasion in Brazil. *J. Econ. Entomol.* **2019**, *112*, 425–433. [CrossRef]
53. Zanuncio-Junior, J.S.; Fornazier, M.J.; Andreazza, F.; Culik, M.P.; Mendonça, L.D.P.; Oliveira, E.E.; Martins, D.D.S.; Fornazier, M.L.; Costa, H.; Ventura, J.A. Spread of Two Invasive Flies (Diptera: Drosophilidae) Infesting Commercial Fruits in Southeastern Brazil. *Fla. Entomol.* **2018**, *101*, 522–525. [CrossRef]
54. De Souza, M.T.; de Souza, M.T.; Morais, M.C.; Oliveira, D.d.C.; de Melo, D.J.; Figueiredo, L.; Zarbin, P.H.G.; Zawadneak, M.A.C.; Bernardi, D. Essential Oils as a Source of Ecofriendly Insecticides for Drosophila Suzukii (Diptera: Drosophilidae) and Their Potential Non-Target Effects. *Molecules* **2022**, *27*, 6215. [CrossRef] [PubMed]
55. Haddi, K.; Turchen, L.M.; Viteri Jumbo, L.O.; Guedes, R.N.C.; Pereira, E.J.G.; Aguiar, R.W.S.; Oliveira, E.E. Re-thinking Biorational Insecticides for Pest Management: Unintended Effects and Consequences. *Pest. Manag. Sci.* **2020**, *76*, 2286–2293. [CrossRef] [PubMed]
56. Vieira, J.G.A.; Krüger, A.P.; Scheuneumann, T.; Morais, M.C.; Speriogin, H.J.; Garcia, F.R.M.; Nava, D.E.; Bernardi, D. Some Aspects of the Biology of Trichopria Anastrephae (Hymenoptera: Diapriidae), a Resident Parasitoid At-tacking Drosophila Suzukii (Diptera: Drosophilidae) in Brazil. *J. Econ. Entomol.* **2020**, *113*, 81–87. [CrossRef] [PubMed]
57. Woltering, S.B.; Romeis, J.; Collatz, J. Influence of the Rearing Host on Biological Parameters of Trichopria Dro-sophilae, a Potential Biological Control Agent of Drosophila Suzukii. *Insects* **2019**, *10*, 183. [CrossRef]
58. Hofmann, T.; Chubanov, V.; Chen, X.; Dietz, A.S.; Gudermann, T.; Montell, C. Drosophila TRPM Channel Is Es-sential for the Control of Extracellular Magnesium Levels. *PLoS ONE* **2010**, *5*, e10519. [CrossRef]
59. Shimomura, K.; Oikawa, H.; Hasobe, M.; Suzuki, N.; Yajima, S.; Tomizawa, M. Contact Repellency by L-Menthol Is Mediated by TRPM Channels in the Red Flour Beetle Tribolium Castaneum. *Pest. Manag. Sci.* **2021**, *77*, 1422–1427. [CrossRef]
60. Turner, H.N.; Armengol, K.; Patel, A.A.; Himmel, N.J.; Sullivan, L.; Iyer, S.C.; Bhattacharya, S.; Iyer, E.P.R.; Landry, C.; Galko, M.J.; et al. The TRP Channels Pkd2, NompC, and Trpm Act in Cold-Sensing Neurons to Mediate Unique Aversive Behaviors to Noxious Cold in Drosophila. *Curr. Biol.* **2016**, *26*, 3116–3128. [CrossRef]

 *insects*

Article

New Insights on Antennal Sensilla of *Anastrepha ludens* (Diptera: Tephritidae) Using Advanced Microscopy Techniques

Larissa Guillén [1,*], Lorena López-Sánchez [2], Olinda Velázquez [2], Greta Rosas-Saito [2], Alma Altúzar-Molina [1], John G. Stoffolano, Jr. [3], Mónica Ramírez-Vázquez [2] and Martín Aluja [1,*]

[1]   Red de Manejo Biorracional de Plagas y Vectores, Instituto de Ecología, A.C.—INECOL, Clúster Científico y Tecnológico BioMimic, Carretera antigua a Coatepec 351, El Haya, Xalapa 91073, Ver., Mexico; alma.altuzar@inecol.mx

[2]   Red de Estudios Moleculares Avanzados, Instituto de Ecología, A.C.—INECOL, Clúster Científico y Tecnológico BioMimic®, Carretera antigua a Coatepec 351, El Haya, Xalapa 91073, Ver., Mexico; lorena.lopez@inecol.mx (L.L.-S.); olinda.velazquez@inecol.mx (O.V.); greta.rosas@inecol.mx (G.R.-S.); m.ramirez@ciencias.unam.mx (M.R.-V.)

[3]   Stockbridge School of Agriculture, University of Massachusetts, Amherst, MA 01003, USA; stoff@umass.edu

*   Correspondence: larissa.guillen@inecol.mx (L.G.); martin.aluja@inecol.mx (M.A.); Tel.: +52-(228)-8421841 (L.G. & M.A.)

Simple Summary: In insects, including tephritid fruit flies, some of which are notorious pests of commercially grown fruit, the antenna harbors the sensilla responsible for the perception of odors (chemicals carried by air), temperature, humidity, and movement. As one of the methods used to monitor and control these agricultural pests is using traps baited with attractive odors, or toxic bait sprays, both of which an adult fly detects through the antenna, the study of this organ is crucial in understanding the behavior of the insect and applying this information in its environmentally friendly control/management. In this study, we detected up to 16 different subtypes of sensilla and various other hitherto unknown structures with the help of various types of microscopes in the antenna of the Mexican Fruit Fly, *Anastrepha ludens*, a pest of citrus and mango. We describe these sensilla/structures and suggest possible functions. As other researchers have previously worked on this topic, we made a special effort to uniformize the criteria used to classify these key structures, update the terminology, and better describe each sensilla with the help of detailed photographs.

Abstract: Using light, transmission, scanning electron, and confocal microscopy, we carried out a morphological study of antennal sensilla and their ultrastructures of the Mexican Fruit Fly *Anastrepha ludens* (Loew), an economically important species that is a pest of mangos and citrus in Mexico and Central America. Our goal was to update the known information on the various sensilla in the antennae of *A. ludens*, involved in the perception of odors, temperature, humidity, and movement. Based on their external shape, size, cuticle-thickness, and presence of pores, we identified six types of sensilla with 16 subtypes (one chaetica in the pedicel, four clavate, two trichoid, four basiconic, one styloconic, and one campaniform-like in the flagellum, and three additional ones in the two chambers of the sensory pit (pit-basiconic I and II, and pit-styloconic)), some of them described for the first time in *A. ludens*. We also report, for the first time, two types of pores in the sensilla (hourglass and wedge shapes) that helped classify the sensilla. Additionally, we report a campaniform-like sensillum only observed by transmission electronic microscopy on the flagellum, styloconic and basiconic variants inside the sensory pit, and an "hourglass-shaped" pore in six sensilla types. We discuss and suggest the possible function of each sensillum according to their characteristics and unify previously used criteria in the only previous study on the topic.

Keywords: *Anastrepha ludens*; Tephritidae; olfaction; antennal ultrastructures; sensilla; sensory pit; hourglass-pore

Citation: Guillén, L.; López-Sánchez, L.; Velázquez, O.; Rosas-Saito, G.; Altúzar-Molina, A.; Stoffolano, J.G., Jr.; Ramírez-Vázquez, M.; Aluja, M. New Insights on Antennal Sensilla of *Anastrepha ludens* (Diptera: Tephritidae) Using Advanced Microscopy Techniques. *Insects* **2023**, *14*, 652. https://doi.org/10.3390/insects14070652

Academic Editor: Peter H. Adler

Received: 1 June 2023
Revised: 4 July 2023
Accepted: 7 July 2023
Published: 20 July 2023

1. Introduction

The antenna is considered the major sensory organ of insects [1] because it contains the main receptors involved in perceiving odors, movement, temperature, and humidity [2]. The receptors are located in sensilla, which have been classified according to their function (i.e., olfactory, mechanoreceptors, or thermohygroreceptors), external morphology (e.g., clavate, trichoid, basiconic, campaniform, and styloconic), cuticle texture (i.e., multiporous pitted sensilla (MPS), no-pore sensilla (NPS), multiporous grooved sensilla (MPGS)), size, thickness-walled cuticle (i.e., thick-walled and thin-walled) and a combination of some of these characteristics [3–7].

In the case of fruit flies (Diptera: Tephritidae), different authors have reported distinct types of sensilla in various economically important species. For example, for *Ceratitis capitata* (Wiedemann), Levinson et al. [8] reported three types of sensilla, whereas Mayo et al. [9], Dickens et al. [5], and Bigiani et al. [10] reported four. Four types were reported in *Anastrepha* (formerly *Toxotrypana*) *curvicauda* (Gerstaecker) [11], *Anastrepha ludens* (Loew), *Bactrocera cucurbitae* (Coquillett) and *Bactrocera dorsalis* (Hendel) [5]. Recently, Perre et al. [12] also reported four types of olfactory sensilla in *Anastrepha obliqua* (Macquart), *Anastrepha bistrigata* Bezzi, *Anastrepha grandis* (Macquart), *Anastrepha serpentina* Wiedemann, *Anastrepha* sp.2 aff. *fraterculus* (s. Selivon), *Anastrepha sororcula* Zucchi, *Anastrepha montei* Lima, and *Anastrepha pickeli* Lima. Additionally, Hu et al. [13] reported six types of sensilla in *B. cucurbitae* and *B. dorsalis*, and other authors reported five in *Eurosta solidaginis* Fitch [14], six in *Bactrocera oleae* (Rossi) [15], *Bactrocera tryoni* Froggatt [13,16], *Bactrocera tau* (Walker), *Bactrocera minax* (Enderlein), *Bactrocera diaphora* (Hendel), *Bactrocera scutellata* (Hendel) [17], and *Anastrepha fraterculus* (Wiedemann) [7], seven in *Bactrocera zonata* (Saunders) [18,19], and ten types in *A. serpentina* Wiedemann [20]. The challenge one faces with this specialized literature is that the use of different study techniques/methodological approaches for these structures results in different classifications and terminologies for naming them, a fact that can generate confusion.

Anastrepha ludens, the Mexican fruit fly, is an economically important species that attacks citrus and mangos. Despite its significant status as a pest, the antenna have been little studied. The only known study is the one by Dickens et al. [5], who, using Scanning Electronic Microscopy (SEM) and Transmission Electronic Microscopy (TEM), reported, according to the cuticular texture and internal morphology, four types of sensilla (thick-walled MPS, thin-walled MPS, MPGS and NPS) in the antennal flagellum of males and females.

In preliminary observations on the antennae of wild *A. ludens* flies, we recognized some structures that were not mentioned in the work of Dickens et al. [5], which could be important in future electrophysiological studies searching for chemical compounds to develop attractants. We also recently studied the broad morphology and proteomics of the antennae of this pestiferous species, with the aim of better understanding the response to a potent commercial attractant [21]. Considering the above, we report on an in-depth morphological analysis of the sensilla present in the flagellum and sensory pit in the antenna of mature and immature *A. ludens* females and males using light, SEM, TEM, and confocal microscopy techniques. We also update the terminology in the context of the current nomenclature and suggest the types of sensilla that could be associated with the chemical reception of various volatiles.

2. Materials and Methods

2.1. Insects

For SEM and TEM studies, we used wild *A. ludens* flies originating from white sapote fruit (*Casimiroa edulis* La Llave and Lex.), one of the *A. ludens* native hosts collected in Xalapa, Veracruz, Mexico. For confocal microscopy images, we used Laboratory-reared flies maintained at the Red de Manejo Biorracional de Plagas y Vectores at the Instituto de Ecología A.C. in Xalapa, Veracruz [22]. This colony is periodically refreshed with wild material, so we felt justified in using some specimens, as morphological changes in the antenna have not been reported in lab-reared flies.

Newly emerged and 15–20-day old *A. ludens* females and males were used to identify the antennal sensilla using three microscopy techniques (confocal, SEM and TEM). In the case of sexually mature flies (15 days old), they were kept from their emergence as adults (from pupae) until their use in 30 × 30 × 30 cm Plexiglass cages with food ad libitum (mixture 3:1 of sugar and protein) and water in a laboratory at a temperature of $27 \pm 1\,^{\circ}$C and RH of $70 \pm 5\%$. We kept low numbers of flies in these enclosures to avoid damage to the antennae or contamination through excessive dust or other materials.

Considering that there are different terminologies for identifying and classifying fly antennal sensilla in the literature, we reviewed previous publications and summarized them to homologize the terminology/nomenclature being used. We decided to use the classical nomenclature, where the classification scheme is based on the external shape of the sensilla in combination with the cuticular texture terminology used by Giannakakis and Fletcher [16].

2.2. Transmission Electron Microscopy (TEM)

Antennae of five females and males of both ages were fixed over a week in a mixture of 2.5% glutaraldehyde and 2.0% paraformaldehyde in phosphate buffer at pH 7.4 [23]. Samples were then post-fixed in 1.0% OsO4 for 2 h and then dehydrated using a graded ethanol series (30–100%) for 10 min at each concentration. Heads with antennae were mounted in LR–white resin polymerized at 50 °C for 24 h inside jelly capsules (EMS®, Hatfield, UK). Ultrathin longitudinal sections around the antenna (blue peripheral line in Figure 1) of 70 nm were cut with a Leica EMUC7 ultramicrotome; then, basal, medial, and apical sections of the flagellum were analyzed. The slides were placed on a 200-copper mesh (EMS®) and stained with 2% uranyl acetate and lead citrate [24]. Samples were examined with a JEM-1400 PLUS transmission electron microscope (JEOL Ltd., Tokyo, Japan) and photographed using a GAT-830.10U3 camera (GATAN Inc., Pleasanton, CA, USA).

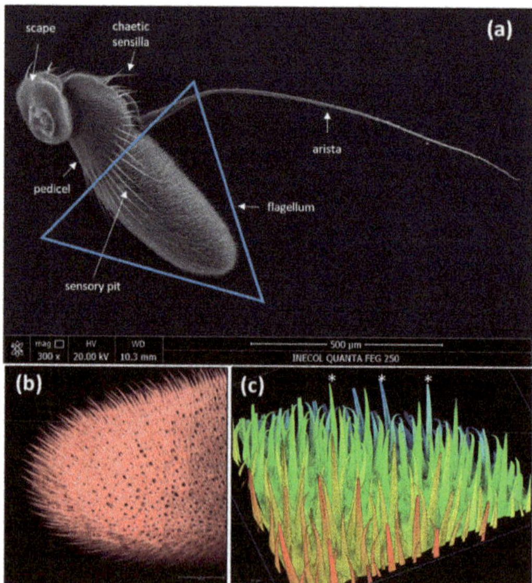

Figure 1. (**a**) Scanning electron micrograph of an overall view of the antennae of *A. ludens*; (**b**) Confocal image of the apical part of the flagellum showing the distribution of different sensilla types represented by black holes of different sizes; (**c**) Confocal image showing different types of sensilla (the longest ones marked with white asterisks are trichoid sensilla); further details in Video S1.

2.3. Scanning Electronic Microscopy (SEM)

Antennae of five females and males of both ages were fixed in a Karnovsky solution [23] for at least a week. Once fixed, specimens were rinsed three times in phosphate buffer at pH 7.2, and then dehydrated using a graded ethanol series (30, 50, 70, and 90%) for 30 min at each concentration and three times with absolute alcohol for 15 min. They were then dried in a critical point dryer (Quorum K850, Quorum technology, UK), followed by attachment to aluminum stubs using a carbon adhesive before coating with gold in a sputtering Quorum Q150 RS [25]. The preparations were studied and photographed with a FEI Quanta 250 FEG scanning electron microscope (FEI Co., Brno, Czech Republic).

2.4. Confocal Microscopy

Antennae of five females and males of both ages were fixed in 4% paraformaldehyde and PBS (0.2 m/7.2 pH). Subsequently, they were placed in a 10% potassium hydroxide solution to remove other tissue, mostly fat bodies. The antennae were stained with Congo Red (Sigma-Aldrich, Steinheim, Germany) dissolved in 70% ethanol and incubated at room temperature for 72 h. The samples were gradually dehydrated in ethanol (70% to 100%). The antennae were mounted with CytosealTM 60 mounting media (Richard-Allan Scientific™ Thermo Scientific™).

Imaging and rendering: Serial optical sections were obtained at 0.2 mm intervals on a TCS-SP8+STED (Leica Microsystems GmbH, Wetzlar, Germany) confocal microscope with an HCX PL APO 40x/1.30 OIL CS2 objective and HCX PL APO 63x/1.40 OIL CS2 objectives. A laser line of 488 nm was used for imaging the Congo-Red-stained cuticle, the laser power was set to 30% and the emitted fluorescent light was detected in the range from 613 nm to 683 nm.

3. Results

As previously reported for *A. ludens* and other tephritid flies [7,8,12,15], the antennae have three segments: scape, pedicel, and flagellum (also called funiculus or post pedicel), covered with different types and subtypes of sensilla and microtrichia (Figure 1, Video S1). The comparison of each antennal-segment size, measured by its length and width, indicates no differences between females and males, except for the width of the flagellum (Table 1). As Dickens et al. [5] originally reported, the antenna also has an arista inserted on the dorsal–proximal end of the flagellum and a sensory pit (also named olfactory pit) present on the dorsal–proximal surface of the flagellum (Figure 1).

Table 1. Mean (± SE) length and width of different antennal segments of *A. ludens* females and males (n = 50).

Segment	Length (μm)		T-Value	Width (μm)		T-Value
	Female	Male	(p-Value)	Female	Male	(p-Value)
Scape	97.3 ± 1.8	96.6 ± 1.5	0.31 (0.76)	180.9 ± 3.2	178.4 ± 2.0	0.66 (0.51)
Pedicel	169.5 ± 2.8	166.0 ± 2.4	0.94 (0.35)	193.2 ± 1.9	195.6 ± 1.5	−0.98 (0.33)
Flagellum	432.3 ± 3.3	430.5 ± 2.7	0.44 (0.66)	225.2 ± 4.5	214.8 ± 2.2	2.08 (**0.04**)
Arista	995.3 ± 4.7	998.8 ± 4.0	−0.55 (0.58)	38.36 ± 1.3	39.0 ± 1.6	−0.33 (0.74)

(p-Value) in bold numbers are significantly different ($p < 0.05$).

Based on the shape, length, cuticle thickness, pore density in the cuticle (i.e., multiple, few, one, or none), pore shape, and if the sensillum is socket-based, we identified a total of 16 different sensilla subtypes (Figures 2–12) (including the three sensilla in the sensory pit) and microtrichia (mi), mainly distributed on the flagellum (Figure 1b,c, Video S1; Table 2) of the antennae of *A. ludens* females and males.

Table 2. Equivalencies between the names of the sensilla in the *A. ludens* flagellum used by us and the ones by Dickens et al. [5], including their putative/potential function.

No.	Sensillum Name Used Here	Sensillum Name *sensu* Dickens et al. [5]	Putative Function and (Location)
1	Basiconic I (rounded tip, the longest with socket, thin-walled multipore (MPS) with hourglass-like porous)	Thin-walled-MPS	Chemoreception (flagellum)
2	Basiconic II (rounded tip, shorter and with a bigger diameter than b-I with socket, thin-walled MPS with hourglass-like porous)	Not reported	Chemoreception (flagellum)
3	Basiconic III (thick-walled with few wedge-like pores)	Not reported	Chemoreception (flagellum)
4	Basiconic IV (smallest, in socket, thick-walled, with wedge-like porous	Not reported	Chemoreception (flagellum)
5	Pit-basiconic I (nutshell-like cuticle texture with tip protuberance)	Not reported	Contact chemoreception (Internal chamber of the sensory pit)
6	Pit-basiconic II (scaly-like cuticle texture with a tip rosette)	Not reported	Chemoreception/Thermoreception (Internal chamber of the sensory pit)
7	Campaniform-like	Not reported	(flagellum)
8	Chaetica	No-pore (NPS)	Mechanoreception (scape and pedicel)
9	Clavate I (thick-walled, wedge-like pore shape)	Not reported	Chemoreception (flagellum)
10	Clavate II (thin-walled, MP with hourglass-like pore shape and with socket)	Not reported	Chemoreception (flagellum)
11	Clavate III (the shortest, thick-walled, and without pores)	Not reported	Mechanoreception (flagellum)
12	Clavate IV (similar to C-I shape but thin-walled and hourglass-like pore shape)	Not reported	Chemoreception (flagellum)
13	Styloconic	Multiporous grooved sensilla MPGS	Chemoreception (flagellum)
14	Pit-styloconic	Not reported	(External chamber of the sensory pit)
15	Trichoid I (thin-walled, hourglass-like pores and sharp-tipped, the longest of flagellum)	Thick-walled-MPS	Chemoreception (flagellum)
16	Trichoid II (thin-walled, hourglass-like pores and blunt-tipped)	Not reported	Chemoreception (flagellum)

On the scape, we only detected chaetica sensilla (ch) and microtrichia (Figure 1a). The pedicel has a line of prominent chaetica sensilla in the frontal margin and plenty of microtrichia (Figure 1a).

On the flagellum or funiculus, based on the shape and using TEM and SEM techniques, we identified four main types of sensilla—trichoid (tr), clavate (c), basiconic (b), and styloconic (s)—with different subtypes according to the presence and shape of pores, cuticle width and size (Figures 2–12). The four types were already reported for *A. ludens* by Dickens et al. [5] using different terminology (Table 2). In the TEM study, we also found a different kind of campaniform-like sensilla (cm), which was not previously reported for *A. ludens*. However, since we were unable to conclusively identify it in the SEM study, we handled this finding with caution because it could be an incomplete capture of a sensillum in a bad position. In addition, we report, for the first time, two subtypes of sensilla that differ from all previously described sensilla, inside of the sensory pit of the *A. ludens* flagellum. Below, we provide descriptions of each subtype of sensilla.

3.1. Types, Subtypes, and Descriptions of Sensilla

3.1.1. Basiconic

The basiconic sensillum is digitiform, with a wide base that gradually narrows towards the tip and is shorter than the trichoid sensillum (Figures 2 and 3).

We identified six subtypes of basiconic sensilla. According to their tip shape (sharp or blunt), longitude, the thickness of the cuticular wall, and the presence of hourglass-shaped pores in the flagellum, we recognized four subtypes along the flagellum (Figures 2 and 3) and two inside of the sensory pit of the flagellum (check Section 3.2). Basiconic type I (b-I) are apparently the longest. They have a thin wall with hourglass pores, a rounded tip, and are inserted in a socket (Figures 2a,b and 3a); this sensillum was named "thin-walled multipore pitted sensilla" (MPS) by Dickens et al. [5] and "thin-walled MPS long subtype I" by Castrejón-Gómez and Rojas [20].

Basiconic subtype II (b-II) sensilla are shorter and with a bigger diameter in the tip than b-I, have a thin cuticular wall with hourglass pores, a rounded tip, and are also inserted in a socket (Figures 2a,c and 3b); this sensillum was named "thin-walled MPS short subtype II" by Castrejón-Gómez and Rojas [20].

Basiconic subtype III (b-III) sensilla have a thick cuticular wall with few wedge-shaped pores (Figures 2a,d and 3c). Finally, basiconic subtype IV (b-IV) sensilla are the smallest of our sub-classification, have a socket, a thick cuticular wall, and few wedge-shaped pores (Figures 2a,e and 4e–f).

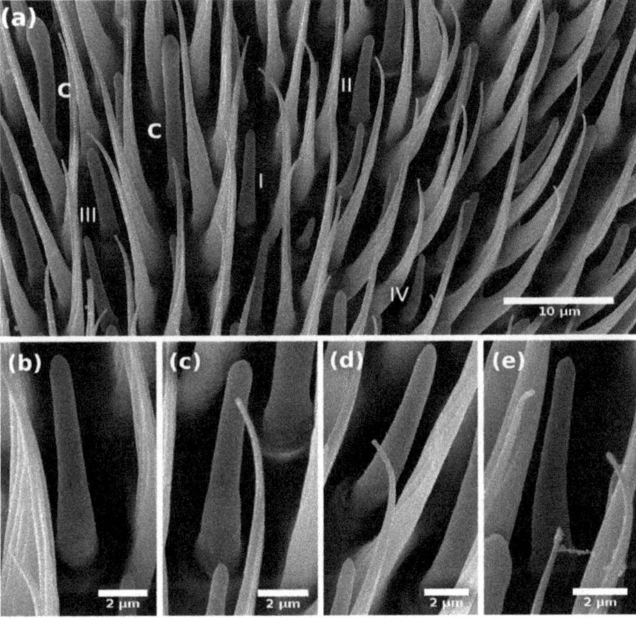

Figure 2. Scanning electron micrographs of the basiconic sensilla on the flagellum (apical segment) of *A. ludens* male: (**a**) Segment showing basiconic sensilla subtypes I–IV; (**b**) Subtype I; (**c**) Subtype II; (**d**) Subtype III; (**e**) Subtype IV.

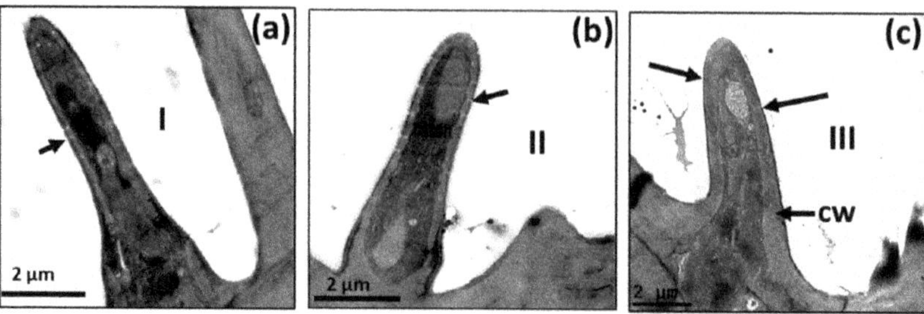

Figure 3. Transmission electron micrographs of a longitudinal section of the basiconic sensilla on the flagellum of *A. ludens* antenna showing cuticle wall [cw] and pores [arrows]: (**a**) Subtype I with thin-cuticle wall and pores of 15-day-old male; (**b**) Subtype II with thin cuticle wall of 15-day-old male; (**c**) Subtype III with thick cuticle wall of 15-day-old female.

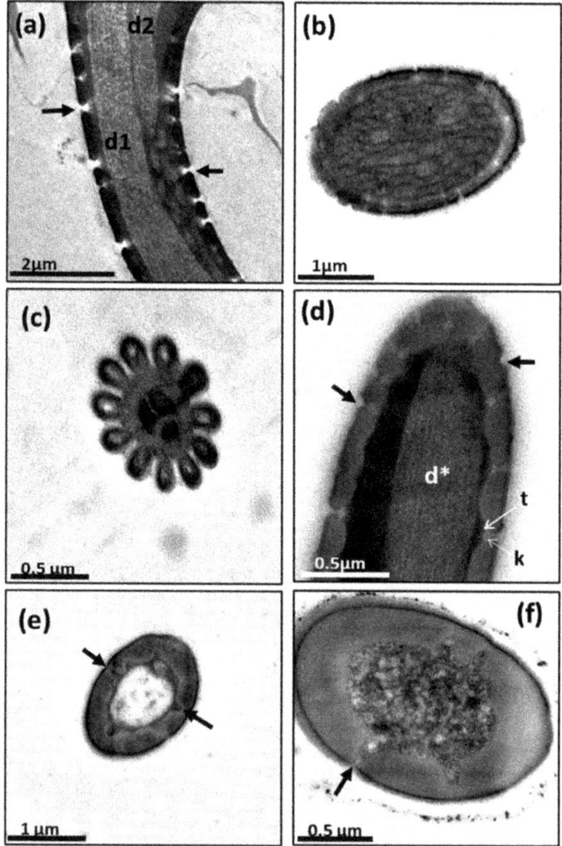

Figure 4. Transmission electron micrographs of sensilla in the antenna of *A. ludens* individuals: (**a**) Close-up of the longitudinal section of a trichoid sensillum showing in its internal middle part two dendrites [d1 and d2]. Note the hourglass pores (arrows), the kettle or pore pot, and the pore tubules through which the volatile molecules disperse towards the sensilla lymph; (**b**) Cross-section of the distal part of a clavate sensillum showing the lamellar dendrites inside and its cuticle interrupted

by pores; (**c**) Cross-section of a styloconic sensillum showing double-wall, three well-defined dendrites in the central part and 11 digitiform projections similar to what Dickens et al. [5] reported; (**d**) Longitudinal section of a basiconic sensillum on its distal part, showing a dendrite (d*) inside and the pores in the cuticle with an hourglass shape (arrows), the kettle (k), and the tubules (t); (**e,f**) Transverse section of two thick-walled sensilla with wedge-shaped pores (arrows) like the one found in the basiconic thick-walled sensillum subtype III and the clavate subtype III.

3.1.2. Chaetica Sensilla

The chaetica sensilla are the longest in the antenna, with a cone shape. They are longitudinally ridged and have a pointed tip (Figure 5). The end of the ridged part of the hair is attached to a socket that is probably suspended in a joint membrane [2], which permits the sensillum's free movement with an apparent mechanoreception function (Figure 5c,d). Interestingly, in the lower outside part of the socket, there is a group of 16–20 tiny pores (Figure 5c,d).

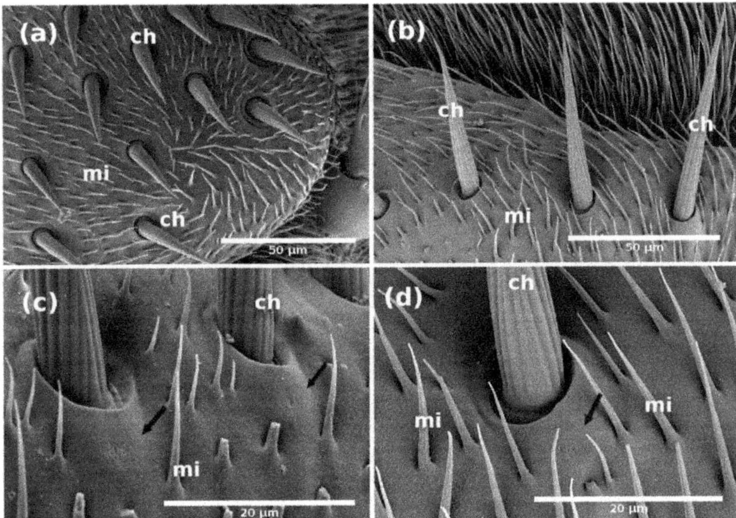

Figure 5. Scanning electron micrographs of chaetica sensilla [ch] and microtrichia [mi] on the pedicel of *A. ludens*: (**a**) Aerial view showing the distribution of chaetica sensilla and microtrichia; (**b**) Lateral view of the marginal part of the pedicel, where it is possible to see the cone shape of the chaetica sensilla; (**c,d**) Close-up of the base of chaetica sensilla showing the longitudinally ridged cuticle, the socket, and the groups of tiny pores in the socket base, denoted by black arrows.

3.1.3. Clavate Sensilla

According to the sensillum shape, the width of the cuticle wall, and the pore shape in the cuticle, we defined four subtypes of clavate sensilla (Figures 6 and 7); one more than reported for other species of tephritids to date.

The clavate type I (c-I) sensilla have a short waist, which widens at the top, and a thick cuticular wall (Figure 7a), probably with wedge-shaped pores, similar to those shown in Figure 4e,f. The clavate-type II (c-II) sensilla, observed on the medium part of the female flagellum, have the smallest diameter in the middle part, are socket-based, and have a thin wall (Figures 6c,d and 7b) with an hourglass-shape and multiple similar to those shown in Figure 4a,d. The clavate type III (c-III) sensilla have the biggest diameter and thickest walls, apparently with pores (Figure 7c). This sensillum, which is reported for the first time in a tephritid fly, has a typical club shape, and a shorter base than subtypes I, II and IV; it was only observed in the distal flagellum of females and males. Finally, we identified a clavate type IV (c-IV) sensillum through TEM with a similar shape to subtype I but exhibiting a thinner wall with multiple hourglass-shaped pores (Figure 7d).

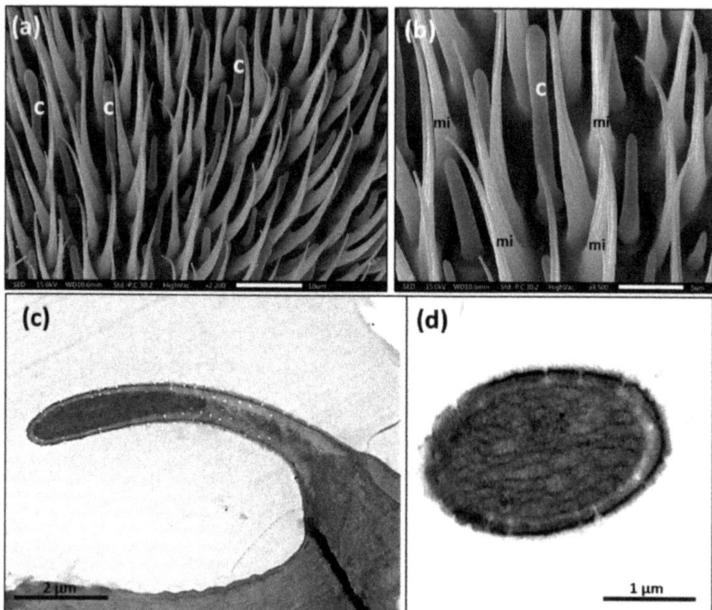

Figure 6. (**a**) Scanning electron micrograph showing the distribution of clavate sensilla (details under 'c') in a flagellum segment of a 15-day-old female *A. ludens*; (**b**) Scanning electron micrograph with a close-up of clavate type sensillum surrounded by microtrichia (mi) and other sensilla types; (**c**) Transmission electron micrograph of a clavate sensilla subtype II characterized by a thin cuticle with multiple pores; (**d**) Transmission electron micrograph showing a transversal section of a thin-walled clavate sensillum in newly emerged males (0 days).

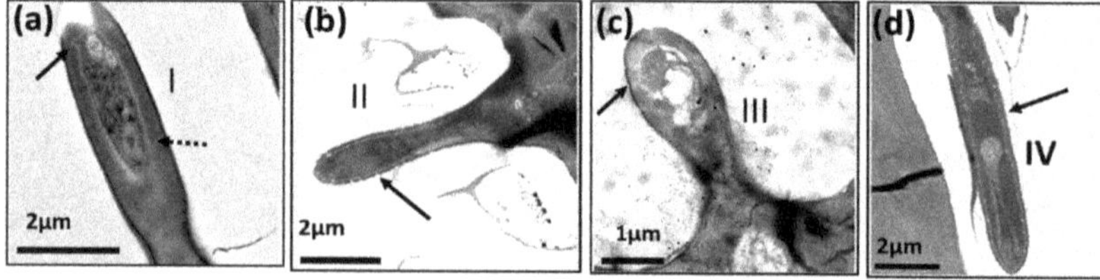

Figure 7. Transmission electron micrographs of longitudinal sections of clavate sensilla on the flagellum of *A. ludens* antenna: (**a**) Subtype I sensillum with thick cuticle wall (dotted arrow) and with wedge-shaped pores (black arrow) of 15-day-old male; (**b**) Subtype II sensillum with thin cuticle wall, hourglass-shape and multiple pores (arrow) of 15-day-old female; (**c**) Subtype III sensillum, the smallest of the subtypes, with thick cuticle wall and wedge-shaped pores (arrow) of 0 day-old female; (**d**) Subtype IV sensillum with thin cuticle wall (arrow) and multiple hourglass-shape pores (arrow) in a 15-day-old male (as shown in Figure 4a).

3.1.4. Styloconic

These sensilla are the smallest ones we identified. They are about 3 µm long and are characterized by grooves and ridges that make them look like a group of digitiform projections (Figure 8). They have a double wall (Figure 4c) that consists of a cuticular sheath that wraps three dendrites in its internal lumen. They have 11-digit type projections with pores between them (Figure 8a,b). We detected them in different areas of the flagellum of males and females.

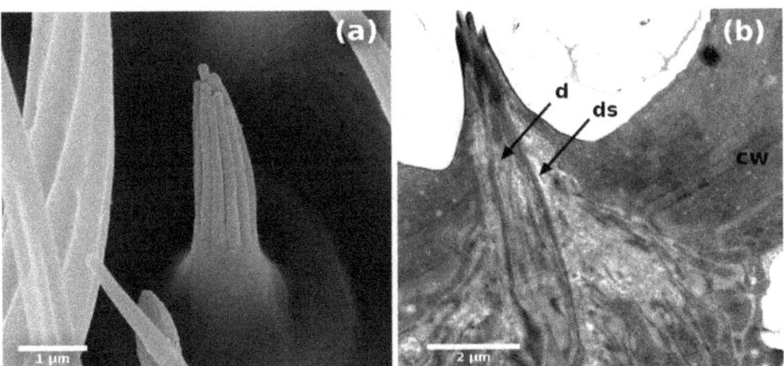

Figure 8. (**a**) Scanning electron micrographs of styloconic sensillum of *A. ludens*; (**b**) Transmission electron micrographs of styloconic sensillum showing an electron-dense dendritic sheath [ds], dendrites [d], and the cuticle wall [cw].

3.1.5. Trichoid

The multipore trichoid-type sensilla we identified had the longest, thinnest, and most conspicuous shape of all sensilla identified in this study (Figures 9 and 10), similar to those previously reported for *B. tryoni* [13,16], *A. curvicauda* [11], and *A. fraterculus* [7]. In our case, however, we identified two subtypes of trichoid sensilla (Figures 9 and 10).

In observations of longitudinal sections via TEM, we detected that within the trichoid sensilla, there were two variants: sharp (Figures 9 and 10a), and blunt-tipped (Figures 9 and 10b), which we named trichoid I (Tr-I) and trichoid II (Tr-II), respectively. Both types are thin-walled with an hourglass shape and multiple pores (Figure 10a,b). In Figure 4a, a close-up of the trichoid sensillum of the proximal flagellum, it is possible to perceive the hourglass-shaped pores and the tubule projections that run into the dendrite branches. Tr-I are the most abundant and longest in the flagellum and are longer than Tr-II (Figure 9a).

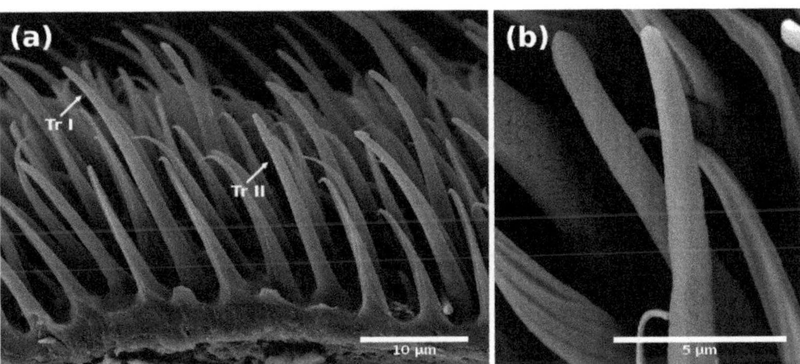

Figure 9. Scanning electron micrographs showing: (**a**) Trichoid sensilla among microtrichia; subtype I (Tr-I) sharply tipped and subtype II (Tr-II) blunt tipped on the flagellum of a 15-day-old *A. ludens* female; (**b**) Close-up of trichoid sensilla showing multiple pores on the surface.

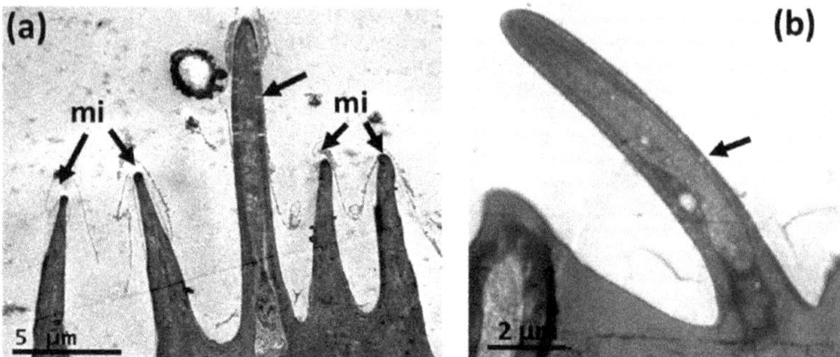

Figure 10. Transmission electron micrographs of a longitudinal section of the trichoid sensilla on the flagellum of an *A. ludens* adult antenna: (**a**) Subtype I with sharp-tip among microtrichia (mi) in a newly emerged female (black arrow points to hourglass-shaped pores); (**b**) Subtype II with a blunt tip (black arrow points to hourglass-shaped pores) in a 15-day-old female.

3.1.6. Campaniform-like Sensillum and Glands

In the flagellum of an *A. ludens* male, we identified, with the help of TEM images, campaniform-like sensilla (Figure 11). In this case, three long cells were observed behind the cuticula, clustered very close to the sheath surrounding the sensilla's dendrite, like a campaniform sensillum, which could possibly be a secretory cell associated with this sensillum (Figure 11a).

In the TEM images, we also observed campaniform-like sensilla in females, which have a flattened external cuticular area (there is no hair as such) and are apparently innervated by two sensitive cells. Moreover, this type of sensillum is found very close to a group of secretory cells directly in contact with the cuticula, where numerous channels can be observed (Figure 11b–d). Unfortunately, we were not able to identify these sensilla with SEM images.

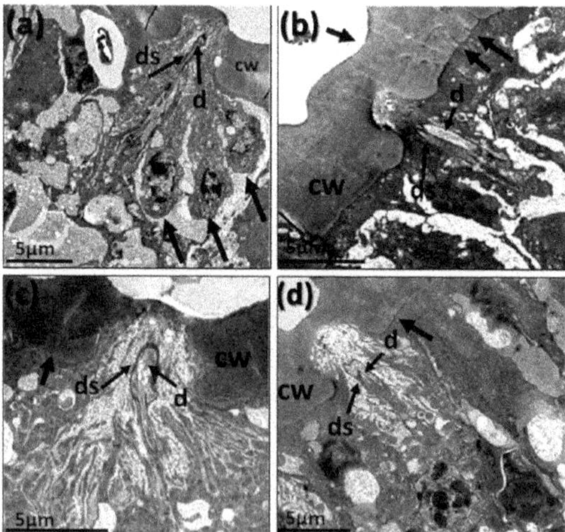

Figure 11. Transmission electron micrographs of the ultrastructure of a campaniform-like sensillum in *A. ludens*: (**a**) Longitudinal section of male sensilla showing three cells (black arrows) close to the sheath surrounding the sensilla's dendrite and cuticle wall [cw]; (**b**–**d**) Longitudinal section of like-campaniform sensillum of females showing two sensory dendrites [d] with dendrite sheath [ds] and numerous channels (black arrows).

3.2. Sensory Pit

The olfactory or sensory pit is located in the dorsobasal part of the antenna (Figure 12a, Video S2), and it is composed of two chambers (internal and external) where we found three subtypes of sensilla (Figure 12b). Chambers are physically semi-separated by a bridge of structures that look like modified microtrichia of different sizes and shapes, as well as modifications of the cuticular floor (Figure 12b,c).

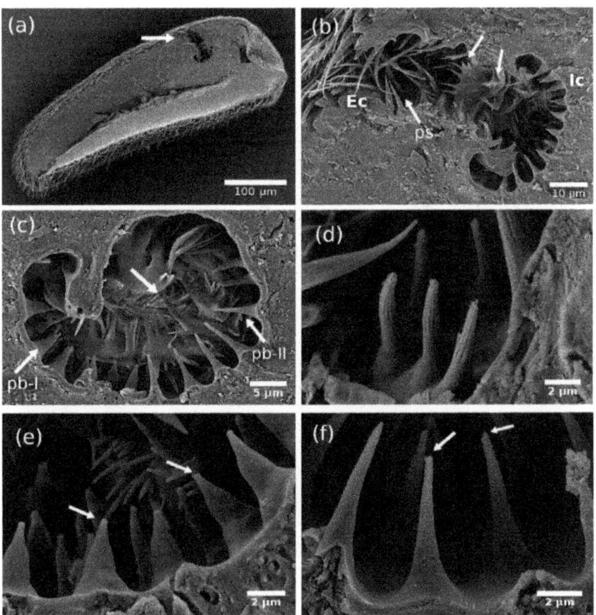

Figure 12. Scanning electron micrographs of the structures in the sensory pit located in the flagellum of the *A. ludens* antenna: (**a**) Internal part of the antenna showing the two chambers of the sensory pit; (**b**) Close-up of the chambers showing the structures (non-uniform wall and modified microtrichia pointed with white arrows) between the chambers. The distribution of the three subtypes of pit-sensilla is also discernible; (**c**) Internal chamber showing the distribution of pit-basiconic sensillum subtype I [pb-I] and II [pb-II]; White arrows point to non-uniform wall and modified microtrichia; (**d**) Close-up of pit-styloconic sensilla; (**e**) Close-up of pit-basiconic subtype I sensilla [pb-I] which have nutshell-like cuticle texture, nipple-like shape, and a tip protuberance (pointed by white arrows); (**f**) Close-up of pit-basiconic subtype II sensilla [pb-II], which have scaly-like cuticle texture with a rosette (pointed by white arrows) at the tip that appears to open and close.

External chamber (Ec): This chamber is the outermost in the pit and the smallest. It has at least eight styloconic sensilla (Figure 12b), which have, from the middle towards the tip, the characteristic finger-like form of the styloconic sensillae with different longitudinal fingers, and have a smooth cuticle from the middle towards the base (Figure 12d). We named them pit-styloconic sensilla (ps). They are longer (ca. 7 µm long) than the styloconic sensilla found in the rest of the flagellum (ca. 3.76 µm long) and are inserted into sockets, either alone or in pairs. This type of sensillum is like the one reported as "grooved sensillum" by Honda et al. [26] in the "large olfactory pit" of the onion fly, *Hylemya antiqua* Meigen (Diptera: Anthomyiidae). Although these authors suggested an olfactory function, we detected no apparent pores in the cuticle, so their function remains uncertain.

Internal chamber (Ic): This chamber is bigger than the external one and has two subtypes of basiconic sensilla, which are different from all sensilla in the rest of the flagellum. We named them pit-basiconic sensilla type I (pb-I) (Figure 12e) and type II (pb-II) (Figure 12f).

There are about 20 pit-basiconic-type I sensilla (with a nipple-like shape) (Figure S1), mainly located on the proximal side to the base of the antenna (Figure 12b). They are approximately 3–4 µm long, have a nutshell-like cuticle texture in two-thirds of the sensillum cuticle from the tip to the base, are socketed, and end with a spherical protuberance or a kind of porous plug (Figures 12e and 13).

The pit-basiconic type II (pb-II) sensilla are about 6 µm long. Approximately 13–15 pb-II are located on the side facing the apical end of the antenna (Figure 12b). They have a scaly-like texture with a rosette at the tip that appears to open and close (Figure 12f). This sensillum is similar in shape to the "striated pit sensillum" reported by Honda et al. [26] in the "large olfactory pit" of the onion fly. These authors reported, for the "striated pit sensillum", the presence of two sensory neurons that extend their dendrites to the sensillum tip, but they did not observe any pores or opening in the tip that could have suggested a gustatory or olfactory function. In our case, we do not have an internal image of this sensillum showing the dendrites, but we have images suggesting that the rosette in the tip could be a type of mouth that opens and closes. Considering the similitude of the pit-basiconic type II with the "striated pit sensillum" reported by Honda et al. [26], we suggest that this sensillum has an olfactory function.

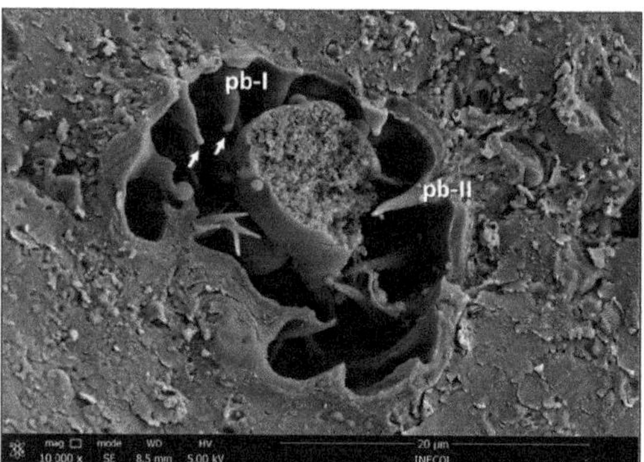

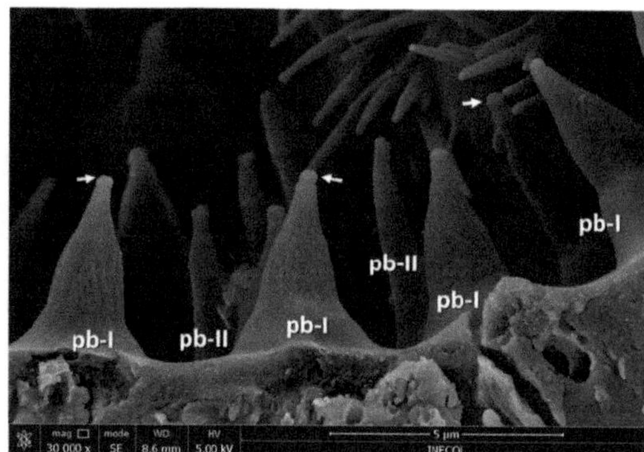

Figure 13. Scanning electron micrographs of the internal chamber of the sensory pit with pb-I and pb-II sensilla. Short arrows show the tip of pit-basiconic subtype I [pb-I] with a kind of porous plug, which could also be a secretion of viscous fluid containing a mucopolysaccharide that usually covers the tips of chemoreceptor dendrites and sometimes is exuded through the terminal pore [1].

3.3. Microtrichia (mi)

These microstructures ("mi" in Figures 5b, 6b, 10a and 14) distributed along the antenna are curved, grooved, long, and thin projections that narrow at their apical part, ending in a sharp point. These projections are non-innervated, as becomes apparent in Figures 10a and 14a–c), and in line with what other authors have reported [5,16]. However, in a transversal cut made in the middle of a microtrichium, it is possible to observe what appears to be a dendrite (Figure 14b). We propose that microtrichia are likely more associated with a protective function of the sensilla in the antenna. We also suggest that the longitudinal ridges of microtrichia could help conduct some substances by runoff to the pores on the antenna cuticle (Figure 14d). Some of the microtrichia inside the sensory pit are modified and partially separate the two chambers (Figure 12b,c).

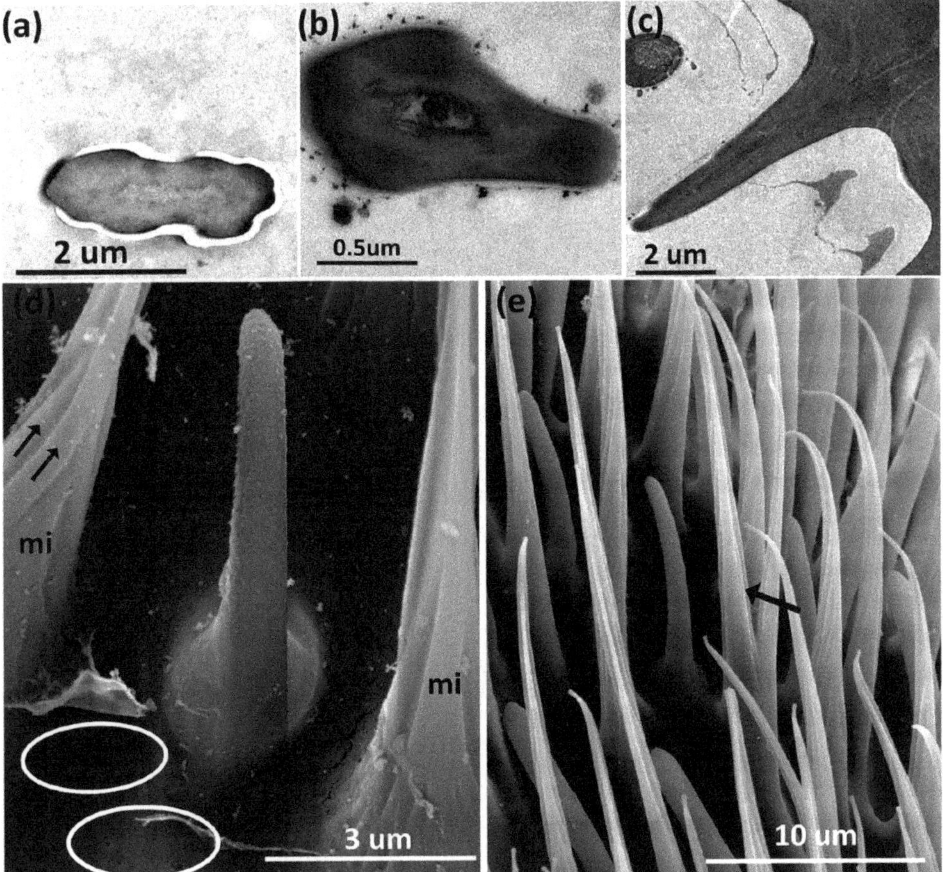

Figure 14. TEM and SEM of microtrichia: (**a**) TEM image of a transversal section of a microtrichium showing the lumen in the center; (**b**) TEM image of a transversal section of a microtrichium showing, in the center, a probable dendrite; (**c**) TEM of a microtrichium longitudinal section showing the lumen; (**d**) SEM of basiconic sensillum and microtrichia [mi] with some pores (inside white circles) in the antenna cuticle near their bases; black arrows point to the ridges of microtrichia; (**e**) Microtrichia (black arrow) surrounding sensilla.

3.4. Other Structures

In addition to sensilla, we report on other structures discovered during the preparation of the samples for SEM and TEM studies. For example, when the antenna was cut longitudinally, we found a tracheal tube crossing the medial–internal part (Figure 15a–c). Tracheae are part of the insect's air supply system, and their abundance in specific body parts or tissues reflects the demand for oxygen in those parts [1]. Other structures that we found included several rough spherical structures distributed along the deeper medium part of the flagellum (Figure 15d). We also identified small pores on the antenna cuticle on the base of some sensilla and microtrichia (Figures 5c,d and 14d).

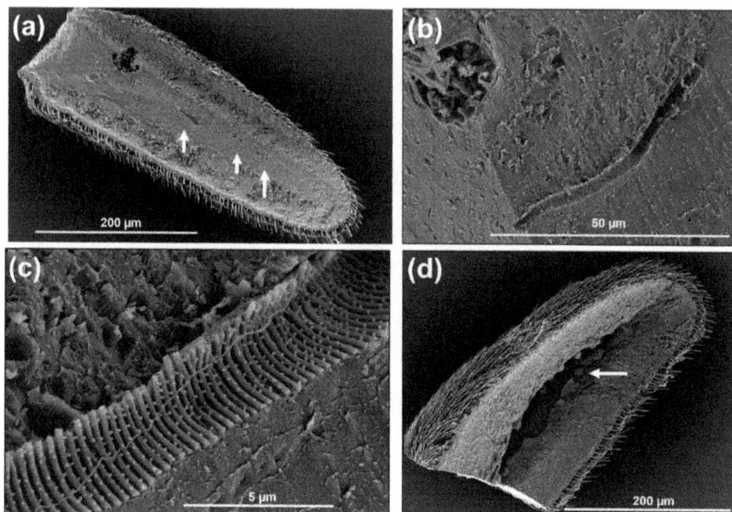

Figure 15. (**a**) The internal part of the antenna showing sections of the tracheal tube that passes through the antenna longitudinally (white arrows); (**b**) Cose-up of the trachea behind the sensory pit; (**c**) Close-up of the trachea formed by tubes (tubes were cut during sample preparation) with small pedicelled spherical protuberances and a duct crossing the antenna; (**d**) Internal part of the antenna showing several rough spherical structures.

4. Discussion

Considering the shape, size, wall thickness, and presence of pores in the cuticle and the location (flagellum and sensory pit), we were able to characterize and suggest the function of 16 subtypes of sensilla (13 in the flagellum and three in the sensory pit) of the *A. ludens* antennae. That is, we described 12 additional subtypes to those reported by Dickens et al. [5] who only characterized four types (Table 2). We report, for the first time in *A. ludens*: (a) two types of pores in the sensilla (hourglass and wedge shapes) that we also used to classify the sensilla; (b) the description of the sensory pit and their associated sensilla, classifying them according to their shape (to name them, we add the prefix "pit"); (c) the presence of two chambers (external and internal chambers); (d) pit-styloconic (ps) sensilla in the external chamber, and pit-basiconic subtype I (pb-I) and subtype II (pb-II) sensilla in the internal chamber (Figure 12); (e) a porous plug or secretion that apparently flows from the sensilla pit-basiconic subtype I (pb-I) (Figure 13); (f) a campaniform-like sensillum in the flagellum (although we could not confirm its presence with SEM images); (g) the presence of pores in the antenna cuticle (Figures 5c,d and 14); and (h) the presence of a tracheal tube crossing the internal part of the antenna longitudinally (Figure 15).

The difference in the number of sensilla subtypes with respect to the ones reported by Dickens et al. [5] is partly because these authors based their classification on the terminology proposed by Altner [3], and thus only considered the presence/absence of pores and the thickness of the cuticle, classifying all types of sensilla as no-pore sensilla (NPS) and multiple pitted sensilla (MPS), and in the case of subtypes, thick (Thick-Walled MPS), thin (Thin-Walled MPS) and Multiporous Grooved Sensilla (MPGS) (Table 2). However, the complex repertoire of behaviors that *A. ludens* and other fruit flies exhibit during host and food location, as well as courtship and mating, suggests the existence of a more sophisticated group of sensilla than just four types.

The terminology used to describe and classify sensilla has been changing according to the development of microscopy techniques that now allow for us to detect/describe internal details that are helpful in better characterizing a sensillum. In our case, we partially used the old system of Schenk [27], based on the shape and mode of insertion in the antenna-cuticular wall (e.g., the presence of a socket), which is still practical when distinguishing one type from another with a light microscope. We also used a system used by other authors e.g., [3,5,28–31] based on the presence or absence of cuticular pores, a single (thin) or double (thick) cuticular-wall, as well as the study of other internal structures discovered with SEM and TEM techniques. These two approaches helped us considerably refine the classification used in the only other study on the antenna of *A. ludens* [5].

In the case of true fruit flies (Diptera: Tephritidae), most investigations have been restricted to the study/description of morphological structures in the sensilla of the antenna, and very few have tested the functionality of these structures. Considering that adult fruit flies follow odors as cues to find food, mates, and hosts, the great advantage and contribution of the work by Dickens and collaborators [5] was the inclusion of some electrophysiological tests with *C. capitata* that confirmed that sensilla with pores were related to chemoreception and those without pores were related to mechanoreception.

We found abundant microtrichia in the pedicel and flagellum of the *A. ludens* antenna, which coincides with the findings of Dickens et al. [5], and other authors such as Giannakakis and Fletcher [16], Bisotto de Oliveira et al. [7], Hu et al. [17], and Perre et al. [12], among others, who, working with other species of fruit flies, all reported that microtrichia constitute the major cuticular structures in the antenna compared with sensilla. However, most authors describe microtrichia as non-innervated setae, curved and longitudinally ridged, omitting any mention of their function, except for Hu et al. [17], who named them "microtrichial sensilla" inferring a mechanoreception function for six species of *Bactrocera*. However, in the latter study, it is not possible to observe the elastic membrane (the joining or socket membrane observed in the chaetica sensilla in Figure 5c,d of our study) that, according to Keil and Steinbrech [2], permits the sensillum movement and, with it, the stimulation of the outer dendritic segment that in mechanosensitive sensilla is only located in the internal base of the sensillum. In our case, we show innervated microtrichia without the elastic membrane, socket, and pores (Figures 10a and 14a,c), which indicates that they do not have mechanical or chemoreceptive functions. However, we found probable evidence that microtrichia may have what appears to be a dendrite in the center, although we could not see the other usually related ultrastructures (Figure 14b). Since we could not see a well-developed dendrite in all samples, we suggest that the microtrichia could be vestiges of true sensilla that gradually lost their main function in the evolutionary specialization process of this group of flies and that, since they are very abundant and surround the true sensilla, they could possibly work as physical protectors of sensilla. We surmise that they could also possibly capture chemicals to avoid chemoreceptor sensilla saturation and conduct the captured chemicals through their longitudinal ridges to the base of the antenna cuticle, where some pores are present (Figures 5c–d and 14). Also, since we did not observe pores in the microtrichia, we suppose that they do not have a chemoreceptive function.

The trichoid sensilla we found in *A. ludens* are similar to those reported in other fruit flies such as *B. tryoni* [13,16], *A. curvicauda* [11], *A. serpentina* [20], *A. fraterculus* [7,16], and the other eight species of *Anastrepha* [12]. Notably, the trichoid sensilla we observed were not reported by Dickens et al. [5], who mention "longitudinally ridged trichoid mechanosensory sensilla along the distal margins of both the scape and the pedicel" that they classified as "No-Pore Sensilla" but that we classified as Chaetica sensilla. The same authors refer to the arista as "an elongated trichoid arista". In our study, Tr-I and Tr-II were found to be thin-walled with multiple hourglass-like pores, a detail not reported before; Tr-I are sharp, and Tr-II blunt-tipped (Figures 9 and 10). Pore presence and the shape of Tr-I coincide with the trichoid type I reported for *B. tryoni* by Giannakakis and Fletcher [16]; the difference is that, in our case, we report that trichoid subtype I (Tr-I) have a thin wall and sharp tip (Figures 9a and 10a), while those authors report a thick wall for trichoid type I. Trichoid subtype II is also similar to the trichoid type II reported by Giannakakis and Fletcher [16]. Trichoid sensilla are considered chemoreceptors and are associated with the behavior of orientation and intraspecific communication [2], specifically pheromone recognition [32,33]. In fruit flies, Levinson et al. [8] and Dickens et al. [5] reported that trichoid sensilla (= thick-walled-[MPS] of *C. capitata*) respond to sex pheromone and trimedlure extracts (an attractant based on sexual pheromones), respectively.

In the case of basiconic and clavate sensilla, there is great diversity in the *A. ludens* flagellum we studied; therefore, it is sometimes complicated to discern one from the other through SEM. These types of sensilla have been also reported in several other tephritid fruit flies such as *A. fraterculus* [7], *A. curvicauda* [11], *C. capitata* [8], *A. serpentina* [20], *B. tryoni* [13,16] and, very recently, in eight species of *Anastrepha* present in Brazil [12]. In the case of *B. dorsalis*, Liu et al. [34] did not find clavate sensilla in the antenna of this species. Despite the above, no specific studies on the function of these sensilla have been performed, although most of the previously cited authors suggest that basiconic and clavate are chemosensilla, mainly based on the presence of pores. Keil and Steinbrecht [2] mention that basiconic sensilla in *Bombyx mori* L. have a thin cuticular wall, higher pore density, a higher numbers of pore tubules per pore, and a greater number of dendrites than trichoid sensilla. Although these authors did not identify a functional role in the studied structures, based on the fact that *B. mori* basiconic sensilla respond to fatty acids and alcohols, and considering that other insects have basiconic sensilla with similar features, they suggest a possible function in food finding and selection.

In our observations, the hourglass pores were more related to thin cuticular-wall sensilla, and wedge or funnel-like pores were more related to thick cuticular walls, with both pore types presenting several tubules, which coincides with that reported by Keil and Steinbrecht [2]. These authors report that the pheromone-sensitive trichoid sensilla of *B. mori* have thick walls and funnel or wedge-shaped pores, which have a narrow channel with tubules running from the channel to a broader fluid-filled canal to contact the dendrites. Considering this, we surmise that the pores in the trichoid sensilla, the longest ones (Figures 1b,c and 9a) in the flagellum, could help transport pheromones to the dendrites.

With the help of TEM, we observed probable secretory epithelial cells contiguous to the cuticle (class 1 glands) associated with campaniform-like sensilla, which had not been described in the antenna of *A. ludens* and other fruit flies, except in *B. zonata*. In this case, Awad et al. [18] reported the presence of campaniform sensilla on the pedicel of males and suggested that they are mechanoreceptors. The glands on antennae in both males and females, first discovered via TEM, have been widely described in egg parasitoids associating Type 1 glands with campaniform sensilla [35]. Usually, campaniform sensilla are located in structures where a mechanical deformation occurs on the cuticle [2,36]. In our case, we could not identify these glands with the help of SEM as campaniform sensilla in the flagellum are surrounded by many microtrichia and other sensilla, which made it difficult to find them.

We report a styloconic sensilla (Figure 8) distributed along the pedicel of females and males. This sensillum type is similar to those reported with the same name by Giannakakis and Fletcher [16], Arzuffi et al. [11], Bissotto de Oliveira et al. [7], and referred to as multiporous grooved sensilla (MPGS) by Dickens et al. [5] and Castrejón and Rojas [20], grooved sensilla by Mayo et al. [9] and Bigiani et al. [10], or coeloconic sensilla by Keil and Steinbrech [2] and Awad et al. [18], among other authors. In other insect species, in the case of this sensillum type, chemoreception [13,16], higroreception and thermoreception functions have been reported [3].

We found that the sensory pit in the *A. ludens* antenna has two chambers, the external (most outer) and the internal, with distinct types of sensilla in each one (Figure 12). The external chamber has a group of pit-styloconic sensilla, which coincides with the only sensory pit chamber reported in *B. zonata* [18], where only styloconic sensilla are found. Styloconic sensilla are also found in the "large pit" of the onion fly (*H. antiqua*), with the difference that, in that chamber, there are two subtypes of sensilla [26]. In both cases, the authors propose an olfactory function for these sensilla.

The internal chamber of *A. ludens* is like "Chamber III" of *Drosophila melanogaster* Meigen (its sensory tip has three chambers), which has two types of sensilla [37]. However, the sensilla are quite different. In *A. ludens*, the pit basiconic type I (pb-I) sensillum is similar to the "no-pore coeloconica sensilla (np-CS)" in "Chamber II" of the *D. melanogaster* sensory pit because both have a kind of protuberance at the tip (Figures 12e and 13) and a conical shape; however, they differ in the cuticular wall, which is smooth in the np-SC in *D. melanogaster* and nutshell-like in the pb-I (Figures 12e and 13) of *A. ludens*. Shanbhag et al. [37] indicate that the protuberance of np-SC is the molting pore of the sensillum and that the lumen of the peg is filled with the dendritic outer segment of two sensory neurons and with electron-dense material. In our case, the protuberance in the pb-I (Figures 12e and 13) could be a porous plug separating the dendrite ends from the environment, or it could also be part of a secretion of viscous fluid containing a mucopolysaccharide, which covers the tips of the contact chemoreceptor dendrites and is sometimes exuded through the terminal pore of the sensillum [1]. Honda et al. [26] reported a similar sensillum with a protuberance at the tip and elongated pores in the "small olfactory pit" of the onion fly, but this fly species has one large sensory pit and several (8–10) small olfactory pits.

The sensillum pit basiconic type II (pb-II) we describe is similar to the "striated pit sensillum" of the *H. antiqua* "large chamber" [26] and to the "grooved sensilla 1 and 2 (GS1 and GS2)" of *D. melanogaster* in the "Chamber III" [37]. These sensilla have an open slit channel system that permits access to the external environment, as is probably the case in the sensilla pb-II of *A. ludens* (Figure 12f). In *D. melanogaster*, Shanbhag et al. [37], considering the internal structure of this sensillum type, suggested a combination of olfactory and thermoreceptive functions.

Finally, we report some structures observed in the internal part of the antenna (Figures 14 and 15) that will need to be studied in more detail to discover their function.

5. Conclusions

In conclusion, our data suggest that the antenna of *A. ludens* contains a complex group of chaetica, trichoid, clavate, basiconic, styloconic, and campaniform-like sensilla that likely participate in the perception of volatiles originating from congeners, host plants and food sources, as well as mechanoreception, thermoreception, and hygroreception. These functions need to be confirmed via electrophysiological, neurological, and behavioral studies, but an important step towards updating the knowledge on the antenna of *A. ludens*, a key pest of fruit in the Americas, has been achieved here. We also need to confirm if the various subtypes of sensilla identified here have different functions or simply represent natural variabilty in shape.

Supplementary Materials: The following supporting information can be downloaded at: https://www.mdpi.com/article/10.3390/insects14070652/s1, Figure S1: Scanning electron micrograph of the internal chamber of the sensory pit showing the distribution of at least 20 pit-basiconic type I (pb-I) sensilla, identified by white asterisks; Video S1: Three-dimensional view using confocal microscopy of the sensilla and microtrichia of *Anastrepha ludens*; several sizes of sensilla and microtrichia with forked tips are shown (depth coding mode); Video S2: Three-dimensional rotational view of the sensory pit of *Anastrepha ludens*; green color represents autofluorescence of the cuticle (447–543 nm) with 405 nm laser excitation, and red color represents chitin stained with Congo Red.

Author Contributions: Conceptualization, M.A. and L.G.; methodology, M.A., L.G., L.L.-S., O.V., G.R.-S. and A.A-M.; resources, M.A.; Data curation, M.R.-V.; writing—original draft preparation, L.G. and L.L.-S.; writing—review and editing, M.A., L.G. and J.G.S.J.; supervision, M.A.; project administration, M.A. and A.A.-M.; funding acquisition, M.A. All authors have read and agreed to the published version of the manuscript.

Funding: This research was funded by Programa Nacional de Moscas de la Fruta (DGSV-SENASICA-SADER) via the Consejo Nacional Consultivo Fitosanitario (CONACOFI), and by the Instituto de Ecología, A.C. (INECOL).

Data Availability Statement: Data are contained within the article or Supplementary Material.

Acknowledgments: We thank Lizbeth González-Cobos, Jovita Martínez-Tlapa and Emilio Acosta for technical assistance. We are thankful to two anonymous referees, and the academic editor for their critical and constructive comments that improved the quality of the manuscript.

Conflicts of Interest: The authors declare no conflict of interest.

References

1. Chapman, R.F. *The Insects: Structure and Function*, 4th ed.; Cambridge University Press: New York, NY, USA, 1998; pp. 1–770.
2. Keil, T.A.; Steinbrecht, R.A. Mechanosensitive and olfactory sensilla of insects. In *Insect Ultrastructure*; King, R.C., Akai, H., Eds.; Plenum Press: New York, NY, USA, 1984; pp. 477–516. [CrossRef]
3. Altner, H. Insect sensillum specificity and structure: An approach to a new typology. In *Olfaction and Taste VI*; Le Magnen, J., Mac Leod, P., Eds.; Information Retrieval: London, UK, 1977; pp. 295–303.
4. Altner, H.; Prillinger, L. Ultrastructure of invertebrate chemo-, thermo- and hygroreceptors and its functional significance. *Int. Rev. Cytol.* **1980**, *67*, 69–139. [CrossRef]
5. Dickens, J.C.; Hart, W.G.; Light, D.M.; Jang, E.B. Tephritid olfaction: Morphology of the antennae of four tropical species of economic importance (Diptera: Tephritidae). *Ann. Entomol. Soc. Am.* **1988**, *81*, 325–331. [CrossRef]
6. Hartenstein, V. Development of insect sensilla. In *Comprehensive Molecular Insect Science*; Gilbert, L.I., Iatrou, K., Gill, S.S., Eds.; Elsevier: Oxford, UK, 2005; Volume 1, pp. 379–419.
7. Bisotto-de-Oliveira, R.; Redaelli, L.R.; Santana, J. Morphometry and distribution of sensilla on the antennae of *Anastrepha fraterculus* (Wiedemann) (Diptera: Tephritidae). *Neotrop. Entomol.* **2011**, *40*, 212–216. [CrossRef] [PubMed]
8. Levinson, H.Z.; Levinson, A.R.; Schäfer, K. Pheromone biology of the Mediterranean fruit fly (*Ceratitis capitata* Weid.) with the emphasis on the functional anatomy of the pheromone glands and antennae as well as mating behaviour. *J. Appl. Entomol.* **1987**, *104*, 448–461. [CrossRef]
9. Mayo, I.; Anderson, M.; Burguete, J.; Robles Chillida, E.M. Structure of superficial chemoreceptive sensilla on the third antennal segment of *Ceratitis capitata* (Wiedemann) (Diptera: Tephritidae). *Int. J. Insect Morphol. Embryol.* **1987**, *16*, 131–141. [CrossRef]
10. Bigiani, A.; Scalera, G.; Crnjar, R.; Barbarossa, I.T.; Magherini, P.C.; Pietra, P. Distribution and function of the antennal olfactory sensilla in *Ceratitis capitata* Wied. (Diptera, Trypetidae). *Boll. Zool.* **1989**, *56*, 305–311. [CrossRef]
11. Arzuffi, R.; Robledo, N.; Valdez, J. Antennal sensilla of *Toxotrypana curvicauda* (Diptera, Tephritidae). *Fla. Entomol.* **2008**, *91*, 669–673. [CrossRef]
12. Perre, P.; Bassan, M.; Perondini, A.L.P.; Selivon, D. Antennal sensilla of specialist and generalist *Anastrepha* species. *Bull. Insectology* **2023**, *76*, 63–71.
13. Hull, C.D.; Cribb, B.W. Ultrastructure of the antennal sensilla of Queensland fruit fly, *Bactrocera tryoni* (Froggatt) (Diptera: Tephritidae). *Int. J. Insect Morphol. Embryol.* **1997**, *26*, 27–34. [CrossRef]
14. Vasey, C.E.; Ritter, E. Antennal sensilla and setal patterns of the goldenrod gall fly, *Eurosta solidaginis* (Fitch) (Diptera: Tephritidae). *J. N. Y. Entomol. Soc.* **1987**, *95*, 452–455.
15. Hallberg, E.; Van Der Pers, J.N.C.; Haniotakis, G.E. Funicular sensilla of *Dacus oleae*: Fine structural characteristics. *Entomol. Hell.* **1984**, *2*, 41–46. [CrossRef]
16. Giannakakis, A.; Fletcher, B.S. Morphology and distribution of antennal sensilla of *Dacus tryoni* (Froggatt) (Diptera: Tephritidae). *Aust. J. Entomol.* **1985**, *24*, 31–35. [CrossRef]
17. Hu, F.; Zhang, G.N.; Jia, F.X.; Dou, W.; Wang, J.J. Morphological characterization and distribution of antennal sensilla of six fruit flies (Diptera: Tephritidae). *Ann. Entomol. Soc. Am.* **2010**, *103*, 661–670. [CrossRef]
18. Awad, A.A.; Ali, N.A.; Mohamed, H.O. Ultrastructure of the antennal sensillae of male and female peach fruit fly, *Bactrocera zonata*. *J. Insect Sci.* **2014**, *14*, 45. [CrossRef]
19. Awad, A.A.; Mohamed, H.O.; Ali, N.A. Differences in antennal sensillae of male and female peach fruit flies in relation to hosts. *J. Insect Sci.* **2015**, *15*, 8. [CrossRef]
20. Castrejón-Gómez, V.R.; Rojas, J.C. Antennal sensilla of *Anastrepha serpentina* (Diptera: Tephritidae). *Ann. Entomol. Soc. Am.* **2009**, *102*, 310–316. [CrossRef]
21. Ruiz-May, E.; Altúzar-Molina, A.; Elizalde-Contreras, J.M.; Santos, J.A.; Monribot-Villanueva, J.L.; Guillén, L.; Vázquez-Rosas-Landa, M.; Ibarra-Laclette, E.; Ramírez-Vázquez, M.; Ortega, R.; et al. A first glimpse of the Mexican fruit fly *Anastrepha ludens* (Diptera: Tephritidae) antenna morphology and proteome in response to a proteinaceous attractant. *Int. J. Mol. Sci.* **2020**, *21*, 8086. [CrossRef]
22. Aluja, M.; Sivinski, J.; Ovruski, S.; Guillen, L.; Lopez, M.; Cancino, J.; Torres-Anaya, A.; Gallegos-Chan, G.; Ruíz, L. Colonization and domestication of seven species of native New World hymenopterous larval-prepupal and pupal fruit fly (Diptera: Tephritidae) parasitoids. *Biocontrol Sci. Technol.* **2009**, *19*, 49–79. [CrossRef]
23. Karnovsky, M.J. A formaldehydeglutaraldehyde fixative of high osmolality for use in electron microscopy. *J. Cell Biol.* **1965**, *27*, 137–138.
24. Reynolds, E.S. The use of lead citrate at high pH as an electron-opaque stain in electron microscopy. *J. Cell Biol.* **1963**, *17*, 208. [CrossRef]

25. Bozzola, J.J.; Russell, L.D. *Electron Microscopy–Principles and Techniques for Biologists*; Jones and Bartlett Publishers: Boston, MA, USA, 1992; pp. 1–542.
26. Honda, I.; Ishikawa, Y.; Matsumoto, Y. Morphological studies on the antennal sensilla of the onion fly, *Hylemya antiqua* Meigen (Diptera: Anthomyiidae). *Appl. Entomol. Zool.* **1983**, *18*, 170–181. [CrossRef]
27. Schenk, O. Die antennalen Hautsinnesorgane einiger Lepidopteren und Hymenopteren. *Zool. Jb. Anat.* **1903**, *17*, 573–618.
28. Schneider, D. Specialized odor receptors of insects. In *Gustation and Olfaction*; Ohlolf, G., Thomes, A.F., Eds.; Academic Press: New York, NY, USA, 1971; pp. 45–60.
29. Keil, T.A. Morphology and development of the peripheral olfactory organs. In *Insect Olfaction*; Hansson, B.S., Ed.; Springer: Berlin/Heidelberg, Germany, 1999; pp. 5–47. [CrossRef]
30. Zacharuck, R.Y. Ultrastructure and function of insect chemosensilla. *Ann. Rev. Entomol.* **1980**, *25*, 27–47. [CrossRef]
31. Steinbrecht, R.A. Pore structures in insect olfactory sensilla: A review of data and concepts. *Int. J. Insect Morphol. Embryol.* **1997**, *26*, 229–245. [CrossRef]
32. Steinbrecht, R.A.; Schneider, D. Pheromone communication in moths: Sensory physiology and behaviour. In *Insect Biology in the Future*; Locke, M., Smith, D.S., Eds.; Academic Press: New York, NY, USA, 1980; pp. 685–703. [CrossRef]
33. Steinbrecht, R.A.; Gnatzy, W. Pheromone receptors in *Bombyx mori* and *Antheraea pernyi*. I. Reconstruction of the cellular organization of the sensilla trichodea. *Cell Tissue Res.* **1984**, *235*, 25–34. [CrossRef]
34. Liu, Z.; Hu, T.; Guo, H.-W.; Liang, X.-F.; Cheng, Y.-Q. Ultrastructure of the olfactory sensilla across the antennae and maxillary palps of *Bactrocera dorsalis* (Diptera: Tephritidae). *Insects* **2021**, *12*, 289. [CrossRef]
35. Romani, R.; Isidoro, N.; Bin, F. Antennal structures usen in communication by egg parasitoids. In *Egg Parasitoids in Agroecosystems with Emphasis on Trichogramma*; Consoli, F.L., Parra, J.R.P., Zucchi, R.A., Eds.; Progress in Biological Control; Springer: Dordrecht, Germany, 2009; Volume 9, pp. 57–96. [CrossRef]
36. Gnatzy, W.; Grünert, U.; Bender, M. Campaniform sensilla of *Calliphora vicina* (Insecta, Diptera). *Zoomorphology* **1987**, *106*, 312–319. [CrossRef]
37. Shanbhag, S.R.; Singh, K.; Singh, R.N. Fine structures and primary sensory projections of sensilla located in the sacculus of the antenna of *Drosophila melanogaster*. *Cell Tissue Res.* **1995**, *282*, 237–249. [CrossRef]

Article

Provisioning Australian Seed Carrot Agroecosystems with Non-Floral Habitat Provides Oviposition Sites for Crop-Pollinating Diptera

Abby E. Davis [1,*], Lena Alice Schmidt [1], Samantha Harrington [2], Cameron Spurr [3] and Romina Rader [1]

[1] Department of Environmental and Rural Science, University of New England, Armidale, NSW 2351, Australia; lschmidt@une.edu.au (L.A.S.); rrader@une.edu.au (R.R.)
[2] South Pacific Seeds, Griffith, NSW 2680, Australia
[3] SeedPurity Pty Ltd., Margate, TAS 7054, Australia; cspurr@seedpurity.com
* Correspondence: aed236@cornell.edu

Simple Summary: Planting a diverse array of flowers in crop field settings can support insect crop pollinators, as many pollinating insects feed on floral pollen and nectar as adults. Although adult pollinating flies that feed on floral resources will also be supported by flower plantings, fly larvae rarely feed on floral nectar and pollen. Here, we deployed pools filled with habitats (decaying plant materials, soil, water) that pollinating fly larvae are known to feed on in seed carrot fields in an attempt to attract flies to lay eggs. We found many fly eggs and larvae within the pools after 12 to 21 days. More eggs were laid on decaying plant stems and carrot roots, compared to other locations (e.g., on decaying carrot flowers, leaves, etc.) within the pools. The habitat pools we trialed within the seed carrot fields could be a quick and easy way to support the reproduction of beneficial fly pollinators. These results can be used to support future studies to examine the effect of habitat pools in crop fields on the number of flies that visit crop flowers.

Abstract: The addition of floral resources is a common intervention to support the adult life stages of key crop pollinators. Fly (Diptera) crop pollinators, however, typically do not require floral resources in their immature life stages and are likely not supported by this management intervention. Here, we deployed portable pools filled with habitat (decaying plant materials, soil, water) in seed carrot agroecosystems with the intention of providing reproduction sites for beneficial syrphid (tribe *Eristalini*) fly pollinators. Within 12 to 21 days after the pools were deployed, we found that the habitat pools supported the oviposition and larval development of two species of eristaline syrphid flies, *Eristalis tenax* (Linnaeus, 1758) and *Eristalinus punctulatus* (Macquart, 1847). Each habitat pool contained an average (±S.E.) of 547 ± 117 eristaline fly eggs and 50 ± 17 eristaline fly larvae. Additionally, we found significantly more eggs were laid on decaying plant stems and carrot roots compared to other locations within the pool habitat (e.g., on decaying carrot umbels, leaves, etc.). These results suggest that deploying habitat pools in agroecosystems can be a successful management intervention that rapidly facilitates fly pollinator reproduction. This method can be used to support future studies to determine if the addition of habitat resources on intensively cultivated farms increases flower visitation and crop pollination success by flies.

Keywords: non-bee pollinators; Syrphidae; pollinator management interventions; fly reproduction

Citation: Davis, A.E.; Schmidt, L.A.; Harrington, S.; Spurr, C.; Rader, R. Provisioning Australian Seed Carrot Agroecosystems with Non-Floral Habitat Provides Oviposition Sites for Crop-Pollinating Diptera. *Insects* 2023, 14, 439. https://doi.org/10.3390/insects14050439

Academic Editors: Sérgio M. Ovruski and Flávio Roberto Mello Garcia

Received: 13 April 2023
Revised: 24 April 2023
Accepted: 28 April 2023
Published: 4 May 2023

1. Introduction

The abundance and diversity of insects that provide pollination services within agroecosystems often depend on suitable habitat options for the insects to complete their life cycles [1,2]. These habitats (e.g., remnant vegetation, semi-natural landscape features) are typically not within the crop system itself, but nearby, and provide food, reproduction sites, overwintering resources, and shelter from agricultural management practices such as tilling, harvesting, and pesticide application [3–7]. When non-crop habitat is maintained or restored near intensely managed fields, beneficial insect abundance and diversity generally increase [4,8–12]. Even small patches (e.g., tens of square meters or less) of non-crop habitat can enhance beneficial insect biodiversity in cropping systems [3,4], and result in native species spillover into agroecosystems [4,13,14]. However, most pollinator-friendly habitat enhancements focus on floral interventions, such as floral strips and hedgerows [15–18], to attract adult, wild pollinators, primarily bees. Few studies have focused specifically on interventions to support the habitat needs for non-bee taxa (but see [2,19,20] for exceptions) and their non-floral resource needs (see [21] for an example of bee non-floral resource needs).

While bees are highly dependent on flowers to obtain nutrition for both adults and larvae, non-bee pollinator taxa, such as flies (Diptera), typically do not require floral resources in their larval stages [22]. For example, the larvae of eristaline syrphid flies, which are easily distinguishable due to the siphon-like 'tail' they use to breathe in poorly oxygenated habitats, live in wet substrates commonly found in agricultural environments, including decaying plant materials and manure [23–25]. Adult eristaline syrphid flies, like the cosmopolitan species *Eristalis tenax* (Linnaeus, 1758), have been shown to effectively pollinate crops as they morphologically resemble honeybees (*Apis mellifera* Linnaeus 1758) in size and body hairiness [26–28], despite lacking specialized pollen-collecting structures (e.g.,: scopa, corbicula). In fact, *E. tenax* is already a non-bee pollinator alternative in New Zealand, where the fly is an effective pollinator of multiple crops including seed carrot [29].

Seed carrot is an ideal model crop to study a potential non-bee pollinator since the crop often pollination limited despite attracting high numbers of other insect visitors [30,31]. Honeybees generally find seed carrot flowers unattractive, as the nectar composition is high in ferulic acid, an insect-feeding deterrent, and low in caffeic acid, a bee attractor [32]. As some species of eristaline flies have been shown to be as effective as honey bees at pollinating seed carrot [30,33], we hypothesized that building up populations of these beneficial non-bee pollinators could increase free ecosystem service delivery within seed carrot agroecosystems [29]. Therefore, this study was based in the Riverina region of New South Wales (NSW), Australia (AU), where seed carrot growers plan for the crop to bloom late austral spring and summer (November to December), when almost no other crops are concurrently blooming to best facilitate honeybee pollination services.

In this study, we trialed the deployment of small, portable pools filled with non-floral habitats to support the reproduction of eristaline syrphid flies in seed carrot agroecosystems. Although the life cycle of eristaline syrphid flies is generally well known [34–36], to our knowledge there are no studies that address whether eristaline flies have oviposition preferences within the habitat they utilize to lay eggs in natural field conditions. We, therefore, tested two habitats (soil with decaying carrot plants in water or decaying carrot plants in water only) to determine if existing adult eristaline syrphid flies would utilize the habitat pools as oviposition sites and evaluated where the flies oviposited within the habitat provided. We addressed the following research questions:

1. Will eristaline syrphid flies use provisioned habitat pools as oviposition sites in a commercial field setting otherwise unsuitable for larval development?
2. Which of the two habitats resulted in the greatest number of eggs and larvae?
3. What were the specific features within the habitat pools that resulted in the greatest oviposition?

2. Materials and Methods

2.1. Study Sites

Seven study fields, at a minimum of 315 m apart, of seed carrot monocultures in the Riverina region of NSW, AU, were chosen as sites in four locations comprising three commercial farms and one private farm managed by South Pacific Seeds (Figure 1). The seed carrot plots at the commercial farms varied from 5 to 14 hectares, while the private farm grew commercial-grade seed carrot in small (<500 m) trial rows. Other plant resources flowering nearby included onion (*Allium cepa* L.) at sites six and seven and small patches of native flowering trees, shrubs, and household gardens near sites one, four, and five as these sites were situated near residential areas. No other crops were observed flowering nearby.

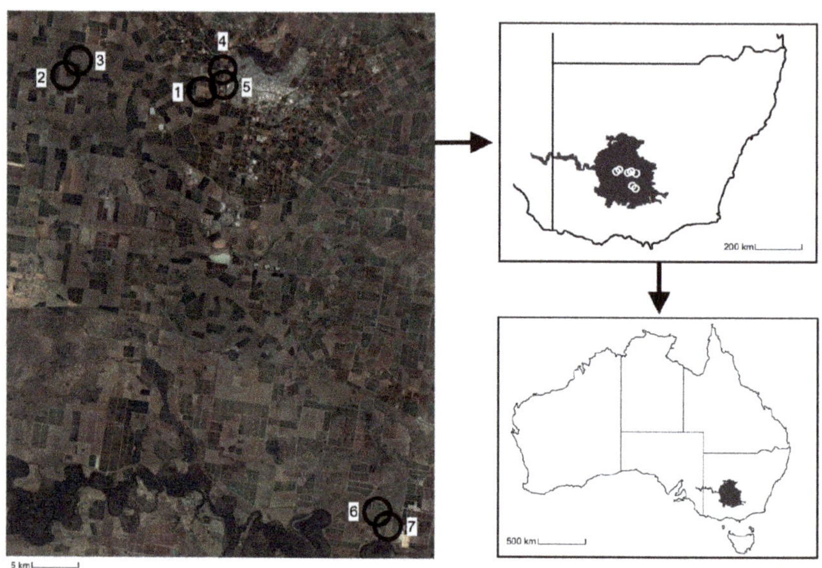

Figure 1. Location of the seed carrot study sites (1–7) where the habitat pools were deployed within the Riverina region of New South Wales, Australia.

2.2. Deploying the Habitat Pools

In preliminary experiments at site one, we placed horse manure and wet, decaying carrot plants within a hybrid seed carrot plot to assess which substrates should be trialed as non-floral fly habitat. The substrates were observed frequently until we observed golden native droneflies, *Eristalinus punctulatus* (Macquart, 1847), oviposit within the wet, decaying carrot plants. As eristaline fly larvae are commonly reared in slurries of manure in laboratory conditions [24,36], this suggests that the larvae are more suited for semi-liquid environments. We did not trial manure at the field sites as golden native droneflies were not observed to oviposit within the manure and some landholders did not want manure placed on their properties; therefore, we chose to trial decaying carrot plants in water as reproduction sites for eristaline flies, with the presence or absence of farm soil. Thus, we hypothesized that more larvae would be found in the semi-liquid decaying carrot plant pool with soil, compared to the treatment with decaying plants and water only.

Pools were deployed during peak bloom (50% flowering) of seed carrot (15 November to 9 December 2021), when adult eristaline flies are most likely to visit seed carrot flowers. Two polypropylene pools (945 mm × 210 mm × 1100 mm each) were placed side by side in a paired experimental design at each site to trial two habitats as eristaline fly oviposition sites (n = 7 per treatment, 14 pools in total). The first habitat consisted of soil, discarded carrot plants and water, while the second habitat consisted of discarded carrot plants and water only. Soil from the farm site was placed within the first pool until the base of the pool was covered, while the second pool contained two bricks to anchor the pool from strong gusts of wind that frequently blow within the region. Three fully grown (150 cm foliage height) male carrot plants were then taken from the site and placed in each pool which was then filled with the same water used to water the seed carrot crop (Figure 2a). We did not include a treatment without water, since eristaline flies cannot survive in habitats devoid of water [35,36]. Likewise, sufficient solid food must be present within the water for eristaline larvae to complete development [30], so we did not include a treatment of water-only pools.

Instead, we conducted preliminary surveys searching for the immature stages of eristaline flies within field sites before pools were deployed, to confirm that no eggs and larvae had been laid in dry soil, dry plant material, or within crop rows. As no eggs or larvae were found in the preliminary field surveys, we excluded them from analyses.

(a)

(b)

Figure 2. Experimental design of the habitat pools deployed to attract eristaline flies: (**a**) a habitat pool yet to be filled with water within a seed carrot field; (**b**) an adult, female *Eristalis tenax* (Linnaeus, 1758) fly within a deployed habitat pool. Arrowheads are pointing to the habitat pool and adult eristaline fly for clarity.

After filling the pools with water, we left them undisturbed to allow the carrot plants within the pools to decay and for eristaline adults to locate the pools (Figure 2b). Due to unprecedented rain events at the time the pools were left undisturbed, site accessibility varied between farms; therefore, the pools were deployed for 12 to 21 days depending on field site accessibility.

2.3. Surveying the Immature Life Stages of Eristaline Flies

Once all farm sites were accessible on the same day, we conducted surveys in each pool to count eristaline syrphid fly egg clutches, a group of eggs laid together in a single oviposition attempt (Figure 3a), individual eggs, and larvae (Figure 3b). All egg clutches and individual eggs were counted on 9 December 2021 (starting at 06:00 and ending at 18:30) and were removed from the pools, so the eggs did not hatch before the larvae were counted. Additionally, we recorded the location of where the eggs were oviposited in the pools (Figure S1). All egg clutches and eggs that were displaced from their original positions in the pools (e.g., due to moving substrates) were counted but not included in statistical analyses.

The day after the eggs were counted and collected, we returned to the pools and counted the eristaline fly larvae over a two-day period (with each pool counted only once for each immature life stage). To count the larvae, all plant material in the pools was thoroughly checked for individuals, and then removed from the pools. Next, we dislodged any larvae within the soil at the bottom of the soil treatment pools, by mixing the water with the soil sediment using a hand-held sieve. We then sifted the soil and water sediment through the sieve five times to determine the total number of larvae in the pools. When larvae were caught in the sieve, they were removed from the pools to avoid duplicate larval counts. For consistency, this procedure was also applied to the carrot and water treatment pools. We did not record where the larvae were found in the pools since we displaced all larvae when mixing the water with the sediment.

(a)

(b)

Figure 3. Immature stages of eristaline syrphid flies: (**a**) a clutch of eggs oviposited on a decaying carrot stem; (**b**) larvae found within a deployed habitat pool. Arrowheads are pointing to the immature stages of eristaline syrphid flies for clarity.

2.4. Rearing Eristaline Flies from Pools

Both eggs and larvae of various growth stages collected from each pool were reared to adulthood in controlled conditions on decaying carrot plants (from inside the pools) or a mixture of decaying carrot plants and sterilized horse manure to confirm species identities. Horse manure was mixed into the habitat as previous studies have successfully reared eristaline syrphid flies from manure [24,37]. Since eristaline syrphid flies have similar morphology in eggs and larvae and are therefore difficult to identify at these stages [23,37–39]; we waited until adults emerged to distinguish species identities using taxonomic keys [23,25].

2.5. Statistical Analyses

Statistical analyses were performed using R version 4.1.2. We created generalized linear mixed-effects models (GLMMs) using the *MASS* package to assess whether the number of eggs and larvae within pools differed based on treatment (two categories) or location (eight categories) [40,41]. To handle overdispersion in the collected count data, all GLMM models were fit to a negative binomial distribution [42]. Additionally, as the number of days the pools were left out in the sites to decay was not standardized as intended due to unprecedented weather conditions, we included the fixed effect 'Day' (continuous: 5 discrete values) in all models. We also included 'Site' as a random factor in all models, to account for the matched pair experimental design.

The *DHARMa* package was employed on all models to perform residual, dispersion, and zero-inflation checks of the data [43]. For all significant models, we performed Tukey pairwise post hoc multiple comparisons tests between fixed effects using the *emmeans* package [44]. All figures were created using the *ggplot2* package [45].

3. Results

Two species of eristaline syrphid flies, the European dronefly, *E. tenax*, and the golden native dronefly, *E. punctulatus*, were reared out of both habitat pools at all seven sites. The fly *E. tenax* was successfully reared from all 14 pools, and *E. punctulatus* was reared from three of the 14 pools. Additionally, eggs and/or larvae of both species were found within all 14 pools (Table 1). The number of eggs within clutches ranged between 10 and 128 eggs (mean ± S.E. 54.7 ± 3.9 eggs/clutch) in the soil, decaying carrot plants, and water habitat and 15 to 125 eggs (mean ± S.E. 54.4 ± 3.6 eggs/clutch) in the decaying carrot plants and water only habitat.

Table 1. Developmental stages of eristaline syrphid flies, *Eristalis tenax* (Linnaeus, 1758) and *Eristalinus punctulatus* (Macquart, 1847) found in two habitats (1 = soil, carrot plants, and water and 2 = carrot plants and water) at seven sites in the Riverina region of New South Wales, Australia. Both habitats were left to decay for a minimum of 12 days before surveying for fly egg clutches, eggs, and larvae.

Habitat	Site	Days	Clutches	Eggs	Larvae	Species
1	Site 1	21	16	910	0	*E. tenax*
1	Site 2	14	7	494	26	*E. tenax*
1	Site 3	13	6	296	41	*E. tenax*
1	Site 4	12	9	694	117	*E. tenax, E. punctulatus*
1	Site 5	12	22	1355	107	*E. tenax*
1	Site 6	19	9	382	9	*E. tenax*
1	Site 7	19	4	113	3	*E. tenax*
2	Site 1	21	9	476	0	*E. tenax*
2	Site 2	14	2	233	16	*E. tenax*
2	Site 3	13	0	0	41	*E. tenax*
2	Site 4	12	4	258	201	*E. tenax, E. punctulatus*
2	Site 5	12	21	1497	137	*E. tenax, E. punctulatus*
2	Site 6	19	6	548	4	*E. tenax*
2	Site 7	19	8	401	0	*E. tenax*

Location within the habitat pools also influenced how many eggs were oviposited by female eristaline flies. We found significantly more eggs were oviposited within decaying carrot plant stems and decaying carrot vegetables compared to all other locations (Figure 4). There were no significant differences in the number of eggs laid within the pools based on habitat ($p > 0.05$ for both, Table S1). Additionally, the number of days the pools were left out to decay did not significantly impact the number of eggs laid within habitat pools ($z_{1,4} = -0.012, p = 0.91$).

First, second, and third instar eristaline fly larvae were found within both habitat pools across all sites (Table 2). Significantly more living larvae were found compared to dead larvae ($z_{1,1} = 6.13, p < 0.001$); however, the longer the habitat pools were left out to decay, the fewer larvae of all three instars were found in the pools (Figure 5; see Table S2 for statistics). There were no significant differences in the number of larvae found within the pools based on habitat ($z_{1,1} = -0.468, p = 0.64$). Additionally, there were no larval instars more abundant than others within the habitat pools ($p > 0.05$ for all, Table S3).

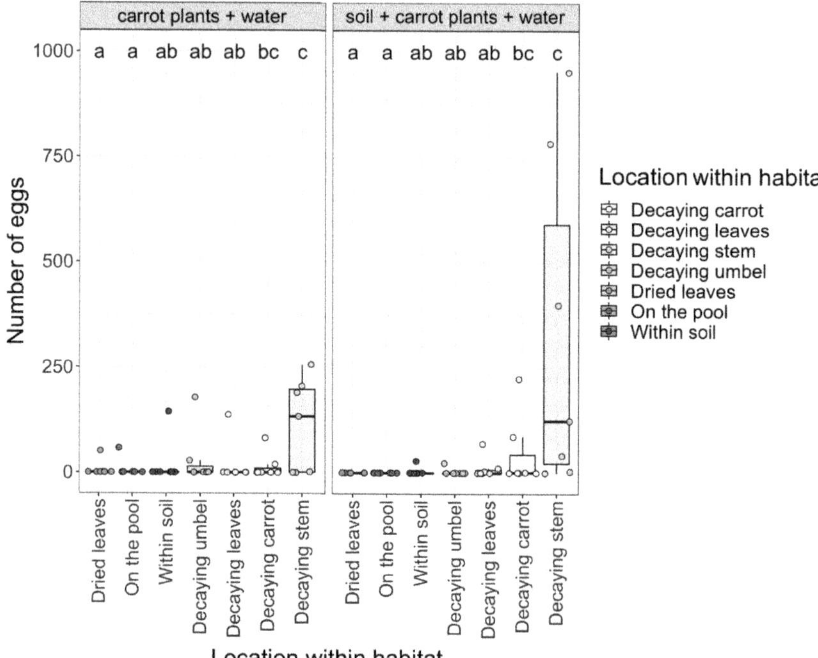

Figure 4. The number of eggs oviposited by female eristaline syrphid flies within the deployed pools based on habitat (carrot plants + water only and the soil + carrot plants + water) and the location where the eggs were laid. Letters indicate significant differences between locations ($p < 0.05$). Individual data points representing each habitat pool ($n = 14$ in total) are jittered onto the figure for clarity. Lower to upper box boundaries indicate the inter-quartile range (IQR). Whiskers are extended to the furthest data point within 1.5x the IQR from each box end.

Table 2. Total number of larvae found within habitat pools (1 = soil, carrot plants, and water and 2 = carrot plants and water) deployed at seven seed carrot sites in the Riverina region of New South Wales, Australia.

Habitat	Site	1st Instar	2nd Instar	3rd Instar	Dead
1	Site 1	0	0	0	0
1	Site 2	17	4	3	2
1	Site 3	35	1	0	5
1	Site 4	54	50	12	1
1	Site 5	100	0	0	7
1	Site 6	1	4	4	0
1	Site 7	3	0	0	0
2	Site 1	0	0	0	0
2	Site 2	2	4	4	6
2	Site 3	5	19	13	4
2	Site 4	79	90	32	0
2	Site 5	135	2	0	0
2	Site 6	0	0	0	4
2	Site 7	0	0	0	0

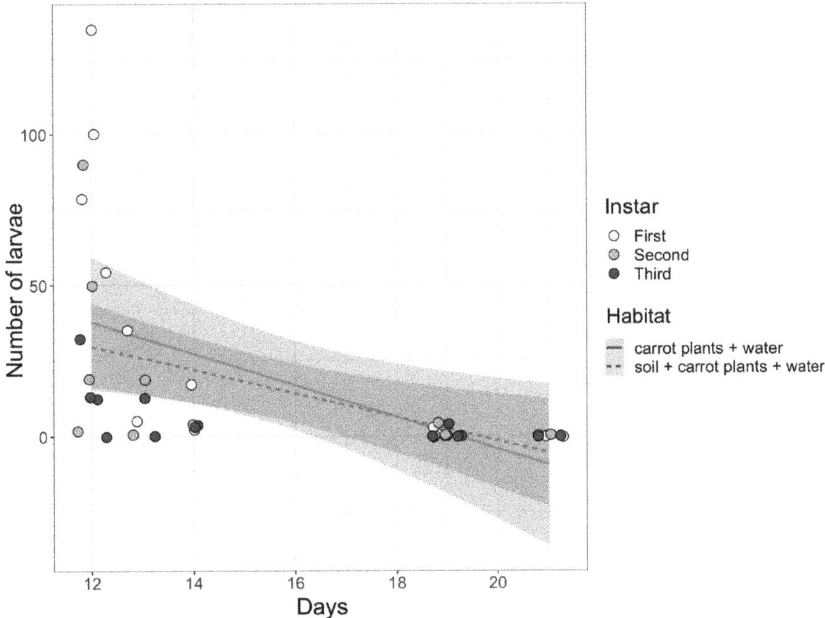

Figure 5. The number of larvae (first, second and third instar) found within the pools based on habitat (soil, carrot plants, and water or carrot plants and water only) and the number of days the habitat pools were left to decay. Individual data points representing larval instars found per pool (*n* = 14 per instar; 42 in total) are jittered onto the figure for clarity. The shaded outline of the linear regression lines indicates standard error.

4. Discussion

In this study, we demonstrate that beneficial fly pollinators can utilize small, portable pools filled with locally available, cheap substrates (habitat) in seed carrot agroecosystems as oviposition sites. To our knowledge, no other study has deployed non-floral habitats to provide oviposition sites for pollinating eristaline flies. We found that all the habitat pools contained eggs or larvae of *E. tenax*, a cosmopolitan fly species that is an effective pollinator of carrot, onion, canola, and other cropping systems [28,46]. We also showed that the habitat pools encouraged oviposition by *E. punctulatus*, an eristaline fly endemic to the Australasian region [23]; however, other fly species in the genus *Eristalinus* Rondani, 1845 are known to be effective pollinators of other cropping systems including celery and fennel (Apiaceae), which are close relatives of seed carrot [47]. Species within the *Eristalinus* genus are found globally and have similar larval habitat and diet requirements [23,25,39]. Therefore, we predict that closely related flies from diverse biogeographical regions will be attracted to the habitat additions tested as well. Both fly species demonstrated oviposition preferences within the habitat, as more eggs were laid on decaying carrot stems, likely because this location within the pool offered protection for the eggs from the sun, preventing the eggs from desiccation, or from predators.

We conducted this study to determine whether habitat pools could host eristaline fly reproduction. Thus, while it has been demonstrated that these pools act as oviposition sites for resident populations of flies, it is unclear how many pools are required to impact pollination services within different-sized fields. While we did observe *E. tenax* and *E. punctulatus* flies visiting seed carrot flowers in low numbers before these habitat pools were deployed within fields, it was beyond the scope of the study to compare the effect of habitat pool presence and absence on crop yield.

It is well-known that many species of flies can develop in large numbers from small, transient resources [48–50]. The habitat pools tested were proven to be effective and were a quick and easy way to attract flies since both eggs and larvae were found within the habitat after a minimum of 12 days when the seed carrot crop was at peak bloom (50% flowering). For both fly species reared, the time needed to undergo different developmental stages is similar, as eggs hatch after 48 hours, and in optimal conditions, the larvae take an average of 12 days to develop before pupation [23,39]. Although the number of larval instars found did not differ significantly between pools, the majority had recently hatched and were in the first instar of development, and thus likely oviposited 48 to 96 hours previously. Hence, to best facilitate fly pollination during peak crop bloom, we suggest placing the pools nearby a different flowering crop or a small planting of flowers, 12 to 15 days before the flowering onset of the desired crop to ensure that adult eristaline flies locate the pools and two to three generations of syrphids emerge by the time the desired crop reaches peak bloom.

Environmental conditions in the region at the time the habitat pools were deployed significantly influenced pool management. While these results suggest that the deployed habitat pools were low maintenance, we suspect that, under more average (i.e., drier, hotter) environmental conditions at this time of year within the region, this may not be the case. As the Riverina region is typically hot and dry in austral summer, we anticipated refilling the pools with water at least once or moving the pools to a shaded location to ensure the deployed habitat remained a suitable oviposition site for eristaline flies; however, rain events were common when we performed this experiment, so farms became inaccessible to check on the progress of the pools. As the mean temperature between November to December 2021 in the region was 21 °C to 24 °C, none of the habitat pools dried out completely; however, the pools were shallow and not completely shaded, so the sun could have heated up the habitat pools, which could have negatively affected egg and larvae survival. Therefore, we recommend deploying pools in completely shaded environments and monitoring the water level within pools, to ensure that the pools do not become ecological traps for eristaline flies [51].

Similarly, fewer larvae were found within the pools the longer the pools were left out to decay, suggesting the larvae within the pools left undisturbed for longer had either crawled out of the pools to pupate, died competing for food resources, or had been predated upon. As large amounts of decaying carrot plant debris were found within all pools, it is unlikely that the larvae died competing for food resources. The only observed predator within the pools was the rove beetle *Creophilus erythrocephalus* (Fabricius, 1775) which was present in two pools at the same farm. These rove beetles are known predators of fly larvae [52,53], although they have not been recorded feeding on rat-tailed maggots, specifically. Further research is required to better understand how to scale up these habitats to meet pollination service needs, the length of time the portable habitat pools should be placed on farms, the water conditions that eristaline syrphid fly larvae require to survive, the potential predators of the fly larvae, and whether these pools attract non-target or potential pest species to crop fields.

5. Conclusions

In this study, we successfully trialed a non-floral resource habitat intervention which acted as oviposition sites for beneficial fly pollinators. These habitat pools are a unique, yet cheap, alternative for landholders who want to support pollinating fly reproduction but may not be able to set aside arable land for non-crop habitats. The flies oviposited within decaying carrot plant habitat, and larvae of all instars were found in pools within 12 days. Eristaline flies were found to preferentially oviposit underneath decaying plant stems, likely to protect eggs from predation or adverse environmental conditions. The substrates placed within the habitat pools (soil, discarded carrot plants, and water) are locally available, cheap, and the pools are small and portable, enabling placement and removal at key flowering times. This approach may increase the natural population of flies that provide critical pollination services to crops in intensely managed agricultural systems.

Supplementary Materials: The following supporting information can be downloaded at: https://www.mdpi.com/article/10.3390/insects14050439/s1, Figure S1: Location of eggs within habitat pools [22]; Table S1: Egg results [22]; Table S2: Larval results [22]; Table S3: Larval instar pairwise comparisons [22].

Author Contributions: Conceptualization (A.E.D., C.S. and R.R.), methodology (A.E.D., L.A.S., S.H., C.S. and R.R.), validation (S.H., C.S. and R.R.), formal analysis (A.E.D.), investigation (A.E.D. and L.A.S.), resources (A.E.D., L.A.S., S.H. and R.R.), data curation (A.E.D.), writing—original draft preparation (A.E.D.), writing—review and editing (A.E.D., L.A.S., S.H., C.S. and R.R.), visualization (A.E.D., L.A.S. and R.R.), supervision (S.H., C.S. and R.R.), project administration (R.R.), and funding acquisition (R.R.). All authors have read and agreed to the published version of the manuscript.

Funding: The research was funded by the Horticulture Innovation Australia Limited Project Managing Flies for Crop Pollination (PH16002) and an Australian Research Council Future Fellowship (FT210100851) awarded to R. Rader.

Data Availability Statement: The data presented in this study are available within the paper and its Supplementary Materials.

Acknowledgments: We thank the growers of the sites for allowing us to conduct this study on their properties; Mark Hancock, Mike Tarr and John Hall at South Pacific Seeds for discussion about the study; Karen Santos for help with initial set-up of the habitat pools; Blake Dawson for feedback on the final manuscript; Julius Roberts for help designing Figure 1; several anonymous reviewers for comments and suggestions that have improved the manuscript; and we acknowledge the Wiradjuri people as the traditional owners of the land this research was conducted on and pay respect to Wiradjuri Elders past, present, and future.

Conflicts of Interest: Author S.H. was employed by the company South Pacific Seeds. Author C.S. was employed by the company SeedPurity. The remaining authors declare that the research was conducted in the absence of any commercial or financial relationships that could be construed as a potential conflict of interest.

References

1. Fijen, T.P.M.; Read, S.F.J.; Walker, M.K.; Gee, M.; Nelson, W.R.; Howlett, B.G. Different landscape features within a simplified agroecosystem support diverse pollinators and their service to crop plants. *Landsc. Ecol.* **2022**, *37*, 1787–1799. [CrossRef]
2. Finch, J.; Gilpin, A.-M.; Cook, J. Fishing for flies: Testing the efficacy of "stink stations" for promoting blow flies as pollinators in mango orchards. *J. Pollinat. Ecol.* **2023**, *32*, 79–100. [CrossRef]
3. Carvalheiro, L.G.; Seymour, C.L.; Nicolson, S.W.; Veldtman, R. Creating patches of native flowers facilitates crop pollination in large agricultural fields: Mango as a case study. *J. Appl. Ecol.* **2012**, *49*, 1373–1383. [CrossRef]
4. Knapp, M.; Rezac, M. Even the smallest non-crop habitat islands could be beneficial: Distribution of carabid beetles and spiders in agricultural landscape. *PLoS ONE* **2015**, *10*, e0123052. [CrossRef] [PubMed]
5. Altieri, M.A. The ecological role of biodiversity in agroecosystems. *Agric. Ecosyst. Environ.* **1999**, *74*, 19–31. [CrossRef]
6. Fahrig, L.; Baudry, J.; Brotons, L.; Burel, F.G.; Crist, T.O.; Fuller, R.J.; Sirami, C.; Siriwardena, G.M.; Martin, J.L. Functional landscape heterogeneity and animal biodiversity in agricultural landscapes. *Ecol. Lett.* **2011**, *14*, 101–112. [CrossRef]
7. Tscharntke, T.; Clough, Y.; Wanger, T.C.; Jackson, L.; Motzke, I.; Perfecto, I.; Vandermeer, J.; Whitbread, A. Global food security, biodiversity conservation and the future of agricultural intensification. *Biol. Conserv.* **2012**, *151*, 53–59. [CrossRef]
8. Sunderland, K.; Samu, F. Effects of agricultural diversification on the abundance, distribution, and pest control potential of spiders: A review. *Entomol. Exp. Appl.* **2000**, *95*, 1–13. [CrossRef]
9. Wilson, J.D.; Morris, A.J.; Arroyo, B.E.; Clark, S.C.; Bradbury, R.B. A review of the abundance and diversity of invertebrate and plant foods of granivorous birds in northern Europe in relation to agricultural change. *Agric. Ecosyst. Environ.* **1999**, *75*, 13–30. [CrossRef]
10. Rischen, T.; Frenzel, T.; Fischer, K. Biodiversity in agricultural landscapes: Different non-crop habitats increase diversity of ground-dwelling beetles (Coleoptera) but support different communities. *Biodivers. Conserv.* **2021**, *30*, 3965–3981. [CrossRef]
11. González, E.; Štrobl, M.; Janšta, P.; Hovorka, T.; Kadlec, T.; Knapp, M. Artificial temporary non-crop habitats support parasitoids on arable land. *Biol. Conserv.* **2022**, *265*, 109409. [CrossRef]
12. Landis, D.A. Designing agricultural landscapes for biodiversity-based ecosystem services. *Basic Appl. Ecol.* **2017**, *18*, 1–12. [CrossRef]

13. Schellhorn, N.A.; Bianchi, F.; Hsu, C.L. Movement of Entomophagous Arthropods in Agricultural Landscapes: Links to Pest Suppression. *Annu. Rev. Entomol.* **2014**, *59*, 559–581. [CrossRef] [PubMed]

14. Saunders, M.E.; Peisley, R.K.; Rader, R.; Luck, G.W. Pollinators, pests, and predators: Recognizing ecological trade-offs in agroecosystems. *Ambio* **2016**, *45*, 4–14. [CrossRef]

15. Albrecht, M.; Kleijn, D.; Williams, N.M.; Tschumi, M.; Blaauw, B.R.; Bommarco, R.; Campbell, A.J.; Dainese, M.; Drummond, F.A.; Entling, M.H.; et al. The effectiveness of flower strips and hedgerows on pest control, pollination services and crop yield: A quantitative synthesis. *Ecol. Lett.* **2020**, *23*, 1488–1498. [CrossRef]

16. Scheper, J.; Bommarco, R.; Holzschuh, A.; Potts, S.G.; Riedinger, V.; Roberts, S.P.M.; Rundlof, M.; Smith, H.G.; Steffan-Dewenter, I.; Wickens, J.B.; et al. Local and landscape-level floral resources explain effects of wildflower strips on wild bees across four European countries. *J. Appl. Ecol.* **2015**, *52*, 1165–1175. [CrossRef]

17. Jonsson, A.M.; Ekroos, J.; Danhardt, J.; Andersson, G.K.S.; Olsson, O.; Smith, H.G. Sown flower strips in southern Sweden increase abundances of wild bees and hoverflies in the wider landscape. *Biol. Conserv.* **2015**, *184*, 51–58. [CrossRef]

18. Haenke, S.; Scheid, B.; Schaefer, M.; Tscharntke, T.; Thies, C. Increasing syrphid fly diversity and density in sown flower strips within simple vs. complex landscapes. *J. Appl. Ecol.* **2009**, *46*, 1106–1114. [CrossRef]

19. Gillespie, M.; Wratten, S.; Sedcole, R.; Colfer, R. Manipulating floral resources dispersion for hoverflies (Diptera: Syrphidae) in a California lettuce agro-ecosystem. *Biol. Control* **2011**, *59*, 215–220. [CrossRef]

20. Munoz, A.E.; Plantegenest, M.; Amouroux, P.; Zaviezo, T. Native flower strips increase visitation by non-bee insects to avocado flowers and promote yield. *Basic Appl. Ecol.* **2021**, *56*, 369–378. [CrossRef]

21. Requier, F.; Leonhardt, S.D. Beyond flowers: Including non-floral resources in bee conservation schemes. *J. Insect Conserv.* **2020**, *24*, 5–16. [CrossRef]

22. Davis, A.E.; Bickel, D.; Saunders, M.E.; Rader, R. Crop-pollinating Diptera have diverse diet and habitat needs in both larval and adult stages. *Ecol. Appl.* **2023**, e2859. [CrossRef]

23. Campoy, A.; Pérez-Bañón, C.; Aznar, D.; Rojo, S. Description of the preimaginal stages of the golden native dronefly from Australia, *Eristalinus punctulatus* (Macquart, 1847) (Diptera: Syrphidae). *Austral Entomol.* **2020**, *59*, 784–793. [CrossRef]

24. Davis, A.E.; Deutsch, K.R.; Torres, A.M.; Loya, M.J.M.; Cody, L.V.; Harte, E.; Sossa, D.; Muniz, P.A.; Ng, W.H.; McArt, S.H. *Eristalis* flower flies can be mechanical vectors of the common trypanosome bee parasite, *Crithidia bombi. Sci. Rep.* **2021**, *11*, 15852. [CrossRef] [PubMed]

25. Young, A.; Marshall, S.; Locke, M.; Moran, K.; Crins, W.; Skevington, J. *Field Guide to the Flower Flies of Northeastern North America*; Princeton University Press: Princeton, NJ, USA, 2019.

26. Jauker, F.; Bondarenko, B.; Becker, H.C.; Steffan-Dewenter, I. Pollination efficiency of wild bees and hoverflies provided to oilseed rape. *Agric. For. Entomol.* **2012**, *14*, 81–87. [CrossRef]

27. Jarlan, A.; De Oliveira, D.; Gingras, J. Effects of *Eristalis tenax* (Diptera: Syrphidae) pollination on characteristics of greenhouse sweet pepper fruits. *J. Econ. Entomol.* **1997**, *90*, 1650–1654. [CrossRef]

28. Cook, D.; Voss, S.; Finch, J.; Rader, R.; Cook, J.; Spurr, C. The role of flies as pollinators of horticultural crops: An Australian case study with worldwide relevance. *Insects* **2020**, *11*, 341. [CrossRef]

29. Howlett, B.G.; Gee, M. The potential management of the drone fly (*Eristalis tenax*) as a crop pollinator in New Zealand. *N. Z. Plant Prot.* **2019**, *72*, 221–230. [CrossRef]

30. Gaffney, A.; Allen, G.R.; Brown, P.H. Insect visitation to flowering hybrid carrot seed crops. *N. Z. J. Crop Hortic. Sci.* **2011**, *39*, 79–93. [CrossRef]

31. Gaffney, A.; Bohman, B.; Quarrell, S.R.; Brown, P.H.; Allen, G.R. Frequent insect visitors are not always pollen carriers in hybrid carrot pollination. *Insects* **2018**, *9*, 61. [CrossRef]

32. Broussard, M.A.; Mas, F.; Howlett, B.; Pattemore, D.; Tylianakis, J.M. Possible mechanisms of pollination failure in hybrid carrot seed and implications for industry in a changing climate. *PLoS ONE* **2017**, *12*, e0180215. [CrossRef] [PubMed]

33. Perez-Banon, C.; Petanidou, T.; Marcos-Garcia, M.A. Pollination in small islands by occasional visitors: The case of *Daucus carota* subsp *commutatus* (Apiaceae) in the Columbretes archipelago, Spain. *Plant Ecol.* **2007**, *192*, 133–151. [CrossRef]

34. Reilly, J.; Artz, D.; Biddinger, D.; Bobiwash, K.; Boyle, N.; Brittain, C.; Brokaw, J.; Campbell, J.; Daniels, J.; Elle, E.; et al. Crop production in the USA is frequently limited by a lack of pollinators. *Proc. R. Soc. B-Biol. Sci.* **2020**, *287*, 20200922. [CrossRef] [PubMed]

35. Dolley, W.L., Jr.; Hassett, C.C.; Bowen, W.J. Laboratory culture of the drone fly, *Eristalis tenax. Science* **1933**, *78*, 313–314. [CrossRef] [PubMed]

36. Nicholas, S.; Thyselius, M.; Holden, M.; Nordstrom, K. Rearing and long-Term maintenance of *Eristalis tenax* hoverflies for research studies. *J. Vis. Exp.* **2018**, *135*, 57711. [CrossRef]

37. Campoy, A.; Sáez, L.; Pérez-Bañón, C.; Rojo, S. Demography and population parameters of two species of eristaline flower flies (Diptera, Syrphidae, Eristalini). *J. Appl. Entomol.* **2020**, *144*, 133–143. [CrossRef]

38. Campoy, A.; Aracil, A.; Pérez-Bañón, C.; Rojo, S. An in-depth study of the larval head skeleton and the external feeding structures related with the ingestion of food particles by the eristaline flower flies *Eristalis tenax* and *Eristalinus aeneus. Entomol. Exp. Appl.* **2020**, *168*, 783–798. [CrossRef]

39. Campoy, A.; Pérez-Bañón, C.; Rojo, S. Intra-puparial development in the hoverflies *Eristalinus aeneus* and *Eristalis tenax* (Diptera: Syrphidae). *J. Morphol.* **2020**, *281*, 1436–1445. [CrossRef]

40. Venables, W.N.; Ripley, B.D. *Modern Applied Statistics with S*, 4th ed.; Springer: New York, NY, USA, 2002.

41. Bates, D.; Machler, M.; Bolker, B.M.; Walker, S.C. Fitting linear mixed-effects models using lme4. *J. Stat. Softw.* **2015**, *67*, 1–48. [CrossRef]

42. O'Hara, R.B.; Kotze, D.J. Do not log-transform count data. *Methods Ecol. Evol.* **2010**, *1*, 118–122. [CrossRef]

43. Hartig, F. DHARMa: Residual Diagnostics for Hierarchical (Multi-Level/Mixed) Regression Models. 2022. Available online: https://cran.r-project.org/web/packages/DHARMa/vignettes/DHARMa.html (accessed on 10 October 2022).

44. Hothorn, T.; Bretz, F.; Westfall, P. Simultaneous inference in general parametric models. *Biom. J.* **2008**, *50*, 346–363. [CrossRef] [PubMed]

45. Wickham, H. *ggplot2: Elegant Graphics for Data Analysis*; Springer: New York, NY, USA, 2016.

46. Rader, R.; Bartomeus, I.; Garibaldi, L.; Garratt, M.; Howlett, B.; Winfree, R.; Cunningham, S.; Mayfield, M.; Arthur, A.; Andersson, G.; et al. Non-bee insects are important contributors to global crop pollination. *Proc. Natl. Acad. Sci. USA* **2016**, *113*, 146–151. [CrossRef] [PubMed]

47. Sánchez, M.; Belliure, B.; Montserrat, M.; Gil, J.; Velásquez, Y. Pollination by the hoverfly *Eristalinus aeneus* (Diptera: Syrphidae) in two hybrid seed crops: Celery and fennel (Apiaceae). *J. Agric. Sci.* **2022**, *160*, 1–13. [CrossRef]

48. Levot, G.W.; Brown, K.R.; Shipp, E. Larval growth of some calliphorid and sarcophagid Diptera. *Bull. Entomol. Res.* **1979**, *69*, 469–475. [CrossRef]

49. Carey, J.R.; Liedo, P.; Muller, H.G.; Wang, J.L.; Chiou, J.M. Relationship of age patterns of fecundity to mortality, longevity, and lifetime reproduction in a large cohort of Mediterranean fruit fly females. *J. Gerontol. Ser. A-Biol. Sci. Med. Sci.* **1998**, *53*, B245–B251. [CrossRef]

50. Vasconcelos, S.D.; Cruz, T.M.; Salgado, R.L.; Thyssen, P.J. Dipterans associated with a decomposing animal carcass in a rainforest fragment in Brazil: Notes on the early arrival and colonization by necrophagous species. *J. Insect Sci.* **2013**, *13*, 145. [CrossRef]

51. Cox, M.E.; Johnstone, R.; Robinson, J. Relationships between perceived coastal waterway condition and social aspects of quality of life. *Ecol. Soc.* **2006**, *11*, 35. Available online: https://www.jstor.org/stable/26267792 (accessed on 15 October 2022). [CrossRef]

52. Greene, G.L. Rearing techniques for *Creophilus maxillosus* (Coleoptera: Staphylinidae), a predator of fly larvae in cattle feedlots. *J. Econ. Entomol.* **1996**, *89*, 848–851. [CrossRef]

53. Bornemissza, G.F. An analysis of arthropod succession in carrion and the effect of its decomposition on the soil fauna. *Aust. J. Zool.* **1957**, *5*, 1–12. [CrossRef]

Communication

Rescuing the Inhibitory Effect of the Salivary Gland Hypertrophy Virus of *Musca domestica* on Mating Behavior

Marissa Gallagher [1], Arianna Ramirez [2], Christopher J. Geden [3,*] and John G. Stoffolano, Jr. [4]

[1] Neuroscience Department, University of Massachusetts, Amherst, MA 01003, USA
[2] Biology Department, University of Massachusetts, Amherst, MA 01003, USA
[3] Center for Medical, Agricultural and Veterinary Entomology, USDA, Agricultural Research Service, Gainesville, FL 32608, USA
[4] Stockbridge School of Agriculture, University of Massachusetts, Amherst, MA 01003, USA; stoff@umass.edu
*   Correspondence: chris.geden@usda.gov

Simple Summary: House flies have been global pests of humans and animals since antiquity, and are notoriously difficult to control. Flies in nature are sometimes infected with salivary gland hypertrophy virus (*Md*SGHV), which prevents them from mating or laying eggs. A better understanding of how the virus works could be helpful for through use as a fly management tool. In this study, we found that infected female flies, which normally do not mate, could be induced to mate by treating them with hormones that are involved in normal fly reproduction. The results provide insight into the mechanisms by which the virus tricks the fly into being unresponsive to male suitors.

Abstract: Infection with salivary gland hypertrophy virus (*Md*SGHV) of *Musca domestica* prevents female flies from accepting copulation attempts by healthy or virus-infected males. This study focused on supplemental hormonal rescue therapy for mating behavior in virus-infected female house flies. The inhibitory effect of the virus on mating behavior in females injected with *Md*SGHV was reversed by hormonal therapy in the form of octopamine injections, topical application of methoprene, or both therapies combined along with 20-hydroxyecdysone. Infected females whose mating responsiveness had been restored continued to have other viral pathologies associated with infection such as hypertrophy of the salivary glands and a lack of ovarian development.

Keywords: juvenile hormone; octopamine; methoprene; corpus allatum; sesquiterpenoids; hormone supplemental rescue therapy; mating receptivity

Citation: Gallagher, M.; Ramirez, A.; Geden, C.J.; Stoffolano, J.G., Jr. Rescuing the Inhibitory Effect of the Salivary Gland Hypertrophy Virus of *Musca domestica* on Mating Behavior. *Insects* **2023**, *14*, 416. https://doi.org/10.3390/insects14050416

Academic Editors: Sérgio M. Ovruski and Flávio Roberto Mello Garcia

Received: 14 March 2023
Revised: 24 April 2023
Accepted: 25 April 2023
Published: 27 April 2023

1. Introduction

Mating behavior is essential for those insects that rely on the successful transfer of both viable sperm and female egg development. Without either, individuals have wasted gametes. Various factors have been shown to influence normal mating in insects. One factor that is currently under investigation is the effect of viruses on either sperm/egg production or on mating behavior. Studies on the nematode, *Caenorhabditis elegans*, demonstrated that virus infection somehow changes male mating choice [1]. One of the most complete studies showing the effect of a virus on insect mating behavior is that of Burand et al. [2], who showed that the virus Hz-2v altered mating behavior and pheromone production in female moths. The review by Kariithi et al. [3], in addition to focusing on tsetse flies, provides information that diverse viruses of insects, including dipterans, affect both male and female reproductive systems.

Hytrosaviruses are a relatively recently discovered group of viruses that are mostly known from forms that infect house flies and *Glossina* species [4]. They are double-stranded DNA, enveloped viruses that are characterized by causing hypertrophy of the salivary glands and effects on the reproductive system. The virus infecting house flies (*Md*SGHV) is thought to be transmitted per os when infected flies deposit the virus on food and has been regarded as a potential biological control agent [5,6]. In contrast, the viruses infecting *Glossina* spp. are mainly viewed as an impediment to tsetse mass-rearing efforts for releases in sterile insect technique programs [7].

Coler et al. [8] first reported that *Md*SGHV shuts down ovarian development in house flies. They did not, however, mention the effect of the virus on mating behavior for either sex. Later, Leitze et al. [9] reported on mating trials using different combinations of healthy versus infected males and females at different times post-infection. They demonstrated that females virally infected for 72 h, post-eclosion at the previtellogenic stage, had almost zero percentage of copulation when paired with healthy males. They suggested that the virus somehow influenced the central nervous system, thus shutting down mating receptivity.

To explain the effect of the virus on mating receptivity, Kariithi et al. [10] provided evidence that low hemolymph sesquiterpenoid levels may account for the female's refusal to mate. They reported that "*Md*SGHV replication in the CA/CC [corpus allatum/corpus cardiacum] complex potentially explains the significant reduction of hemolymph sesquiter-penoid levels, the refusal to mate, and the complete shutdown of ovarian development in viremic females." They did not, however, examine the effects of biogenic amines or (S)-methoprene, a juvenile hormone (JH) mimic, both of which have previously been used by researchers to study mating behavior in flies [11,12]. In their review paper, Kariithi et al. [4] reported that hytrosavirus replicates within the CA and suggested that it disrupts JH hormone biosynthesis.

Because our laboratory has previously studied mating behavior in flies [13,14], we decided to see if we could reverse the effect that salivary gland hypertrophy virus (*Md*SGHV) has on mating responsiveness in house flies. Compared to previous studies, a different approach for rescuing mating behavior in infected females was used here. We treated infected female house flies with two chemicals—octopamine (OA) and JH [i.e., (S)-methoprene]. OA, a biogenic amine, is a neurohormone in insects known for its involvement in fly mating [15]. (S)-methoprene, a synthetic analog of juvenile hormone (JH), has been previously shown to influence the mating behavior of flies [16]. The effects of OA and (S)-methoprene were examined separately on infected females. Our hypothesis was that mating responsiveness could be rescued in virus-infected females if they were given hormone therapy that could counteract the effects of infection.

2. Materials and Methods

2.1. Maintaining Flies

Flies were from the WTF strain maintained at USDA-ARS-CMAVE in Gainesville, FL. Adults were separated upon emergence and put into separate cages based on sex. Cages (20 × 20 × 20 cm) were provided with two 30 mL plastic containers of water with saturated Absorbal© wicks and one 30 mL plastic container with a 1:1 mixture of dry granulated sugar and powdered milk. Cages were held at 24–25 °C in incubators.

The WTF house fly colony includes a small but variable proportion of females that are autogenous (do not require protein for mating receptiveness or ovarian development) in each generation. To eliminate autogenous flies from the assays, females were pre-screened for signs of autogeny by placing them for 1 h with active males (1:1 females:males) ready to mate and removing any females or males that mated from the study. Only non-autogenous females were used for the mating studies. After removing all flies suspected of being autogenous, the remaining flies were separated again into groups of males and females.

2.2. Infection with Virus

Female flies were infected within 24 h of emergence with the FL strain of *Md*SGHV as described by Lietze et al. [9] and Shaler et al. [17]. Briefly, frozen virus samples containing a single pair of homogenized/filtered ovaries from infected flies in 50 μL of sterile saline were thawed then serially diluted fourfold (10^{-4} dilution) in PBS. Flies were cold-immobilized and injected in the thorax with 2.5 μL of the diluted virus, resulting in injection of about 8000 viral copies based on Lietze et al. [9]. The 10^{-4} dilution was selected for infection because we had previously determined this to be the best dilution out of a series of 12-fold dilutions to consistently produce 100% infected flies with hypertrophied salivary glands and no ovarian development (unpublished data).

2.3. Hormone Treatments

Octopamaine (OA) treatments were administered to the females via the same injection used to deliver the virus to avoid mortality from multiple injections. Because it is soluble in PBS, OA (6 mg) was directly dissolved into the *Md*SGHV inoculum (200 µL), producing a final diluted concentration of OA (30 µg/µL). When the 10^{-4} diluted virus inoculum was injected (2.5 µL) into each cold-immobilized female, final dosages of 75 µg of OA [15] were administered per fly.

Methoprene was applied topically. A stock solution of (S)-methoprene (5 µg/µL) was prepared by mixing methoprene (5.40 µL), density of 926.1 µg/µL, with acetone (994.60 µL). Cold-immobilized flies were treated by applying 1 µL of this solution (5 µg (S)-methoprene) to the thoracic surface of each female at 48 and 72 h after infection, resulting in a final dosage of 10 µg (S)-methoprene per fly.

A final experimental condition was a combination of: (1) topical application of methoprene as before; (2) injection with octopamine as before; and (3) inclusion of 2.5 µg of 20-hydroxyecdysone (20E) in the initial injection along with the virus and OA.

Several sets of control flies were set up as well: (1) uninfected, untreated flies; (2) uninfected flies injected with 2.5 µL PBS; and (3) uninfected flies treated topically with 1 µL acetone. Finally, uninfected control females were also set up that were denied protein (sugar-fed only) and either left untreated or treated topically with methoprene as described previously.

For each bioassay, cages of 50 healthy male flies and 15 flies from each of the treatment or control groups were set up and provided with food and water. An additional sample of five females injected with the virus were set aside from each batch of virus-injected flies to provide a virus quality control check before mating bioassays. These flies were dissected 72 h after viral injection and examined to confirm both hypertrophy of the salivary glands and lack of ovarian development. Mating bioassays were only conducted if all of the injected flies in a batch were symptomatic for infection.

2.4. Mating Timeline in House Flies

To determine an appropriate timeline of mating behavior for our assays, preliminary tests were first conducted to determine when females were optimally receptive to mating attempts. To do this, 24 h-old healthy females were placed into 7 separate, 16 oz plastic containers, with water, granulated sugar, and powdered milk, and 24 h-old males were added to each cup for 7 consecutive days. Each day, when males were added, they were observed for mating behavior for 1 h. Males showed clear mating behavior attempts beginning when females were 48 h old, but females did not accept male attempts until after 120 h post-eclosion. Based on these observations, mating behavior observations were done with females that were 120 h old at the time of bioassays.

2.5. Observation of Mating Behavior

Females from treatment or control groups were removed from their holding cages and transferred individually to 30 mL cups, each with a ventilated cap. Three healthy males of the same age were cold-immobilized and introduced into each cup containing 1 female. Sexual behaviors, and especially copulation, were observed for 2.5 h, as previously reported by Lietze et al. [9]. If a copulating pair included a hormone-treated female, she was saved for later dissection to confirm viral symptomology. Successful mating acceptance was defined as when females extended their ovipositor and contacted the claspers and aedeagus of the male [13]. Copulation acceptance from the female was indicated when a male and female fly embraced in mating for at least 30 min and did not unclasp from each other. Mating acceptance data were analyzed by G-tests of independence (chi square estimate, [18]) comparing (1) uninfected untreated controls with others; and (2) infected untreated females with others.

3. Results

Successful copulation was observed between infected females treated with hormone therapy and untreated, healthy males (Figure 1).

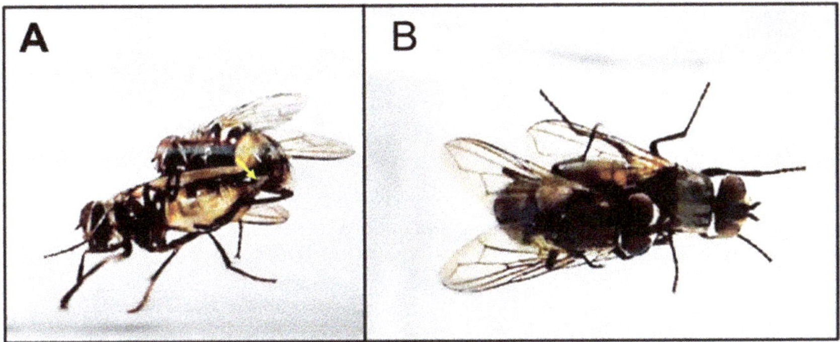

Figure 1. A *Md*SGHV-infected female treated with octopamine, (S)-methoprene, and 20-hydroxyecdysone mated on day 5 with an uninfected, untreated male (**A**,**B**). The copulatory position of the pair is correct, with the male on top (**A**,**B**). The yellow arrow (**A**) points to the female ovipositor that is being drawn into the male's genital area by his claspers (**A**) while, in (**B**), the normal mating positions is shown, with the male on top.

Healthy, untreated control females showed a 65% copulation rate for a duration longer than 30 min (Table 1). The mating success of PBS-injected females (80.4%) did not differ significantly from the untreated controls. No copulation was observed in virus-injected flies that were given no hormone therapy. The copulation rates of infected flies that were given octopamine (23%) or methoprene (27.8%) alone were significantly higher than for untreated infected flies, although somewhat lower than for uninfected females. Infected flies that were given both OA and methoprene had copulation success rates (88.9%) that did not differ from uninfected flies. The uninfected flies that fed only on sugar (no protein) did not mate at all, but treatment of these sugar-fed flies with methoprene resulted in mating success (50%) that did not differ significantly from flies that were provided with both sugar and protein (65%).

Table 1. The effect of various treatments on female house fly mating behavior/copulation. Adult, anautogenous females subjected to various treatments [a] and who then mated with uninfected, active mating males. Inf. = females infected with salivary gland hypertrophy and no ovarian development.

Treatment	Dose	# Mated/N [b]	%Mated [c]	Chi-Square [d]	
				Uninfected	Infected
				vs	vs
No injection/no treatment	N/A	17/30	65.0	-	38.154 **
PBS-injected control	2.5 μL	45/70	80.4	0.513	61.564 **
Infected with *Md*SGHV	2.5 μL	0/45	0	33.010 **	-
Inf. + Octopamine (OA)	75 μg	3/20	23.0	9.339 **	6.605 **

Table 1. *Cont.*

Treatment	Dose	# Mated/N [b]	%Mated [c]	Chi-Square [d]	
				Uninfected	Infected
				vs	vs
Inf. + Methoprene (Meth)	2×5 µg	10/60	27.8	14.834 **	11.043 **
Inf. + *Md*SGHV + acetone	2.5 µL + 1 µL	0/10	0	12.652 **	0.102 ns
Inf. + OA, Meth. + 20E	5, 2.5 + 2.5 µg	8/20	88.9	1.340 ns	20.592 **
Sugar-fed only	N/A	0/10	0	12.652 **	0.102 ns
Sugar-fed only + Meth	1 µL	5/10	50.0	0.134	18.722 **

[a] All treatments were injected except methoprene, which was applied topically. [b] Number of females that mated over the total number of females of all trials. #, Number of females that accepted copulation attempts by male flies. [c] Percentage of females that copulated with healthy males for a duration longer than 30 min. [d] Mating success of either uninfected controls or infected untreated flies compared to others; **, $p < 0.01$, ns, $p > 0.05$.

4. Discussion

Our results on the effect of viral infection on female mating receptivity are in broad agreement with those of Leitze et al. [9], who also found that early infection with *Md*SGHV causes females to be refractory to mating attempts by males. The two studies differ somewhat in regard to defining successful copulation. We used the same behavioral observations for successful copulation as described by Tobin and Stoffolano [4] for house flies, which includes the female voluntarily everting her ovipositor so that the male can grasp it and pull it into his genital opening. This contrasts somewhat with Leitze et al. [13], who stated that the female "extended her ovipositor into the genital opening" and that this constituted a successful copulation. In fact, the male grabs and pulls the ovipositor into his genital opening [13].

Manning [19] appears to be the first to have demonstrated the importance of JH in the mating receptivity of females in the Diptera, in this case *D. melanogaster*. Adams and Hintz [20] subsequently discussed how JH stimulates mating in female house flies, while Barth and Lester [21] and Ringo [11,22] later discussed the various factors influencing receptivity in insects and provided references demonstrating that JH is essential for receptivity in many insects (i.e., including flies), as well as that the JH analogue, (S)-methoprene, can induce or restore mating receptivity when given as a hormone replacement therapy. Another important physiological event that can influence mating with respect to JH in flies is when they enter adult diapause [23]. Stoffolano [24] examined the spermathecae of female *Phormia regina* and found that, based on the absence of sperm, females in diapause failed to mate, while non-diapausing females successfully copulated. During their adult diapause, *Phormia regina* and *Protophormia terraenovae* adults refuse to mate [24,25] and, presumably, this is related to the diapause syndrome, which is due to an insufficient amount of JH. Tanigawa et al. [25] were able to rescue CA ablation in *Protophormia* that prevented mating by using a topical application of methoprene. In another study, Teal et al. [12] demonstrated that JH was essential for mating in the Caribbean fruit fly *Anastrepha suspensa* (Loew).

We found that the JH analog methoprene and octopamine were both effective at rescuing mating receptivity in infected females. In contrast, Kariithi et al. [10] attempted to rescue mating behavior in virus-infected house flies by injecting them with ecdysone, commercial JH-III, or methyl farnesoate, and were unsuccessful in their attempt to produce hormonal therapy. Differences in methodology between the two studies include our use of lower viral doses, the topical application of JH (methoprene), and the use of more than one application of methoprene.

Adams and Hintz [20] demonstrated that JH was essential for female house flies to accept mating attempts by males, and Yin et al. [14,26] showed that removal of the CA in *P. regina* females significantly reduced receptivity, which could be reversed if they topically applied methoprene. In their paper, Kariithi et al. [10] noted that "*Md*SGHV replication in CA/CC potentially explains the significant reduction of hemolymph sesquiterpenoids levels, the refusal to mate, and the complete shutdown of egg development in viremic females". The involvement of the CA/CC complex suggests that low or no JH is involved in the lack of mating receptivity in virus-infected female house flies. Evans et al. [15] showed that two applications of 5 μg of methoprene or one 75 μg dose of OA can significantly increase the percentage of insemination in *P. regina* that were fed only sugar, which normally do not mate. OA is a neurohormone that regulates the reproductive function of *Drosophila melanogaster* by controlling the metabolism of JH directly and 20E indirectly [27–30]. Our results indicate that either JH or OA therapy alone was sufficient to partially restore mating acceptance (23–28%) in virus-infected flies, whereas flies that received both therapies plus 20-hydroxyecdysone (20E) copulated at the same rates (88.9%) as uninfected controls. Further research would be needed to determine whether 20E contributed to the effectiveness of the combination of JH and OA.

The topical application of various juvenoids has been shown to rescue mating receptivity in flies, per the following examples: methoprene for *Protophormia terraenovae* [25], methoprene or fenoxycarb for *Anastrepha suspensa* [12], trans, trans-10, ll-epoxy farnesenic acid, methyl ester for house fly [20], methoprene for *Drosophila melanogaster* [28], OA for *Phormia regina* [15], and now methoprene and octopamine for virus-infected house flies.

We were able to rescue mating receptivity in virus-infected females. For this study, we applied octopamine by injection, while Barron et al. [31] showed, using honeybees, that various methods of application were suitable. It is possible, as shown by Amsalem et al. [32], that some events in the behavior and physiology of an organism can be rescued by hormone replacement therapy, while other events are unable to be rescued. Hormonal rescue therapy is difficult and can require the application of hormones within a critical window of effectiveness, multiple treatments, an appropriate method of delivery of the treatments, and the tolerance of the study animal to injection. The ability of a therapeutic to rescue a particular pathogen-induced effect may also depend on the dosage of the pathogen or treatment producing the effect.

*Md*SGHV is an attractive biological control agent for managing house flies because of its inhibition of mating behavior and ovarian development. One of the paradoxes of the virus, however, is that flies are only maximally vulnerable to per os infection during a narrow window after adult eclosion, during a time when flies are generally too young to commence feeding [33]. This is thought to be due to development of the peritrophic matrix in the hours after emergence, which prevents the virus from crossing the fly midgut into the hemocoel [34]. House flies are also susceptible to infection by immersion in or sprays with suspensions containing homogenates of virus-infected flies [35]. Although it has limited utility from a fly management standpoint, this viral/house fly system provides a good model to explore the behavioral aspects of how the virus is obtained and spread, the immunity/reproductive tradeoffs, and how it affects mating/copulatory behavior.

5. Conclusions

Injecting octopamine and topically applying methoprene twice, following the injection of the virus into healthy females, resulted in the restoration of mating receptivity of infected females. Treatment with octopamine alone showed a lower percentage of mating behavior than treatment with methoprene alone. We demonstrate that methoprene has the greatest effect on rescuing mating behavior in house flies when the treatments are combined. Regardless of hormone treatment, viral injection still resulted in the pathology of the salivary glands and a reduction in ovarian development. The use of the JH-mimic methoprene supports the suggestion that the virus somehow affects sesquiterpenoid production in the corpus allatum or allatotropin from the brain, thus reinforcing JH's long-understood role in mating receptivity in house fly females. Information is now needed as to whether virus-infected males can be hormonally rescued to mate, and it remains to be determined whether either sex is able to detect virus-infected mates, which might determine mate choice.

Author Contributions: Conceptualization, J.G.S.J.; methodology, J.G.S.J., M.G. and A.R.; formal analysis, C.J.G.; investigation, M.G. and A.R.; resources, J.G.S.J.; writing—original draft preparation, A.R.; writing—review and editing, C.J.G.; supervision, J.G.S.J.; project administration, J.G.S.J.; funding acquisition, J.G.S.J. All authors have read and agreed to the published version of the manuscript.

Funding: This research was partially supported by the National Institute of Food and Agriculture, U.S. Department of Agriculture, the Massachusetts Agricultural Experiment Station and the Stockbridge School of Agriculture at the Univ. of Massachusetts, Amherst, under project number MAS00527/S51076 to J.G.S.J.

Data Availability Statement: Data available upon request from J.G.S.J. or C.J.G

Acknowledgments: Thanks to Jennifer Schaler and Shelby Tonelli for assistance in handling and injection procedures of the virus. Thanks to Douglas VanGundy of Central Life Sciences, Dallas, TX 75234, USA, for his generous supply of methoprene. Thanks also to Alexandra Pagac for rearing and shipping the flies used in the project. The contents are solely the responsibility of the authors and do not necessarily represent the official views of the USDA or NIFA.

Conflicts of Interest: The authors declare no conflict of interest. The funders had no role in the design of the study; in the collection, analyses, or interpretation of data; in the writing of the manuscript; or in the decision to publish the results.

References

1. van Sluijs, L.; Liu, J.; Schrama, M.; van Hamond, S.; Vromans, S.P.J.M.; Scholten, M.H.; Žibrat, N.; Riksen, J.A.G.; Pijlman, G.P.; Sterken, M.G.; et al. Virus infection modulates male sexual behaviour in *Caenorhabditis elegans*. *Mol. Ecol.* **2021**, *30*, 6776–6790. [CrossRef] [PubMed]
2. Burand, J.P.; Tan, W.; Kim, W.; Nojima, S.; Roelofs, W. Infection with the insect virus Hz-2v alters mating behavior and pheromone production in female *Helicoverpa zea* moths. *J. Insect Sci.* **2005**, *5*, 6. [CrossRef]
3. Kariithi, H.M.; van Oers, M.M.; Vlak, J.M.; Vreysen, M.J.B.; Parker, A.G.; Abd-Alla, M.M.A. Virology, epidemiology and pathology of *Glossina* hytrosavirus, and its control prospects in laboratory colonies of the tsetse fly, *Glossina pallidipes* (Diptera; Glossinidae). *Insects* **2013**, *4*, 287–319. [CrossRef]
4. Kariithi, H.; Meki, I.; Boucias, D.G.; Abd Alla, A. Hytrosaviruses: Current status and perspective. *Curr. Opin. Ins. Sci.* **2017**, *22*, 71–78. [CrossRef]
5. Geden, C.J.; Lietze, V.-U.; Boucias, D.G. Seasonal prevalence and transmission of salivary gland hypertrophy virus of house flies (Diptera: Muscidae). *J. Med. Entomol.* **2008**, *45*, 42–51. [CrossRef] [PubMed]
6. Weeks, E.N.I.; Machtinger, E.T.; Geden, C.J.; Leemon, D.D. Biological control of livestock pests: Entomopathogens. In *Ecology and Control of Vector-Borne Diseases*; Garros, C., Bouyer, J., Takken, W., Smallegange, R.C., Eds.; Prevention and Control of Pests and Vector-borne Diseases in the Livestock Industry; Wageningen Academic Publishers: Wageningen, The Netherlands, 2018; Volume 5, pp. 337–387.
7. Mutika, G.N.; Marin, C.; Parker, A.G.; Vreysen, M.J.B.; Boucias, D.G. Abd-Alla AMM: Impact of salivary gland hypertrophy virus infection on the mating success of male *Glossina pallidipes*: Consequences for the sterile insect technique. *PLoS ONE* **2012**, *7*, e42188. [CrossRef]
8. Coler, R.R.; Boucias, D.G.; Frank, J.H.; Maruniak, J.E.; Garcia Canedo, A.; Pendland, J.C. Characterization and description of a virus causing salivary gland hyperplasia in the housefly, *Musca domestica*. *Med. Vet. Entomol.* **1993**, *7*, 275–282. [CrossRef]
9. Lietze, V.U.; Geden, C.J.; Blackburn, P.; Boucias, D.G. Effects of salivary gland hypertrophy virus on the reproductive behavior of the housefly, *Musca domestica*. *Appl. Environ. Microbiol.* **2007**, *73*, 6811–6818. [CrossRef]
10. Kariithi, H.M.; Yao, X.; Yu, F.; Teal, P.E.; Verhoeven, C.P.; Boucias, D.G. Responses of the housefly, *Musca domestica*, to the hytrosavirus replication: Impacts on host's vitellogenesis and immunity. *Front. Microbiol.* **2017**, *8*, 583. [CrossRef] [PubMed]
11. Ringo, J.M. Sexual receptivity in insects. *Annu. Rev. Entomol.* **1996**, *41*, 473–494. [CrossRef]
12. Teal, P.E.A.; Gomez-Simuta, Y.; Proveaux, A.T. Mating experience and juvenile hormone enhance sexual signaling and mating in male Caribbean fruit flies. *Proc. Natl. Acad. Sci. USA* **2000**, *97*, 3708–3712. [CrossRef] [PubMed]
13. Tobin, E.N.; Stoffolano, J.G., Jr. The courtship of *Musca* species found in North America, 1. The housefly, *Musca domestica*. *Ann. Entomol. Soc. Am.* **1973**, *66*, 1249–1257. [CrossRef]
14. Yin, C.-M.; Qin, W.-H.; Stoffolano, J.G., Jr. Regulation of mating behavior by nutrition and the corpus allatum in both male and female *Phormia regina* (Meigen). *J. Insect Physiol.* **1999**, *45*, 815–822. [CrossRef]
15. Evans, B.P.; Stoffolano, J.G., Jr.; Yin, C.-M.; Meyer, J.S. A pharmacological and endocrinological study of female insemination in *Phormia regina* (Diptera: Calliphoridae). *J. Insect Behav.* **1997**, *10*, 493–508. [CrossRef]
16. Evans, P.D. Biogenic amines in the insect nervous system. *Adv. Insect Physiol.* **1980**, *15*, 317–473.
17. Schaler, J.; Stoffolano, J.G., Jr.; Fausto, A.M.; Gambellini, G.; Burand, J. Effect of diet on adult house fly (Diptera: Muscidae) injected with the salivary gland hypertrophy virus (MdSGHV). *J. Insect Sci.* **2018**, *18*, 2. [CrossRef] [PubMed]
18. Sokal, R.R.; Rohlf, F.J. *Biometry*; W. H. Freeman: New York, NY, USA, 1980; 859p.

19. Manning, A. Corpus allatum and sexual receptivity in female *Drosophila melanogaster*. *Nature* **1966**, *211*, 1321–1322. [CrossRef]
20. Adams, T.S.; Hintz, A.M. Relationship of age, ovarian development, and the corpus allatum to mating in the housefly, *Musca domestica*. *J. Insect Physiol.* **1969**, *15*, 201–215. [CrossRef]
21. Barth, R.H.; Lester, L.J. Neuro-hormonal control of sexual behavior in insects. *Ann. Rev. Entomol.* **1973**, *18*, 445–472. [CrossRef]
22. Ringo, J.M. Hormonal regulation of sexual behavior in insects. *Horm. Brain Behav.* **2002**, *3*, 93–114. [CrossRef]
23. Denlinger, D.L. Hormonal control of diapause. In *Comprehensive Insect Physiology, Biochemistry and Pharmacology*; Kerkut, G.A., Gilbert, L.I., Eds.; Pergamon: Oxford, UK, 1985; Volume 8, pp. 353–412. [CrossRef]
24. Stoffolano, J.G., Jr. Influence of diapause and diet on the development of the gonads and accessory reproductive glands of the black blowfly, *Phormia regina* (Meigen). *Canandian J. Zool.* **1974**, *52*, 981–988. [CrossRef]
25. Tanigawa, N.A.; Shiga, S.; Numata, H. Role of the corpus allatum in the control of reproductive diapause in the male blow fly, *Protophormia terraenovae*. *Zool. Sci.* **1999**, *16*, 639–644. [CrossRef]
26. Yin, C.-M.; Zou, B.-X.; Stoffolano, J.G., Jr. Precocene II treatment inhibits terminal oocyte development but not vitellogenin synthesis in the black blowfly *Phormia regina*. *J. Insect Physiol.* **1989**, *35*, 465–474. [CrossRef]
27. Gruntenko, N.E.; Karpova, E.K.; Adonyeva, N.V.; Chentsova, N.A.; Faddeeva, N.V.; Alekseev, A.A.; Rauschenbach, I.Y. Juvenile hormone, 20-hydroxyecdysone and dopamine interaction in *Drosophila virilis* reproduction under normal and nutritional stress conditions. *J. Insect Physiol.* **2005**, *51*, 417–425. [CrossRef]
28. Gruntenko, N.E.; Karpova, E.K.; Alekseev, A.A.; Chentsova, N.A.; Bogomolova, E.V.; Bownes, M.; Yu Rauschenbach, I. Effects of octopamine on reproduction, juvenile hormone metabolism, dopamine, and 20-hydroxyecdysone contents in *Drosophila*. *Arch. Insect Biochem. Physiol.* **2007**, *65*, 85–94. [CrossRef] [PubMed]
29. Rauschenbach, I.Y.; Chentsova, N.A.; Alekseev, A.A.; Gruntenko, N.E.; Karpova, E.K. Dopamine affects the level of 20-hydroxyecdysone in *Drosophila virilis* females. *Dokl. Biol. Sci.* **2006**, *407*, 179–181. [CrossRef]
30. Gruntenko, N.E.; Rauschenbach, I.Y. Interplay of JH, 20E and biogenic amines under normal and stress conditions and its effect on reproduction. *J. Insect Physiol.* **2008**, *54*, 902–908. [CrossRef] [PubMed]
31. Barron, A.B.; Maleszka, J.; Vander Meer, R.K.; Robinson, G.R.; Maleszka, R. Comparing injection, feeding and topical application methods for treatment of honeybees with octopamine. *J. Insect Physiol.* **2007**, *53*, 187–194. [CrossRef] [PubMed]
32. Amsalem, E.; Teal, P.P.; Grozinger, C.M.; Hefetz, A. Precocene-I inhibits juvenile hormone biosynthesis, ovarian activation, aggression and alters sterility signal production in bumble bee (*Bombus terrestris*) workers. *J. Exp. Biol.* **2014**, *217*, 3178–3185. [CrossRef] [PubMed]
33. Prompiboon, P.; Lietze, V.U.; Denton, J.S.; Geden, C.J.; Steenberg, T.; Boucias, D.G. *Musca domestica* salivary gland hypertrophy virus, a globally distributed insect virus that infects and sterilizes female houseflies. *Appl. Environ. Microbiol.* **2010**, *76*, 994–998. [CrossRef]
34. Boucias, D.; Baniszewski, J.; Prompiboon, P.; Lietze, V.; Geden, C. Enhancement of the *Musca domestica* hytrosavirus infection with orally delivered reducing agents. *J. Invert. Pathol.* **2015**, *124*, 35–43. [CrossRef] [PubMed]
35. Geden, C.J.; Steenberg, T.; Lietze, V.U.; Boucias, D.G. Salivary gland hypertrophy virus of house flies in Denmark: Prevalence, host range, and comparison with a Florida isolate. *J. Vector Ecol.* **2011**, *36*, 231–238. [CrossRef]

Article

Host Preferences Shown by Drosophilids (Diptera) in a Commercial Fruit and Vegetable Distribution Center Follow the Wild Neotropical Pattern

Laís Barbosa Ribeiro [1], Carolyn Elinore Barnes Proença [2] and Rosana Tidon [1,3,*

[1] Graduate Program in Ecology, Institute of Biological Sciences, University of Brasilia, Brasilia 70910-900, Brazil; laisribeiro015@hotmail.com

[2] Department of Botany, Institute of Biological Sciences, University of Brasilia, Brasilia 70910-900, Brazil; carolyn.proenca@gmail.com

[3] Department of Genetics and Morphology, Institute of Biological Sciences, University of Brasília, Brasilia 70910-900, Brazil

* Correspondence: rotidon@unb.br

Simple Summary: Drosophilids (fruit flies) are known as study models in several areas of science. Several drosophilid species have recently attracted public attention because they are expanding their geographic distribution and infesting fruit crops. Here, we investigated the relationship between plants and fruit flies in a commercial fruit and vegetable distribution center in Brazil. We accomplished this by collecting 99,478 kg of potential fruit and vegetable hosts from two time periods separated by a decade, representing 48 plant taxa. The 48,894 fruit flies that emerged from these hosts were identified and attributed to 16 fly species. On both collecting occasions, fruit fly assemblages were strongly dominated by basically the same exotic species, which explore a broader range of hosts, especially those of exotic origin, when compared to native neotropical fruit flies. These results are concerning because the studied site, along with other urban markets around the world, might be acting as a source of widespread generalist species that subsequently disperse into surrounding natural vegetation and crops. As these flies are usually superior competitors, they can promote the local extinction of native fruit flies and therefore contribute to the homogenization of fruit fly communities on larger scales. This phenomenon, known as "biotic homogenization" is worrying scientists worldwide.

Abstract: Although drosophilids have been extensively studied in laboratories worldwide, their ecology is still relatively poorly understood. This is unfortunate because some species are currently expanding their geographic distribution and infesting fruit crops. Here, we investigated the relationship between drosophilids and potential plant hosts in a commercial fruit and vegetable distribution center in the Neotropical region. We collected discarded fruits and vegetables from this commercial center during two time periods (2007–2008 and 2017–2018). Resources were weighted and individually monitored in the laboratory. The drosophilids that emerged were identified, and the relationship between them and their resources was explored. From the 99,478 kg of potential hosts collected, we identified 48 plant taxa, from which 48,894 drosophilids of 16 species emerged. On both collecting occasions, drosophilid assemblages were strongly dominated by basically the same exotic species, which explore a broader range of resources, especially those of exotic origin, when compared to neotropical drosophilids. These results are concerning because the studied site, Along with other urban markets around the world, might be acting as sources of generalist widespread species that disperse to surrounding natural vegetation and contribute to biotic homogenization.

Keywords: breeding site; *Drosophila*; fruit markets; invasive species; niche breath; urban ecology; vegetable markets; *Zaprionus*

Citation: Ribeiro, L.B.; Proença, C.E.B.; Tidon, R. Host Preferences Shown by Drosophilids (Diptera) in a Commercial Fruit and Vegetable Distribution Center Follow the Wild Neotropical Pattern. *Insects* **2023**, *14*, 375. https://doi.org/10.3390/insects14040375

Academic Editors: Sérgio M. Ovruski and Flávio Roberto Mello Garcia

Received: 17 March 2023
Revised: 4 April 2023
Accepted: 8 April 2023
Published: 11 April 2023

1. Introduction

The family Drosophilidae includes more than 4600 nominal species [1] that breed preferentially on fermenting substrates such as fruits, flowers, or fungi [2]. While most species are geographically and ecologically restricted, some are generalists and dispersed beyond their native ranges throughout the world [3]. In Brazil, 364 drosophilid species have been recorded, 350 of which are native and 14 of which are exotic to the Neotropical region [4]. Certain exotic species, such as *Drosophila melanogaster* Meigen and *D. simulans* Sturtevant, probably reached Brazil via ships from Africa in the 16th century. Others arrived in the country more recently as a consequence of globalization. From the late 20th century, five new arrivals in the Neotropics were accurately recorded in the earlier stages of invasion: *D. malerkotliana* Parshad and Paika [5], *Zaprionus indianus* Gupta [6], *D. nasuta* Lamb [7], *D. suzukii* Matsumura [8], and *Z. tuberculatus* Malloch [9]. These introductions are especially worrying because some of these species, such as the spotted wing *Drosophila* (*D. suzukii*, see [10,11]) and the African fig fly (*Z. indianus*, see [12,13]), have become invaders and impact agricultural crops.

The establishment of invasive species in new areas also represents an important threat to biodiversity [14]. Widespread species usually present a high climatic tolerance [15] and explore a wider range of resources than narrowly distributed species. As a result, they can outcompete native species. In a comprehensive survey of fruit-breeding drosophilids and their hosts in the Neotropics, Valadão et al. [16] recorded 180 species of plants (representing 50 families) acting as hosts of 100 drosophilid species. These authors also found that exotic drosophilids breed in more plant species and use exotic hosts more frequently than do Neotropical drosophilids. However, Valadão et al. [16] focused primarily on fruits collected near the host plants; fruits from markets and refuse containers were excluded from their analysis. As there is an expressive drosophilid fauna established in urban environments [17–21], it is worth investigating the drosophilid community associated with the resources available in commercial markets.

The Cerrado biome, also known as Brazilian Savanna, spans most of the Central Brazilian highlands [22] and is one of the world's biodiversity hotspots due to its high level of endemism and habitat loss [23]. It comprises a savanna of variable structure on the well-drained interfluves, with gallery forests or other moist vegetation following the watercourses [24]. The climate in the Cerrado is tropical dry winter Aw in 95% of the biome, according to the Koeppen classification, and the precipitation is highly seasonal and concentrated during the rainy season from October to April. Currently, 125 neotropical and 13 exotic species of drosophilid have been recorded in this biome [25]. The drosophilids established in a protected area in the center of the Cerrado biome and monitored since 1998 seem to respond to climate seasonality, vegetation heterogeneity, disturbance (including the arrival of exotic species), resource availability, and parasitoids [26–33]. Given the degree of knowledge of this system, it is relevant to investigate the entry routes and establishment sites of exotic species. In this context, food supply and distribution centers in urban areas, which receive products not only from all over the country but also from abroad, become important places to be explored.

The objective of our study was to investigate the relationship between drosophilids and plant species in a distribution center that supplies many urban markets located in the core area of the Cerrado biome. Our main questions were the following: Does the drosophilid community change over time? How are drosophilid species distributed among plant species? Do exotic drosophilids explore a wider range of resources than neotropical drosophilids?

2. Materials and Methods

2.1. Collections and Taxonomic Determination

Plant resources were collected at the *Centrais de Abastecimento do Distrito Federal* ("Federal District Supply Center" hereafter CEASA-DF), located in the Industry and Supply Sector of Brasília, Brazil. The horticultural products that arrive at CEASA-DF come from different regions of the country and undergo a selection process before being sold. In this process, fruits and vegetables that are deemed unfit for consumption are discarded on the ground, under unloading trucks, and in refuse containers. The collections focused on these decomposing plant resources, which serve as breeding sites for flies and were concentrated over two periods. First, six monthly collections were carried out between August 2007 and January 2008. In the second period, five collections were carried out between October 2018 and May 2019. The collection method in both periods was similar: two collectors randomly collected plant resources. However, in the first period, the collectors spent up to two hours on each collection, while in the second, they spent up to one hour, or until they completed a box of approximately 50 L. The sample units collected (fruits, vegetables, or their fragments) were individually packed and transported to the laboratory.

In the laboratory, each plant sample unit was identified to species (or variety for *Brassica olearacea* L. and *Prunus persica* L.) and classified into types: DF (dry fruits), FF (fleshy fruits), SB (stem bulbs with cataphylls), ST (stem tubers), RT (root tubers), and VL (vegetative leaves). Sample units were then weighed and placed in a transparent plastic container to allow visualization of the hatched flies. In the containers, a thin layer of vermiculite was placed at the bottom to control humidity, and a thin cloth was placed at the opening to trap flies and allow gas exchange. The containers, kept at 25 °C and 12 h:12 h (L:D), were observed every two days. Hatched flies were aspirated and identified by external morphology [34,35] or male terminalia [36,37]. Taxon circumscriptions, names, authors, and geographic distributions of plant and drosophilid species are cited in Valadão et al. [16]. Taxa not included in their study were checked in Taxodros [1] and The World Flora Online [38] for drosophilids and plants, respectively.

2.2. Data Analyses

To assess sampling effort and compare species richness for both collection periods [39], we plotted the drosophilid species accumulation curves using a sample-based rarefaction method (plant taxa) using the function "specaccum" in the Vegan package [40] available in R 4.2.2. We used the Whittaker plot to show the species abundance rank and assess the evenness of the community [41]. To calculate the relative abundance of exotic/neotropical species, we added up the abundances of all the species in each category and divided them by the total abundance. We also assessed the drosophilid density (Nflies/g) per plant species; plant samples without emergencies were not considered.

For each collection period, we built matrices of interactions between drosophilid species and plant species, including all recorded interactions. Then, we generate two bipartite networks to visualize the webs. Moreover, we calculated the Spearman correlation between the mass of each vegetable species and the richness and abundance of flies to understand whether the number of interactions reflected the number of resources. For that, the bipartite package [42] and the function "cor.test," available in R 4.2.2, were used.

The classification of drosophilids as generalists or specialists was based on the criteria established by Magnacca et al. [43]. A species was considered a specialist if two conditions were satisfied: (i) at least two-thirds of its breeding records are associated with a single plant family; and (ii) any other family has <25% of the remaining records, ensuring a clear preference for a single family. For example, a species with 60 breeding records would be considered a specialist if at least 40 records were made in a single plant family and any other family had no more than 15 records. Thus, a drosophilid may be considered a specialist even if it uses alternative plant families as secondary or occasional hosts.

To investigate whether exotic drosophilids explore a wider range of resources than neotropical drosophilids, we calculated the proportion of positive associations observed between Neotropical (N) and exotic (E) hosts (H) and drosophilid (D) species for the four possible pairs: NH × ND, NH × ED, EH × ND, and EH × ED. The expected percentage of associations for each pair was predicted based on the total number of possible associations in the matrix. The adherence between observed and predicted association percentages in each category was tested using the X2 goodness-of-fit test followed by an exact binomial test for each pair as a post hoc test [44]. For this analysis, we only used nominal species.

3. Results

In total, 99.478 kg of plant resources representing 48 species and two varieties (50 taxa in 28 botanical families) were collected and transported to the laboratory (Table 1). From this material, 48,894 drosophilids emerged, representing 16 species (Table 2). Despite the number of sampling units and drosophilids being approximately 50% smaller in the second period compared to the first, the two rarefaction curves stabilized. Thus, in both periods, the sampling effort was sufficient to represent the richness of drosophilids in this urban supply center (Figure 1).

Table 1. Plant families and taxa collected in the Federal District Supply Center in Brasília, Brazil (CEASA).

Family	Taxa	Popular Name	Code	Type	Mass (g)	Empty Mass (%)	Collection Period
Actinidiaceae	*Actinidia chinensis* Planch. [E]	Kiwi; Kiwi	1	FF	592.4	86.26	1', 2
Amaranthaceae	*Beta vulgaris* L. [E]	Beterraba; Beetroot	2	**RT**	482.9	100	2'
Amaryllidaceae	*Allium cepa* L. [E]	Cebola; Onion	3	**SB**	3037.1	57.81	1, 2
Anacardiaceae	*Anacardium occidentale* L. [N]	Caju; Cashew fruit	4	FF	524.0	25.38	1
	Mangifera indica L. [E]	Manga; Mango	5	FF	8595.8	69.42	1, 2
	Spondias mombin L. [N]	Cajá,Cajazinho; Java plum	6	FF	107.2	100	2'
	Spondias purpurea L. [N]	Ciriguela, Seriguela; Gambia plum, Purple mombin	7	FF	8.9	100	2'
Annonaceae	*Annona squamosa* L. [N]	Pinha, Fruta do Conde; Custard apple	8	FF	398.6	100	2'
Apiaceae	*Arracacia xanthorrhiza* Bancr. [N]	Batata baroa, Mandioquinha; Arracache, Peruvian parsnip	9	**RT**	79.7	100	2'
	Daucus carota L. [E]	Cenoura; Carrot	10	**RT**	501.8	100	2'
Araceae	*Colocasia esculenta* (L.) Schott [E]	Inhame; Cocoyam, Taro	11	**ST**	711	73.42	1
Asteraceae	*Lactuca sativa* L. [E]	Alface; Lettuce	12	**VL**	290	0	1
Brassicaceae	*Brassica oleracea* L. var. *acephala* DC. [E]	Couve; Collard greens, Kale	13	**VL**	59.9	100	2'
	Brassica oleracea L. var. *capitata* L. [E]	Repolho; Cabbage	14	**VL**	585	0	1
Bromeliaceae	*Ananas comosus* (L.) Merr. [N]	Abacaxi; Pineapple	15	FF	27,195.5	0	1, 2
Cactaceae	*Selenicereus undatus* (Haw.) D.R.Hunt [N]	Pitaya; Dragon fruit	16	FF	205.9	100	2'
Caricaceae	*Carica papaya* L. [N]	Mamão; Papaya	17	FF	6256.2	87.24	1', 2
Caryocaraceae	*Caryocar brasiliense* Cambess. [N]	Pequi; no English name	18	FF	171	17.54	1
Convolvulaceae	*Ipomoea batatas* (L.) Lam. [N]	Batata doce; Sweet potato	19	**RT**	125.2	100	2'
Cucurbitaceae	*Cucumis anguria* L. [E]	Maxixe; West indian gherkin	20	FF	577.7	78.38	1, 2
	Citrullus lanatus (Thunb.) Matsum. and Nakai [E]	Melancia; Watermelon	21	FF	6071.9	2.86	1, 2
	Cucumis melo L. [E]	Melão; Melon	22	FF	962.9	0	1, 2
	Curcubita moschata Duchesne [N]	Abóbora; Pumpkin, Winter squash	23	FF	1497.3	11.98	1, 2
	Cucumis sativus L. [E]	Pepino; Cucumber	24	FF	844.7	58.01	1, 2
	Sicyos edulis Jacq. [N]	Chuchu; Chayote, Corstophine	25	FF	558.9	31.10	2
Ebenaceae	*Diospyros kaki* L.f. [E]	Caqui; Persimmon	26	FF	282.8	100	2'
Lauraceae	*Persea americana* Mill. [N]	Abacate; Avocado	27	FF	1201.4	100	2'
Malvaceae	*Hibiscus esculentus* L. [N]	Quiabo; Okra, Gumbo, Lady's fingers	28	**DF**	62.9	100	2'
Moraceae	*Artocarpus heterophyllus* Lam. [E]	Jaca; Jackfruit	29	FF	1490	0	1

Table 1. *Cont.*

Family	Taxa	Popular Name	Code	Type	Mass (g)	Empty Mass (%)	Collection Period
Musaceae	*Musa* x *paradisiaca* L. [E]	Banana; Banana	30	FF	5860.3	34.55	1, 2
Myrtaceae	*Psidium guajava* L. [N]	Goiaba; Guava	31	FF	1035.7	54.28	1′, 2
Oxalidaceae	*Averrhoa carambola* L. [E]	Carambola; Star fruit	32	FF	105.3	13.58	1, 2′
Passifloraceae	*Passiflora edulis* Sims [N]	Maracujá; Passion fruit	33	FF	1800.9	43.17	1, 2
Rosaceae	*Fragaria vesca* L. [N]	Morango; Strawberry	34	FF	209.8	64.59	1, 2
	Malus domestica (Suckow) Borkh. [E]	Maçã; Apple	35	FF	5128	94.,44	1, 2′
	Prunus domestica L. [E]	Ameixa; Plum	36	FF	1393	63.03	1, 2
	Prunus persica (L.) Batsch [E]	Pêssego; Peach	37	FF	709	56.56	1, 2′
	Prunus persica var. *nucipersica* (L.)C.K. Schneid. [E]	Nectarina; Nectarine	38	FF	433.8	76.26	1, 2′
	Pyrus communis L. [E]	Pera; Pear	39	FF	1568.4	73.64	1, 2
Rutaceae	*Citrus* x *aurantiifolia* (Christm.) Swingle [E]	Limão; Lime	40	FF	571	80,14	2
	Citrus x *reticulata* Blanco [E]	Mexirica, Bergamota; Tangerine	41	FF	2391	38,62	1, 2
	Citrus sinensis (L.) Osbeck [E]	Laranja; Orange	42	FF	2681.2	47.,78	1, 2
Sapindaceae	*Litchi chinensis* Sonn. [E]	Lichia; Lychee	43	FF	19.9	100	2′
Solanaceae	*Capsicum annuum* L. [N]	Pimentão; Bell pepper	44	FF	2346.5	71.02	1, 2
	Capsicum chinense L. [N]	Pimenta; Chili pepper	45	FF	25.5	100	2′
	Solanum aethiopicum L. [E]	Jiló; Bitterberry	46	FF	239.6	100	2′
	Solanum lycopersicum Lam. [N]	Tomate; Tomato	47	FF	5592	39.33	1, 2
	Solanum melongena L. [E]	Beringela; Eggplant, Aubergine	48	FF	360.5	39.92	2
	Solanum tuberosum L. [N]	Batata; Potato	49	**ST**	3420.4	79.69	1, 2
Vitaceae	*Vitis vinifera* L. × *Vitis labrusca* [EN]	Uva; Grape	50	FF	108.4	40.96	1, 2′

[E]: exotic; [N]: neotropical. Popular names: Brazilian Portuguese; English names. DF: dry fruit; FF: fleshy fruit; RT: root tuber; SB: stem bulb with cataphylls; ST: stem tuber; VL: vegetative leaf. Resources that are not FF in bold. Code: as in Figure 3. Mass: total collected. Empty mass: mass without the emergence of drosophilids. Collection periods 1: 2007–2008; 2: 2018–2019; apostrophe (′): there was no emergence of flies.

Table 2. Genera, subgenera, groups, and species of Drosophilidae recorded on fruits and vegetables collected in the Federal District Supply Center located in Brasília, Brazil, in two collection periods: 2007–2008 (1) and 2018–2019 (2). [E]: exotic; [N]: neotropical; Plant Fam/Spp: number of plant families and plant species with drosophilid records.

Genus	Subgenus	Group	Species	Code	Plant Fam/Spp	Abundance 2007–2008	Abundance 2018–2019
Drosophila	*Dorsilopha*	*busckii*	*D. busckii* Coquillett [E]	A	6/13	1198	1647
	Drosophila	*cardini*	*D. cardini* Sturtevant [N]	B	9/14	121	412
			D. cardinoides Dobzhansky and Pavan [N]	C	1/2	0	11
		immigrans	*D. immigrans* Sturtevant [E]	D	4/4	8	140
			D. nasuta Lamb [E]	E	1/1	0	211
		repleta	*D. hydei* Sturtevant [E]	F	11/14	14,361	1049
			D. mercatorum Patterson and Wheeler [N]	G	9/13	299	472
			D. repleta Wollaston [N]	H	1/1	0	17
		willistoni	*D. nebulosa* Sturtevant [N]	I	4/4	25	0
	Sophophora	*melanogaster*	*D. ananassae* Doleschall [E]	J	10/14	424	2775
			D. kikkawai Burla [E]	K	1/1	2	0
			D. malerkotliana Parshad and Paika [E]	L	6/6	592	0
			D. melanogaster Meigen [E]	M	11/15	6838	3225
			D. simulans Sturtevant [E]	N	15/22	7824	3739
		saltans	*D. sturtevanti* Duda [N]	O	1/1	0	85
Zaprionus			*Z. indianus* Gupta [E]	P	14/20	2154	1265

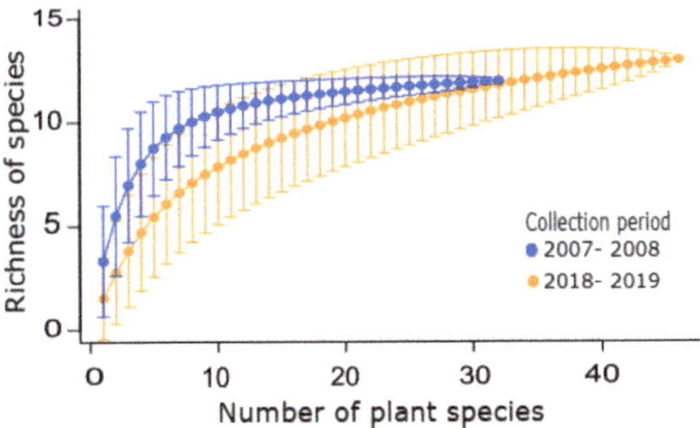

Figure 1. Sample–based rarefaction curves for Drosophilid species were recorded in fruits and vegetables collected in the Federal District Supply Center located in Brasília, Brazil, in two collection periods.

3.1. Temporal Changes

The richness of fly species in both periods was similar: 12 and 13 species, with a strong dominance of exotic species (Figure 2). The two most common species present on the two collection occasions (Table 2) corresponded to 98.7% and 93.4% of the total abundance, respectively. The species composition varied: nine occurred in both periods, three occurred exclusively in 2007–2008, and four occurred only in 2018–2019. The plant species most used by flies in 2007–2008 were pumpkin (1.81 flies/g of resource), melon (1.65 flies/g), and pineapple (1.26 flies/g). In the second period (2018–2019), the plant species with the highest density of drosophilids were potato (4.12 flies/g), mango (1.90 flies/g), and pineapple (1.34 flies/g). Several plant taxa did not register drosophilid emergence (Table 1).

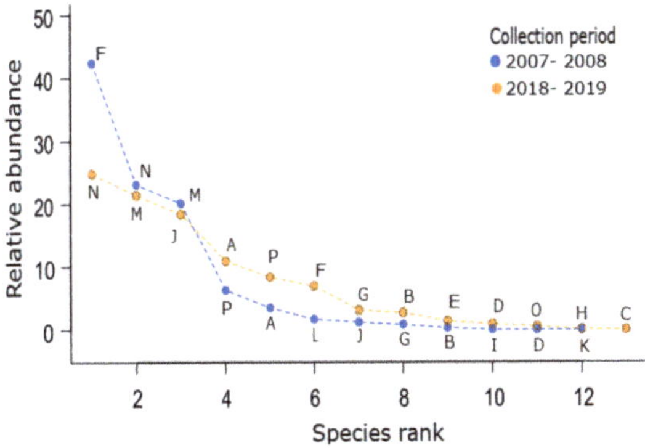

Figure 2. Rank–abundance distribution for drosophilid species recorded in fruits and vegetables collected in the Federal District Supply Center located in Brasília, Brazil, in two collection periods.

3.2. Relationships between Plant and Drosophilid Species

The associations between plant and drosophilid species are shown in Figure 3. The richness of drosophilids in the same plant species varied between 1 and 12, and the most commonly used hosts were pineapple (12 species), banana (11 species), tomato (10 species), melon, and plum (7 species each). Similarly, the number of hosts used by the same species of drosophilid varied between 1 and 22 (Table 2). The drosophilids recorded in most plant species were *Drosophila simulans* (22) and *Zaprionus indianus* (20). However, most species of drosophilids were considered generalists when analyzed using the Magnacca criterion. The exceptions were *D. cardinoides*, *D. nasuta*, *D. repleta*, *D. kikkawai*, and *D. sturtevanti*. The total success rate of fly emergence across the years was 70% of resources (N = 50 species). Fleshy fruits had a success rate of 76.9%; other classes of resources (pooled), such as tubers, bulbs, fruits that are dry at maturity, and leaves, had a lower rate of fly emergence (45.5%).

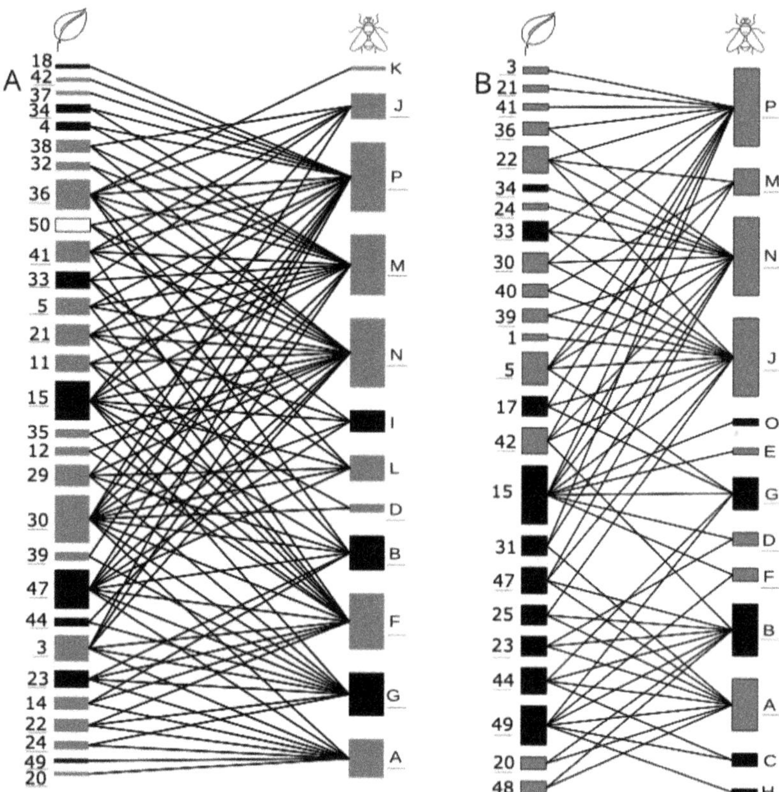

Figure 3. Quantitative food webs for drosophilid species were recorded in fruits and vegetables collected in the Federal District Supply Center located in Brasília, Brazil. (**A**) collection periods 2007–2008 and (**B**) collection periods 2018–2019. For each web, the left bars represent the host plant and the right bars represent drosophilid species. The black lines represent established interactions between plants and drosophilids. The black bars represent native species for the Neotropical region, and the grey bars represent exotic species. The plant and fly species codes are given in Tables 1 and 2, respectively. Underlined codes represent species occurring in both periods, A and B.

The correlation between the drosophilid abundance hatched from each plant species and its mass (weighed in the laboratory) was 0.78 ($p < 0.01$) and 0.76 ($p < 0.01$) in the two collection periods, respectively. The correlation between the drosophilid richness hatched from each plant species and its mass was 0.8 ($p < 0.01$) and 0.72 ($p < 0.01$).

3.3. Neotropical and Exotic Resources Explored by Neotropical and Exotic Drosophilids

Regarding the geographic (native) origin of flies and host species, the matrix between nominal drosophilids and plants showed that 18.11% of all possible associations were recorded. Although the overall chi-square was not significant ($\chi^2 = 7.311$, d.f. $= 3$, $p > 0.05$), the use of exotic hosts was significantly lower than expected for neotropical drosophilids and higher than expected for exotic drosophilids (Table 3). The use of neotropical hosts by neotropical and exotic drosophilids follows the same pattern, except that the p-value ($p < 0.07$) was marginally significant.

Table 3. Percentages of observed interactions (matrix cell occupancy) between drosophilids and host species (i.e., 100% would mean all possible drosophilids using all possible hosts) for each of four possible classes of interaction: (i) Neotropical drosophilid × Neotropical host; (ii) Neotropical drosophilid × exotic host; (iii) exotic drosophilid × Neotropical host; and (iv) exotic drosophilid × exotic host. *p*-values refer to pair Binomial tests.

Drosophilids	Plant Hosts			
	Neotropical (21 Taxa)		Exotic (29 Taxa)	
Neotropical (6 species)	$19/126 = 15.08\%$	$(p = 0.067)$	$15/174 = 8.93\%$	$(p < 0.001)$
Exotic (10 species)	$40/210 = 19.05\%$	$(p = 0.064)$	$68/280 = 24.28\%$	$(p = 0.012)$

4. Discussion

This is the first study to investigate the drosophilid community's association with resources available in a food supply distribution center in the Neotropics. This study demonstrated that the drosophilid communities sampled in the two periods, separated by a decade, were remarkably similar and were mainly composed of generalist species, with most of them being exotic to the Neotropical region. Exotic drosophilid species clearly explored more hosts than their neotropical counterparts, especially exotic host species, supporting the pattern found for wild drosophilids and their fruit hosts [16].

4.1. The Fly Community Remained Relatively Stable after Ten Years

Over time, many similarities have been identified between the two widely isolated samples. Species richness was very similar, and the composition followed the pattern found in natural environments [45]: a few dominant species, generally exotic, in contrast to much more numerous rare species. The relative abundance of species in the community seems to be a predictor of their persistence in the community since the less frequent species (>1.23%) fluctuated the most between the two periods. Our results also support the hypotheses that anthropic environments are favorable for establishing generalist drosophilids [17,19,46]. Markets, especially, provide large amounts and a variety of food resources for drosophilids and are less subject to climate seasonality that affects arthropod communities in tropical savannahs [47–49]. Therefore, distribution centers such as the one studied here may function as reservoirs that support generalist and exotic drosophilid populations.

4.2. Drosophilid Species Are Not Randomly Distributed among Plant Species

The abundance of drosophilids in each plant species was strongly correlated with the mass brought to the laboratory. Previous studies suggest that resource availability is an important predictor of population growth rates [32,50]. However, fly density fluctuated strongly among different hosts, indicating that plant identity also plays an important role in drosophilid abundance. In drosophilids, there are guilds of species associated with flowers, fungi, and fruits [2,51]. Even within the same guild, however, some host species seem to be especially attractive to flies. For the community studied here, pineapple represents an important resource because it is abundant, supports a high density and richness of drosophilids, and all the collected fragments were colonized. The richness and abundance of drosophilids in each host species probably reflect characteristics such as host chemistry, microbial composition, texture, temperature, and the presence or absence of larvae [52]. *Drosophila* females usually explore the substrate with their proboscis and ovipositor to evaluate its quality for oviposition [53]. As the internal microbiome of a single fly represents a highly reduced subset of the external microbial community, the flies might have some level of control over the yeasts and bacteria that inhabit their digestive tracts [54,55].

As might be expected [2], the emergence of Drosophilids from fleshy fruit (FF) resources was higher than from other types of resources, i.e., bulbs (SB), tubers (RT and ST), leaves (VL), and dry fruits (DF). No emergence of flies was detected from any of the root tubers (RT; N = 4), i.e., beetroot, carrot, Peruvian parsnip, or sweet potato; in nature, such resources would rot underground. There was also no emergence from Okra, commercialized as fruit at a very immature stage, that would have naturally matured into a dry, capsular fruit (DF). Stem and leaf resources (SB, ST, and VL), i.e., cabbage, collard greens, lettuce, onions, potato, and taro had varying levels of emergence. Seven species of *Drosophila* emerged in total from these resources, all of which, except for *D. immigrans*, were among the most abundant species in our dataset, with the three most abundant species of flies, *D. hydei*, *D. simulans*, and *D. melanogaster*, respectively colonizing four, three, or two species of these classes of resources.

The insect's choice of the host seems to be based on a decision rule that maximizes the expected production of offspring. Polyphagy is putatively selected for when the chances of an organism finding its preferred resource as well as encountering impalatable, toxic, or poor-quality resources are low, while monophagy is selected for when the chances of it finding its preferred resource, as well as encountering impalatable, toxic, or poor-quality resources are high; both strategies aim at maximizing intake of high-quality resources and avoiding a poor-quality diet [56]. The community studied here is mostly composed of generalist species that oviposit in hosts presenting a wide range of conditions. As urban markets tend to maintain high resource availability throughout the year, even if their composition varies according to seasonality, generalist drosophilids can easily find suitable breeding sites and therefore maintain larger populations in these locations. At the other extreme, four species (*D. kikkawai*, *D. repleta*, *D. sturtevanti*, and *D. nasuta*) were bred from a single host and presented relative abundances lower than 0.05%. Although resource fidelity has been described in drosophilid species [57,58], this is not the case here. *D. kikkawai*, *D. repleta*, and *D. sturtevanti* have been recorded in at least six plant families, and *D. nasuta* was a recent invader in the Neotropics during the period of this study; currently, it is widely established in South America, but its breeding sites are virtually unknown [59].

4.3. Exotic Drosophilids Use More Resources Than Neotropical Drosophilids

The three drosophilid species that exploit the most resources—*Drosophila simulans*, *Zaprionus indianus*, and *D. melanogaster*—are endemic to Africa [3,35], widely distributed throughout the world [1], and abundant in South America [26,32,60]. The success of exotic drosophilids partly stems from their ability to exploit a wider range of plant species. This finding supports Valadão et al. [16], who suggest three complementary hypotheses to explain the broader resource richness used by exotic drosophilids. First, exotic species have survived the trials of introduction, establishment, and dispersion in a new area; thus, they should be adapted to an array of conditions. Second, they may favor their own offspring by inoculating microbes at their breeding sites through their fecal deposits and oviposited eggs [61,62], enhancing the resources available to hatching larvae. Finally, exotic drosophilids can be superior competitors compared to native species: *D. melanogaster* can affect the size of other species when sharing the same resource patch during the larval stage [63], and *Z. indianus* can chase away other species of drosophilids that leave potential breeding sites without laying eggs (see video S1 in Valadão et al. [16]). These processes can promote niche breadth expansion via adaptive evolution [64].

Exotic drosophilids were particularly successful in breeding on exotic host plants, which usually represent much of the resources available in fruit and vegetable distribution centers. Consequently, these markets could provide plentiful breeding sites that act as sources of species that colonize surrounding patches of natural vegetation as well as a possible selection for enhanced generalism. The dispersion of competitive species from urban markets to nature, in turn, could contribute to biotic homogenization by eliminating native species [65,66]. The decline of insect populations and species around the world has been intensely debated, and while there are studies suggesting dramatic rates of extinction of insect species over the next few decades [67,68], there is also some criticism of the methods used to estimate this decline [69].

4.4. Future Research

Drosophilids are good models in many areas of research [70], including conservation and invasion biology [71,72]. In addition to the high-quality, extensively researched publications about nearly all aspects of these flies, there are also complete databanks dealing with *Drosophila* genetics (https://flybase.org/, accessed on 18 January 2023) and taxonomy, including geographical distribution (https://www.taxodros.uzh.ch/, accessed on 18 January 2023). An interesting line of research would be testing the hypothesis that geographically widespread generalists have an apparently greater tendency to use novel, exotic hosts than geographically constrained specialists, as found for butterflies [73,74]. Another promising avenue for research is the standardized approach for systematically monitoring alien species and tracking biological invasions [75]. Considering that exotic species can interact, favoring each other and establishing new ones [76], monitoring becomes especially relevant. In the interval between the two collection periods, the establishment of two exotic species in the Neotropical region was recorded: *Drosophila nasuta* and *D. suzukii*. The first, *D. nasuta*, was registered by us in the second collection period as a rare specialist species (<0.5% of the sample came from a single host), although it was much more abundant (211 records) than the other newly–recorded species of the second period combined (113). This result certainly reflects the short time of introduction at the time of collection, since there are records of *D. nasuta* associated with different trophic levels [77]. Furthermore, given its distribution potential in different Neotropical biomes [78,79], *D. nasuta* possibly uses a variety of resources that should mediate its dispersal. The second, *D. suzukii*, was not found in our samples, although it occurs in natural environments adjacent to the study area [80]. Known as the spotted-winged *Drosophila*, *D. suzukii* has already been found on 64 host plants in 25 families in Latin America, most of which are exotic species [10]. This is an important worldwide pest that infests wild and cultivated small soft-skinned fruits [81], and the rarity of this type of host in our samples might explain its absence in the present study.

5. Conclusions

In short, our study suggests that the drosophilid community established in a fruit and vegetable distribution center located in the core area of South America is relatively stable and dominated by generalist exotic species. Neotropical species were also present, but in general, they were rarer and showed a narrower niche breath. These results are worrying because the studied site, along with other urban markets around the world, might be acting as sources of generalist widespread species that disperse to surrounding natural vegetation and contribute to biotic homogenization.

Author Contributions: Conceptualization, R.T. and L.B.R.; methodology, L.B.R.; formal analysis, L.B.R. and R.T.; investigation, all authors; resources, R.T.; data curation, all authors; writing—original draft preparation, L.B.R. and C.E.B.P.; writing—review and editing, all authors; visualization, all authors; supervision, R.T.; project administration, R.T.; funding acquisition, R.T. All authors have read and agreed to the published version of the manuscript.

Funding: This research was funded by the Conselho Nacional de Desenvolvimento Científico e Tecnológico (CNPq), grant numbers 309973/20171 and 131296/2021-3, Fundação de Apoio à Pesquisa do Distrito Federal (FAPDF), grant number 0193.001710/2017, and Coordenação de Pessoal de Nível Superior (CAPES), code 001.

Data Availability Statement: The data presented in this study are openly available in this manuscript.

Acknowledgments: We are grateful to P.L. Rocha and K. Yuzuki for their technical support in collections and fly identification, and to L.H.P. Vieira and J.A.A. Nunes for their advice in the analyses.

Conflicts of Interest: The authors declare no conflict of interest. The funders had no role in the design of the study, in the collection, analysis, or interpretation of data, in the writing of the manuscript, or in the decision to publish the results.

References

1. TaxoDros: The Database on Taxonomy of Drosophilidae, v.1.04. Database. Available online: https://www.taxodros.uzh.ch/search/class.php (accessed on 18 January 2023).
2. Carson, H.L. *The Ecology of Drosophila Breeding Sites*; Harold L-Lyon Arboretum Lecture; University of Hawaii Foundation Lyon Arboretum Fund: New York, NY, USA, 1971; p. 28.
3. David, J.R.; Tsacas, L. Cosmopolitan, subcosmopolitan and widespread species: Different strategies within the Drosophilid family (Diptera). *C. R. Soc. Biogeogr.* **1981**, *57*, 11–26.
4. Drosophilidae in Catálogo Taxonômico da Fauna do Brasil. PNUD. Available online: http://fauna.jbrj.gov.br/fauna/faunadobrasil/183186 (accessed on 19 January 2023).
5. Val, F.C.; Sene, F.M. Nova Introdução de *Drosophila* No Brasil. *Papéis Avulsos Zool.* **1980**, *33*, 293–298.
6. Vilela, C.R. Is *Zaprionus indianus* Gupta, 1970 (Diptera, Drosophilidae) Currently Colonising the Neotropical Region? *Drosoph. Inf. Serv.* **1999**, *82*, 37–38.
7. Vilela, C.R.; Goñi, B. Is *Drosophila nasuta* Lamb (Diptera, Drosophilidae) Currently Reaching the Status of a Cosmopolitan Species? *Rev. Bras. Entomol.* **2015**, *59*, 346–350. [CrossRef]
8. Deprá, M.; Poppe, J.L.; Schmitz, H.J.; De Toni, D.C.; Valente, V.L.S. The First Records of the Invasive Pest *Drosophila suzukii* in the South American Continent. *J. Pest Sci.* **2014**, *87*, 379–383. [CrossRef]
9. Cavalcanti, F.A.G.S.; Ribeiro, L.B.; Marins, G.; Tonelli, G.S.S.S.; Báo, S.N.; Yassin, A.; Tidon, R. Geographic Expansion of an Invasive Fly: First Record of *Zaprionus tuberculatus* (Diptera: Drosophilidae) in the Americas. *Ann. Entomol. Soc. Am.* **2022**, *115*, 267–274. [CrossRef]
10. Garcia, F.R.M.; Lasa, R.; Funes, C.F.; Buzzetti, K. *Drosophila suzukii* Management in Latin America: Current Status and Perspectives. *J. Econ. Entomol.* **2022**, *115*, 1008–1023. [CrossRef]
11. Lee, J.C.; Rodriguez-Saona, C.R.; Zalom, F.G. Introductory Remarks: Spotlight on Spotted-Wing Drosophila. *J. Econ. Entomol.* **2022**, *115*, 919–921. [CrossRef]
12. Roque, F.; Matavelli, C.; Lopes, P.H.S.; Machida, W.S.; Von Zuben, C.J.; Tidon, R. Brazilian Fig Plantations Are Dominated by Widely Distributed Drosophilid Species (Diptera: Drosophilidae). *Ann. Entomol. Soc. Am.* **2017**, *110*, 521–527. [CrossRef]
13. EFSA Panel on Plant Health (PLH); Bragard, C.; Baptista, P.; Chatzivassiliou, E.; Di Serio, F.; Gonthier, P.; Jaques Miret, J.A.; Justesen, A.F.; Magnusson, C.S.; Milonas, P.; et al. Pest Categorisation of *Zaprionus indianus*. *EFSA J.* **2022**, *20*, e07144. [CrossRef]
14. Mollot, G.; Pantel, J.H.; Romanuk, T.N. Chapter Two—The Effects of Invasive Species on the Decline in Species Richness: A Global Meta-Analysis. In *Advances in Ecological Research*; Bohan, D.A., Dumbrell, A.J., Massol, F., Eds.; Networks of Invasion: A Synthesis of Concepts; Academic Press: Cambridge, MA, USA, 2017; Volume 56, pp. 61–83.
15. da Silva, C.R.B.; Beaman, J.E.; Dorey, J.B.; Barker, S.J.; Congedi, N.C.; Elmer, M.C.; Galvin, S.; Tuiwawa, M.; Stevens, M.I.; Alton, L.A.; et al. Climate Change and Invasive Species: A Physiological Performance Comparison of Invasive and Endemic Bees in Fiji. *J. Exp. Biol.* **2021**, *224*, jeb230326. [CrossRef]
16. Valadão, H.; Proença, C.E.B.; Kuhlmann, M.P.; Harris, S.A.; Tidon, R. Fruit-Breeding Drosophilids (Diptera) in the Neotropics: Playing the Field and Specialising in Generalism? *Ecol. Entomol.* **2019**, *44*, 721–737. [CrossRef]
17. Ferreira, L.B.; Tidon, R. Colonizing Potential of Drosophilidae (Insecta, Diptera) in Environments with Different Grades of Urbanization. *Biodivers. Conserv.* **2005**, *14*, 1809–1821. [CrossRef]
18. Gottschalk, M.S.; De Toni, D.C.; Valente, V.L.S.; Hofmann, P.R.P. Changes in Brazilian Drosophilidae (Diptera) Assemblages across an Urbanisation Gradient. *Neotrop. Entomol.* **2007**, *36*, 848–862. [CrossRef]
19. Garcia, A.C.L.; Valiati, V.H.; Gottschalk, M.S.; Rohde, C.; Valente, V.L.d.S. Two Decades of Colonization of the Urban Environment of Porto Alegre, Southern Brazil, by *Drosophila paulistorum* (Diptera, Drosophilidae). *Iheringia Série Zool.* **2008**, *98*, 329–338. [CrossRef]
20. Bombin, A.; Reed, L.K. The Changing Biodiversity of Alabama *Drosophila*: Important Impacts of Seasonal Variation, Urbanization, and Invasive Species. *Ecol. Evol.* **2016**, *6*, 7057–7069. [CrossRef]
21. Grimaldi, D.; Ginsberg, P.S.; Thayer, L.; McEvey, S.; Hauser, M.; Turelli, M.; Brown, B. Strange Little Flies in the Big City: Exotic Flower-Breeding Drosophilidae (Diptera) in Urban Los Angeles. *PLoS ONE* **2015**, *10*, e0122575. [CrossRef]
22. Eiten, G. The Cerrado Vegetation of Brazil. *Bot. Rev.* **1972**, *38*, 201–341. [CrossRef]
23. Myers, N.; Mittermeier, R.A.; Mittermeier, C.G.; da Fonseca, G.A.B.; Kent, J. Biodiversity Hotspots for Conservation Priorities. *Nature* **2000**, *403*, 853–858. [CrossRef]
24. Oliveira-Filho, A.T.; Ratter, J.A. 6. Vegetation Physiognomies and Woody Flora of the Cerrado Biome. In *The Cerrados of Brazil: Ecology and Natural History of a Neotropical Savanna*; Oliveira, P.S., Marquis, R.J., Eds.; Columbia University Press: New York, NY, USA, 2002; pp. 91–120.
25. Ribeiro, L.B.; Cavalcanti, F.A.G.S.; Viana, J.P.C.; Amaral, J.M.; Costa, S.C.; Tidon, R. Updated List of Drosophilid Species Recorded in the Brazilian Savanna. *Dros. Inf. Serv.* **2022**, *105*, 85–89.
26. Tidon, R.; Leite, D.F.; Leão, B.F.D. Impact of the Colonisation of *Zaprionus* (Diptera, Drosophilidae) in Different Ecosystems of the Neotropical Region: 2 Years after the Invasion. *Biol. Conserv.* **2003**, *112*, 299–305. [CrossRef]
27. Tidon, R. Relationships between Drosophilids (Diptera, Drosophilidae) and the Environment in Two Contrasting Tropical Vegetations. *Biol. J. Linn. Soc.* **2006**, *87*, 233–247. [CrossRef]

28. Roque, F.; Hay, J.D.V.; Tidon, R. Breeding sites of drosophilids (Diptera) in the Brazilian Savanna. I. Fallen fruits of *Emmotum nitens* (Icacinaceae), *Hancornia speciosa* (Apocynaceae) and *Anacardium humile* (Anacardiaceae). *Rev. Bras. Entomol.* **2009**, *53*, 308–313. [CrossRef]

29. Valadão, H.; Hay, J.D.V.; Tidon, R. Temporal Dynamics and Resource Availability for Drosophilid Fruit Flies (Insecta, Diptera) in a Gallery Forest in the Brazilian Savanna. *Int. J. Ecol.* **2010**, *2010*, e152437. [CrossRef]

30. Roque, F.; da Mata, R.A.; Tidon, R. Temporal and Vertical Drosophilid (Insecta; Diptera) Assemblage Fluctuations in a Neotropical Gallery Forest. *Biodivers. Conserv.* **2013**, *22*, 657–672. [CrossRef]

31. Mata, R.A.d.; Valadão, H.; Tidon, R. Spatial and Temporal Dynamics of Drosophilid Larval Assemblages Associated to Fruits. *Rev. Bras. Entomol.* **2015**, *59*, 50–57. [CrossRef]

32. Döge, J.d.S.; de Oliveira, H.V.; Tidon, R. Rapid Response to Abiotic and Biotic Factors Controls Population Growth of Two Invasive Drosophilids (Diptera) in the Brazilian Savanna. *Biol. Invasions* **2015**, *17*, 2461–2474. [CrossRef]

33. Schneider, D.I.; Sujii, E.R.; Laumann, R.A.; Tidon, R. Parasitoids of Drosophilids in the Brazilian Savanna: Spatial–Temporal Distribution and Host Associations with Native and Exotic Species. *Neotrop. Entomol.* **2022**, *51*, 514–525. [CrossRef]

34. Freire-Maia, N.; Pavan, C. Introdução ao Estudo da *Drosophila*. *Rev. Cultus* **1949**, *1*, 3–61.

35. Yuzuki, K.; Tidon, R. Identification Key for Drosophilid Species (Diptera, Drosophilidae) Exotic to the Neotropical Region and Occurring in Brazil. *Rev. Bras. Entomol.* **2020**, *64*, e2019100. [CrossRef]

36. Val, F.C. Male genitalia of some Neotropical *Drosophila*: Notes and illustrations. *Papéis Avulsos Zool.* **1982**, *34*, 309–347.

37. Vilela, C.R. A revision of the *Drosophila repleta* Species Group (Diptera, Drosophilidae). *Rev. Bras. Entomol.* **1983**, *27*, 1–114.

38. The World Flora Online. Available online: http://www.worldfloraonline.org/ (accessed on 19 January 2023).

39. Gotelli, N.J.; Colwell, R.K. Quantifying Biodiversity: Procedures and Pitfalls in the Measurement and Comparison of Species Richness. *Ecol. Lett.* **2001**, *4*, 379–391. [CrossRef]

40. Oksanen, J.; Simpson, G.L.; Blanchet, F.G.; Kindt, R.; Legendre, P.; Minchin, P.R.; O'Hara, R.B.; Solymos, P.; Stevens, M.H.H.; Szoecs, E.; et al. Vegan: Community Ecology Package 2022 (Version 2.6-4). Available online: https://CRAN.R-project.org/package=vegan (accessed on 18 January 2023).

41. Magurran, A.E. The commonness, and rarity, of species. In *Measuring Biological Diversity*, 1st ed.; Wiley-Blackwell: Malden, MA, USA, 2003; pp. 18–71. ISBN 978-0-632-05633-0.

42. Dormann, C.F.; Gruber, B.; Fründ, J. The R Journal: Introducing the Bipartite Package: Analysing Ecological Networks. *R News* **2008**, *8*, 8–11.

43. Magnacca, K.N.; Foote, D.; O'Grady, P.M. A review of the endemic Hawaiian Drosophilidae and their host plants. *Zootaxa* **2008**, *1728*, 1–58. [CrossRef]

44. Zar, J.H. *Biostatistical Analysis*, 4th ed.; Prentice Hall: Hoboken, NJ, USA, 1999; p. 663. ISBN 978-0-13-081542-2.

45. Argemi, M.; Monclús, M.; Mestres, F.; Serra, L. Comparative Analysis of a Community of Drosophilids (Drosophilidae; Diptera) Sampled in Two Periods Widely Separated in Time. *J. Zool. Syst. Evol. Res.* **1999**, *37*, 203–210. [CrossRef]

46. Parsons, P.A. *The Evolutionary Biology of Colonizing Species*; Cambridge University Press: Cambridge, UK, 1983; p. 262. ISBN 0-521-25247-4.

47. Kishimoto-Yamada, K.; Itioka, T. How Much Have We Learned about Seasonality in Tropical Insect Abundance since Wolda (1988)? *Entomol. Sci.* **2015**, *18*, 407–419. [CrossRef]

48. Braga, L.; Diniz, I.R. Trophic Interactions of Caterpillars in the Seasonal Environment of the Brazilian Cerrado and Their Importance in the Face of Climate Change. In *Caterpillars in the Middle: Tritrophic Interactions in a Changing World*; Marquis, R.J., Koptur, S., Eds.; Fascinating Life Sciences; Springer International Publishing: Cham, Switzerland, 2022; pp. 485–508. ISBN 978-3-030-86688-4.

49. Mavasa, R.; Yekwayo, I.; Mwabvu, T.; Tsvuura, Z. Preliminary Patterns of Seasonal Changes in Species Composition of Surface-Active Arthropods in a South African Savannah. *Austral Ecol.* **2022**, *47*, 1222–1231. [CrossRef]

50. Breitmeyer, C.M.; Markow, T.A. Resource Availability and Population Size in Cactophilic *Drosophila*. *Funct. Ecol.* **1998**, *12*, 14–21. [CrossRef]

51. Heed, W.B. A New Cactus-Feeding but Soil-Breeding Species of *Drosophila* (Diptera: Drosophilidae). *Proc. Entomol. Soc. Wash.* **1977**, *79*, 649–654.

52. Markow, T.A.; O'Grady, P. Reproductive Ecology of *Drosophila*. *Funct. Ecol.* **2008**, *22*, 747–759. [CrossRef]

53. Yang, C.; Belawat, P.; Hafen, E.; Jan, L.Y.; Jan, Y.-N. *Drosophila* Egg-Laying Site Selection as a System to Study Simple Decision-Making Processes. *Science* **2008**, *319*, 1679–1683. [CrossRef]

54. Chandler, J.A.; Lang, J.M.; Bhatnagar, S.; Eisen, J.A.; Kopp, A. Bacterial Communities of Diverse *Drosophila* Species: Ecological Context of a Host–Microbe Model System. *PLoS Genet.* **2011**, *7*, e1002272. [CrossRef] [PubMed]

55. Chandler, J.A.; Eisen, J.A.; Kopp, A. Yeast Communities of Diverse *Drosophila* Species: Comparison of Two Symbiont Groups in the Same Hosts. *Appl. Environ. Microbiol.* **2012**, *78*, 7327–7336. [CrossRef]

56. Levins, R.; MacArthur, R. An Hypothesis to Explain the Incidence of Monophagy. *Ecology* **1969**, *50*, 910–911. [CrossRef]

57. Etges, W.J. Evolutionary Genomics of Host Plant Adaptation: Insights from *Drosophila*. *Curr. Opin. Insect Sci.* **2019**, *36*, 96–102. [CrossRef]

58. Auer, T.O.; Shahandeh, M.P.; Benton, R. *Drosophila sechellia*: A Genetic Model for Behavioral Evolution and Neuroecology. *Annu. Rev. Genet.* **2021**, *55*, 527–554. [CrossRef]

59. Lauer Garcia, A.C.; Pessoa Da Silva, F.; Campos Bezerra Neves, C.H.; Montes, M.A. Current and Future Potential Global Distribution of the Invading Species *Drosophila nasuta* (Diptera: Drosophilidae). *Biol. J. Linn. Soc.* **2022**, *135*, 208–221. [CrossRef]
60. Sene, F.M.; Val, F.C.; Vilela, C.R.; Pereira, M.A.Q.R. Preliminary Data on the Geographical Distribution of *Drosophila* Species within Morphoclimatic Domains of Brazil. *Papéis Avulsos Zool.* **1980**, *33*, 315–326.
61. Bakula, M. The Persistence of a Microbial Flora during Postembryogenesis of *Drosophila melanogaster*. *J. Invertebr. Pathol.* **1969**, *14*, 365–374. [CrossRef]
62. Gilbert, D.G. Dispersal of Yeasts, and Bacteria by *Drosophila* in a Temperate Forest. *Oecologia* **1980**, *46*, 135–137. [CrossRef]
63. Atkinson, W.D. A Field Investigation of Larval Competition in Domestic *Drosophila*. *J. Anim. Ecol.* **1979**, *48*, 91–102. [CrossRef]
64. Sexton, J.P.; Montiel, J.; Shay, J.E.; Stephens, M.R.; Slatyer, R.A. Evolution of Ecological Niche Breadth. *Annu. Rev. Ecol. Evol. Syst.* **2017**, *48*, 183–206. [CrossRef]
65. McKinney, M.L.; Lockwood, J.L. Biotic Homogenization: A Few Winners Replacing Many Losers in the next Mass Extinction. *Trends Ecol. Evol.* **1999**, *14*, 450–453. [CrossRef]
66. Olden, J.D.; Comte, L.; Giam, X. The Homogocene: A Research Prospectus for the Study of Biotic Homogenisation. *NeoBiota* **2018**, *36*, 23–36. [CrossRef]
67. Sánchez-Bayo, F.; Wyckhuys, K.A.G. Worldwide Decline of the Entomofauna: A Review of Its Drivers. *Biol. Conserv.* **2019**, *232*, 8–27. [CrossRef]
68. Wagner, D.L. Insect Declines in the Anthropocene. *Annu. Rev. Entomol.* **2020**, *65*, 457–480. [CrossRef]
69. Didham, R.K.; Basset, Y.; Collins, C.M.; Leather, S.R.; Littlewood, N.A.; Menz, M.H.M.; Müller, J.; Packer, L.; Saunders, M.E.; Schönrogge, K.; et al. Interpreting Insect Declines: Seven Challenges and a Way Forward. *Insect Conserv. Divers.* **2020**, *13*, 103–114. [CrossRef]
70. Mohr, S.E. *First in Fly: Drosophila Research and Biologica Discovery*; Harvard University Press: Cambridge, MA, USA, 2018; p. 257. ISBN 978-0-674-97101-1.
71. da Mata, R.A.; McGeoch, M.; Tidon, R. Drosophilids (Insecta, Diptera) as Tools for Conservation Biology. *Nat. Conserv.* **2010**, *8*, 60–65. [CrossRef]
72. Gibert, P.; Hill, M.; Pascual, M.; Plantamp, C.; Terblanche, J.S.; Yassin, A.; Sgrò, C.M. *Drosophila* as Models to Understand the Adaptive Process during Invasion. *Biol. Invasions* **2016**, *18*, 1089–1103. [CrossRef]
73. Slove, J.; Janz, N. The Relationship between Diet Breadth and Geographic Range Size in the Butterfly Subfamily Nymphalinae—A Study of Global Scale. *PLoS ONE* **2011**, *6*, e16057. [CrossRef] [PubMed]
74. Jahner, J.P.; Bonilla, M.M.; Badik, K.J.; Shapiro, A.M.; Forister, M.L. Use of Exotic Hosts by Lepidoptera: Widespread Species Colonize more Novel Hosts. *Evolution* **2011**, *65*, 2719–2724. [CrossRef] [PubMed]
75. Latombe, G.; Pyšek, P.; Jeschke, J.M.; Blackburn, T.M.; Bacher, S.; Capinha, C.; Costello, M.J.; Fernández, M.; Gregory, R.D.; Hobern, D.; et al. A Vision for Global Monitoring of Biological Invasions. *Biol. Conserv.* **2017**, *213*, 295–308. [CrossRef]
76. Simberloff, D.; Von Holle, B. Positive Interactions of Nonindigenous Species. *Biol. Invasions* **1999**, *1*, 21–32. [CrossRef]
77. David, J.R.; McEvey, S.F.; Solignac, M.; Tsacas, L. *Drosophila* Communities on Mauritius and the Ecological Niche of *Drosophila mauritiana* (Diptera, Drosophilidae). *Rev. Zool. Afr.* **1989**, *103*, 107–116.
78. Montes, M.A.; Neves, C.H.C.B.; Ferreira, A.F.; dos Santos, M.d.F.S.; Quintas, J.I.F.P.; Manetta, G.D.Â.; de Oliveira, P.V.; Garcia, A.C.L. Invasion and Spreading of *Drosophila nasuta* (Diptera, Drosophilidae) in the Caatinga Biome, Brazil. *Neotrop. Entomol.* **2021**, *50*, 571–578. [CrossRef]
79. Medeiros, H.F.; Monteiro, M.P.; Caçador, A.W.B.; Pereira, C.M.; de Lurdes Bezerra Praxedes, C.; Martins, M.B.; Montes, M.A.; Garcia, A.C.L. First Records of the Invading Species *Drosophila nasuta* (Diptera: Drosophilidae) in the Amazon. *Neotrop. Entomol.* **2022**, *51*, 493–497. [CrossRef]
80. Paula, M.A.; Lopes, P.H.S.; Tidon, R. First Record of *Drosophila suzukii* in the Brazilian Savanna. *Dros. Inf. Serv.* **2014**, *97*, 113–115.
81. Wollmann, J.; Schlesener, D.C.H.; Mendes, S.R.; Krüger, A.P.; Martins, L.N.; Bernardi, D.; Garcia, M.S.; Garcia, F.R.M. Infestation Index of *Drosophila suzukii* (Diptera: Drosophilidae) in Small Fruit in Southern Brazil. *Arq. Inst. Biológico* **2020**, *87*, e0432018. [CrossRef]

 *insects*

 MDPI

Article

Stage Transitions in *Lucilia sericata* and *Phormia regina* (Diptera: Calliphoridae) and Implications for Forensic Science

Amanda L. Roe [1] and Leon G. Higley [2,*]

[1] Biology Program, College of Saint Mary, Omaha, NE 68106, USA
[2] School of Natural Resources, University of Nebraska-Lincoln, Lincoln, NE 68583, USA
* Correspondence: lhigley1@unl.edu; Tel.: +1-402-560-6684

Simple Summary: Maggot growth is important in estimating the postmortem interval in cases involving decomposing bodies. In turn, the time in which maggots transition from one stage of development to the next is crucial in determining growth rates. In this study, we examined in detail these transition times for two important blow fly species. Species transitioned between stages following a bell-curve pattern which was not previously known. This new information will be valuable for improving maggot growth determinations and postmortem estimates.

Abstract: Blow fly development rates have become a key factor in estimating the postmortem interval where blow flies are among the first decomposers to occur on a body. Because the use of blow fly development requires short time durations and high accuracy, stage transition distributions are essential for proper development modeling. However, detailed examinations of stage transitions are not available for any blow fly species. Consequently, we examined this issue in two blow fly species: *Lucilia sericata* and *Phormia regina*. Transitions for all life stages across all measured temperatures were normally distributed. Use of probit analysis allowed determination of 50% transition points and associated measures of variation (i.e., standard errors). The greatest variation was noted for the L2-L3, L3-L3m, and L3m-P stage transitions. These results invalidate the notion that largest maggots should be preferentially collected for determining current maggot population stage, and further call into question the relationship between intrinsic variation and potential geographic variation in development rates.

Keywords: insect development; postmortem interval; blow fly; maggot development

Citation: Roe, A.L.; Higley, L.G. Stage Transitions in *Lucilia sericata* and *Phormia regina* (Diptera: Calliphoridae) and Implications for Forensic Science. *Insects* **2023**, *14*, 315. https://doi.org/10.3390/insects14040315

Academic Editors: Sérgio M. Ovruski and Iávio Roberto Mello Garcia

Received: 3 March 2023
Revised: 20 March 2023
Accepted: 22 March 2023
Published: 25 March 2023

1. Introduction

The blow flies (family Calliphoridae) have emerged as the most important single group of insects in forensic entomology [1–3]. Their importance arises from their biology, specifically the strong attraction to decay in many species [2,4]. Given their response to decompositional odors coupled with strong flight, it is not surprising that blow flies typically are the first insect colonizers. Indeed, blow flies can arrive at a dead body and lay eggs within literally minutes of death. Consequently, this time of oviposition can be an ideal indicator for estimating time of death. Calculating an estimated time of oviposition itself depends on examining the degree of maggot development at the time of a body's discovery, and on using temperature and development data to determine when egg laying occurred.

Blow fly development is curvilinear, with a linear section in the center and curves at the low and high temperatures [1]. Typically, most research on blow fly development has focused on the linear portion of the development curve, and few of these studies explicitly report on stage transitions, possibly since the assumption is that larval molting events are relatively synchronous and occur over a short time [5]. However, in a study of *Chrysomya megacephala*, Wells and Kurahashi [6] reported that only the molts from egg to first and first to second stages were "highly synchronized", each transition occurring in 6 h or less. Kamal [7] reported large ranges (from hours to days) around his presented modal development times, which suggests large transition times where multiple life stages are found together. Most other studies on blowfly development research either treat transitions as a single point or do not discuss transitions at all.

Because there is very little data on stage transitions of blow flies, the associated transition distribution is unknown, which is integral for determining development time within stages. Here, by "transition distribution" we refer to the temporal frequency (counts or proportions through time) at which a new life stage occurs during the molting from one stage to the next.

Mathematically, the issue of transition distribution is identical to that in which any numerical distribution must be determined. Most commonly, frequency distributions are associated with sampling. With blow fly development, the transition distribution not only represents the temporal pattern at which one stage molts into another, but also is a necessary tool for determining stage duration and estimating variation. Broadly, two types of transition distributions are likely. First is the negative binomial (or similar mathematical) distribution that is asymmetric with a peak to the left. With this distribution, stage duration should be calculated by measuring the time from mode to mode between stages. Second is the normal, or Gaussian distribution. If the frequency distribution for stage transitions is normally distributed, then determination of stage duration requires use of the mean (=the peak of the distribution), and stage duration is calculated as the time from peak to peak between stages.

Debates regarding use of mean or mode routinely occur in the forensic entomology literature [1,8], even though no experimental determination of the actual transition distributions exists for all life stages of any blow fly species. Moreover, transition distributions are necessary for measuring variation and attaching confidence intervals around transitions. Finally, transition distributions are important for determining developmental curves, degrees of uncertainty, and sampling protocols, all of which ultimately influence the accuracy of PMI estimations.

The need to determine transition distribution exists, theoretically, in determining the stage duration for any organism that molts. In practice, if the duration of stage transitions are short relative to the total duration of a stage, using single times to represent a transition is adequate. Similarly, if the time of concern is that of the total larval period, or time from egg to adult (as is often the case with agricultural pests), then details on transition distributions are not needed. In contrast, because the use of blow fly development requires short time durations and high accuracy, transition distributions are essential for proper development modeling. We examined this issue in two blow fly species: *Lucilia sericata* and *Phormia regina*.

Lucilia sericata (Diptera: Calliphoridae) is a ubiquitous fly that belongs to a group of necrophagous insects that are dependent on decomposing flesh to complete their life cycle [4] and has played major roles in sheep strike and other forms of myiasis, maggot therapy, and death investigations. In the Midwest, *L. sericata* often is the first blow fly to oviposit on a dead body, and for this reason, it is potentially one of the most important species in determining the PMI [2]. The age of maggots when a body is discovered provides a starting point from which to count backward to the time of oviposition, providing an estimate of the duration of the exposure of the body [2]. Developmental research on *L. sericata* has focused on linear portions of the development curve (e.g., [7,9,10]) with temperatures between 16 and 35 °C and no explicit consideration of stage transitions.

In the past, much of the research concerning *Phormia regina* (diptera: Calliphoridae) was dedicated to its role in livestock myiasis [11]. However, with its near complete range across the U.S. (except southern Florida) ([4,12]), *P. regina* has steadily increased its role as a colonizer of human and other animal remains. As such, their role in postmortem interval (PMI) estimations has also increased. Like the situation with *L. sericata*, there are insufficient development data for *P. regina* to allow calculation of stage transition distributions. Among existing studies on *P. regina* development (e.g., [7,10,11,13]), issues include insufficient number of replications, inconsistent temperature regimes, variable sampling protocols, and non-specific life stage intervals.

Ideally, blow fly species would have complete developmental data sets [1]. Unfortunately, developmental experiments are expensive, both in labor and materials, and it would be difficult (if not impossible) to capture and maintain colonies of all blow fly species. Designing developmental studies, therefore, requires choosing species that may or may not share certain life history traits but are of similar importance ecologically or legally. For this study, *P. regina* was chosen because of the species' increased geographic area, their increasing role in death investigations, and their placement in a different subfamily (Calliphoridae: *Chrysomyinae*) from *L. sericata* (Calliphoridae: *Luciliinae*).

By examining different subfamilies and comparing the data between *P. regina* and *L. sericata*, the overall goal of having concise development data for the majority of the Calliphorids becomes easier if clear patterns emerge. For example, if stage transition times are consistently normally distributed and variation within life stages is similar between species, then it is not unreasonable that overall development patterns may be similar. Thus, our goals in this study were to establish stage transition distributions with confidence intervals across all life stages (egg to adult) and multiple temperatures for *L. sericata* and *Phomia regina*.

2. Materials and Methods

2.1. Flies

Lucilia sericata were obtained from colonies maintained at the University of Nebraska–Lincoln (Lincoln, Nebraska). The colony was established in October 2010, from field-collected insects provided by Dr. Jeff Wells from Morgantown, West Virginia. At the time of research, the colonies achieved 100 generations without addition of new flies.

Phormia regina were obtained from colonies maintained at the University of Nebraska–Lincoln (Lincoln, Nebraska). The colony was established in 2011 from field-collected insects in Lancaster County, Nebraska. At the time of research, the colonies achieved 100 generations without addition of new flies. For specific rearing protocols, see Roe and Higley [14].

Adult flies were maintained in screen cages (46 cm × 46 cm × 46 cm) (Bioquip Products, Claremont, CA, USA) in a rearing room at 27.5 °C (±3 °C), with a 16:8 (L:D) photoperiod. Multiple generations were maintained in a single cage, and ca. 1000 adult flies were introduced every 1–2 weeks. Adults had access to granulated sugar and water ad libitum, and raw beef liver for protein and as an ovipositional substrate. After egg laying, eggs and liver were placed in an 89 mL plastic cup which was surrounded by pine shavings in a 1.7 L plastic box. The pine shavings served as a pupation substrate. The 1.7 L box was placed in a I30-BLL Percival biological incubator (Percival Scientific, Inc., Perry, IA, USA) set at 26 °C (±1.5 °C). After eclosion, adults were released into the screened cages.

2.2. Incubators

Incubator information has been previously discussed in Authement et al. [14]; pertinent information was revisited here. Incubators were customized, model SMY04-1 DigiTherm® CirKinetics Incubators (TriTech Research, Inc., Los Angeles, CA, USA). The DigiTherm® CirKinetics Incubators have microprocessor -ontrolled temperature regulation, internal lighting, a recirculating air system (to help maintain humidity), and a thermoelectric heat pump (rather than coolant and condenser as is typical with larger incubators and growth chambers). Customizations included the addition of a data port, vertical lighting (so all shelves were illuminated), and an additional internal fan. The manufacturer's specifications indicate an operational range of 10–60 °C ± 0.1 °C. It is worth noting that a range of ±0.1 °C is an order of magnitude more precise than is possible in conventional growth chambers. Although growth chambers have been shown to display substantial differences between programmed temperatures and actual internal temperatures [15], incubators tested with internal thermocouples in a replicated study showed internal temperatures on all shelves within incubators never varied by more than 0.1 °C from the programmed temperature, in agreement with the manufacturer's specifications. Given the high level of measured accuracy with programmed temperatures, we were able to use incubators for temperature treatments, which improved our experimental efficiency and helped reduce experimental error.

2.3. Experimental Design

The studies with *Lucilia sericata* and *Phormia regina* used the same experimental design. For each species, treatments comprised eleven temperatures (7.5, 10, 12.5, 15, 17.5, 20, 22.5, 25, 27.5, 30, and 32.5 °C) with a light: dark cycle of 16:8. Twenty eggs (collected within 30 min of oviposition) were counted onto a moist black filter paper triangle and placed in direct contact with 10 g of beef liver in a 29.5 mL plastic cup. The cup was placed in a 7 cm × 7 cm × 10 cm plastic container that had 2.5 cm of wood shavings in the bottom for L. sericata, and 2.5 cm of damp sand for *P. regina*. The container was then placed randomly in an incubator. A total of 22 incubators were available for use: the experimental unit was an incubator, temperature treatments were randomized by incubator, and replications were conducted through time to provide sufficient incubators for all treatments. Each life stage (egg–1st stage, 1st–2nd stage, 2nd–3rd stage, 3rd–3rd migratory, 3rd migratory–pupation, pupation–adult) was calculated using Kamal's [4] data which were converted to accumulated degree hours (ADH) and divided equally into five sampling times (specific times are listed in Table 1 for *Lucillia sericata* and Table 2 for *Phormia regina*). Each sample was replicated 4 times, for a total of 20 samples per life stage (4 replications times 5 sampling periods). In total, there were 120 samples per treatment (4 replications times 5 sampling periods per stage transition times 6 life stage transitions). During each sample time, a container was pulled from each of the four incubators and the stage of each maggot was documented morphologically using the posterior spiracular slits.

Table 1. Sample times for *Lucilia sericata* were calculated by converting the minimum and maximum data reported in Kamal [4] into accumulated degree hours (ADH). The ADHs were calculated for each life stage and sampling temperature, converted back into hours and divided into 5 equal sample times.

Life Stage	Temperature °C										
	7.5	10.0	12.5	15.0	17.5	20.0	22.5	25.0	27.5	30.0	32.5
Egg–1st	35	35	35	17	12	9	7	6	5	4	4
1st–2nd	56	56	56	28	19	14	11	9	8	7	6
2nd–3f	79	79	79	39	26	20	16	13	11	10	9
3f–3m	143	143	143	71	48	36	29	24	20	18	16
3m–Pupal	335	335	335	167	112	84	67	56	48	42	37
Pupal–Adult	527	527	527	263	176	132	105	88	75	66	59

Table 2. Sample times for *Phormia regina* were calculated by converting the minimum and maximum data reported in Kamal [4] into accumulated degree hours (ADH). The ADHs were calculated for each life stage and sampling temperature, converted back into hours and divided into 5 equal sample times.

Life Stage	Temperature (°C)										
	7.5	10.0	12.5	15.0	17.5	20.0	22.5	25.0	27.5	30.0	32.5
Egg–1st	16	16	16	8	5	4	3	3	2	2	2
1st–2nd	44	44	44	22	15	11	9	7	6	6	5
2nd–3f	63	63	63	31	21	16	13	10	9	8	7
3f–3m	111	111	111	55	37	28	22	18	16	14	12
3m–Pupal	281	281	281	141	94	70	56	47	40	35	31
Pupal–Adult	441	441	441	221	147	110	88	74	63	55	49

During egg hatch, a larva was recorded as 1st stage if they had broken the egg chorion and were actively emerging. Pharate larvae (larvae that have undergone apolysis but not ecdysis) were recorded as the earlier stage (e.g., 3rd stage spiracular slits can be seen beneath the current spiracular slits would be recorded as 2nd stage), since they had not yet molted. Pupariation started when the larva had a shortened body length and no longer projected its mouth hooks when put in the larval fixative KAAD (kerosene-acetic acid–dioxane). There were times when a larva appeared to be entering the puparium stage but would extend its body length and begin crawling if disturbed or placed in KAAD. These larvae were recorded as 3rd migratory. All life stages were preserved in 70% ethyl alcohol. Third and third–migratory stages were fixed in KAAD for 48 h and transferred to 70% ethyl alcohol.

2.4. Analysis

As previously discussed, the goal of this experiment was to determine the distributions of stage transitions by temperature for each of 6 transitions. With 11 temperatures and 6 transitions we needed to model 66 relationships for each species. We used 2 regression procedures. First, to determine the appropriate transition distributions, we used TableCurve 2d, version 5.01 (SYSTAT Software Inc., San Jose, CA, USA, http://www.sigmaplot.com/products/tablecurve2d/tablecurve2d.php), accessed on 1 July 2022. and Prism, version 6.02 (GraphPad Software, Inc., La Jolla, CA, USA, http://www.graphpad.com/scientific-software/prism/). Here, we fit one of 4 functions (specifically, a regressed proportion (percentage) in stage versus time, at each temperature tested). Variables were y = proportion in stage, x = time (degree days), and a, b, c, and d = regression coefficients. The equations used were the following:

1. A Gaussian equation (a standard normal curve):

$$y = a \exp\left[-\frac{1}{2}\left(\frac{x-b}{c}\right)^2\right].$$

2. A modified Gaussian equation (a form of Gaussian curve with a plateau at 100%):

$$y = a \exp\left[-\frac{1}{2}\left(\frac{|x-b|}{c}\right)^d\right].$$

3. A cumulative Gaussian equation (a form of the Gaussian curve used for adults to model a sigmoidal increase to a plateau):

$$y = \frac{a}{2}\left[1 + erf\left(\frac{x-b}{\sqrt{2}c}\right)\right].$$

4. A reversed cumulative Gaussian equation (a form of the cumulative Gaussian equation used for eggs, to model a sigmoidal decrease from a plateau):

$$y = \frac{a}{2}\left[1 - erf\left(\frac{x-b}{\sqrt{2}c}\right)\right].$$

Cumulative forms of the equations were needed to model the transitions from egg or to adult. For the larval and pupal stages, the distinction between fitting a Gaussian or modified Gaussian equation usually depended on length of time in stage. Because longer lasting stages often had a plateau when all individuals were in the same stage between transitions, the modified Gaussian relationship was more appropriate. Fitting these relationships provided evidence for the mathematical distribution of individuals during stage transitions.

A different regression procedure was needed to determine the duration of individual stages (50% of L1 to L2, for example). Various approaches could be used, for example, determining the time from peak of one stage to peak of the next. However, we used the time between 50% transition into a stage to 50% transition out of a stage. We made this choice because we can determine a standard error in the 50% transition point, which is not always possible with determining peaks. Determination of the 50% transition point itself is straightforward using a probit model, with the probit choices of being in the first stage or the next. Through probit modelling, it is possible to determine any desired % transition and the associated variation. Probit models were constructed with Prism 6.02.

For all regression analyses, the data were examined closely to determine their propriety for inclusion in analysis. In a few instances, individuals were sampled with extraordinarily extended durations. These were treated as outliers and excluded from analysis (indicated in Tables S1 and S4).

3. Results

3.1. Lucilia sericata

Data for *Lucilia sericata* can be found in Table S1 and all calculations are contained in Tables S2 and S3. Data from 7.5 °C were not analyzed because there was very little egg development and no larval eclosion. In all 66 stage transitions examined, transition frequencies were normally distributed (Figure 1 shows the normal distribution of the transitions periods). Gaussian curves were used on life stages that had little or no plateauing, whereas modified Gaussian curves were a better fit for plateaued data (i.e., pupation). There were large spreads in transition times, particularly during the later life stages (third, 3m, and pupation) for all temperatures reported (Figure 1), covering a period of hours (egg, first, second stages) to days (third, 3m, pupation).

Probit analysis indicated 50% transition times (Figure 2). There was large variation at 10.0 and 12.5 °C, the majority due to high mortality rates (Figure 2). The effect of mortality can also be seen at 30.0 and 32.5 °C (Figure 2), with few individuals reaching 100% of the early life stages. However, there was still enough resolution to determine the means for those temperatures (Table 3).

Table 3. Mean time (hours) for *Lucilia sericata* to reach 50% of the population for each life stage and temperature.

Life Stage	Temperature °C										
	7.5	10.0	12.5	15.0	17.5	20.0	22.5	25.0	27.5	30.0	32.5
Egg–1st	N/A	186	41	52	47	28	21	17	18	12	14
1st–2nd	N/A	407	136	131	79	59	46	36	31	24	24
2nd–3f	N/A	463	231	179	137	92	78	63	49	42	36
3f–3m	N/A	683	431	314	237	193	141	111	124	85	95
3m–Pupal	N/A	2179	1935	1206	326	217	202	168	158	139	180
Pupal–Adult	N/A	4011	2896	1710	760	554	424	371	344	297	353

3.2. Phormia regina

Data for *Phormia regina* can be found Table S4 and calculations are contained in Tables S5 and S6. Data from 7.5 °C were not used because there was no egg development. Experiments at 10 °C were stopped at 400 h (16.7 days) due to complete egg and first stage mortality. The remaining stage transitions were all normally distributed (Figure 3 shows the normal distribution of the transitions periods). The later life stages, particularly L3m and pupation, had the largest variability, with large error bars on most of the samples. Starting at 12.5 °C, the pupation stage was broad, covering hundreds of hours, which was also true at 32.5 °C, where the entire life cycle was complete in 250 h.

The method of calculating sampling times was not as accurate in this species as compared to *L. sericata* for the L3m stage (Figure 4). Of all the L3m samples taken, few were at the required time within the stage, with 6 of the 10 temperatures never sampling at 100%. The most pronounced data reduction can be observed at 15.0 °C. Figure 4 illustrates this issue more clearly by the almost vertical slope and <100% in stage observed at the L3m locations. However, there were enough data to determine the means for all temperatures (Table 4).

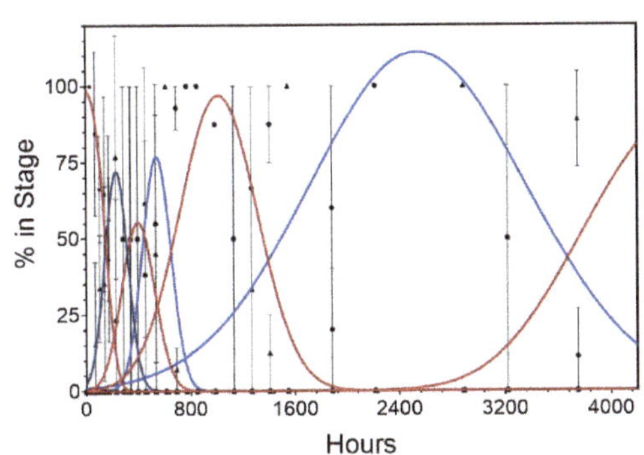

A. 10.0°C *Lucilia sericata* E-A

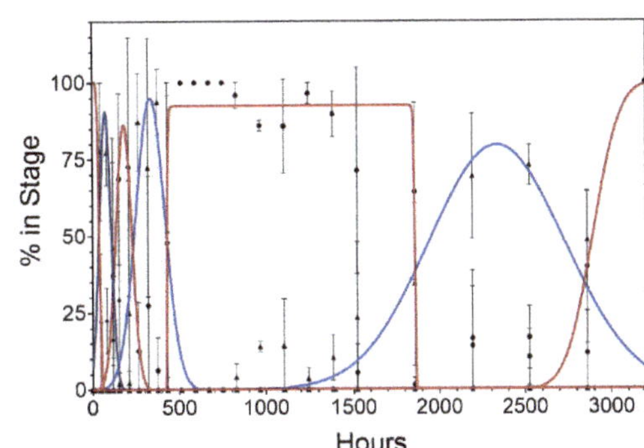

B. 12.5°C *Lucilia sericata* E-A

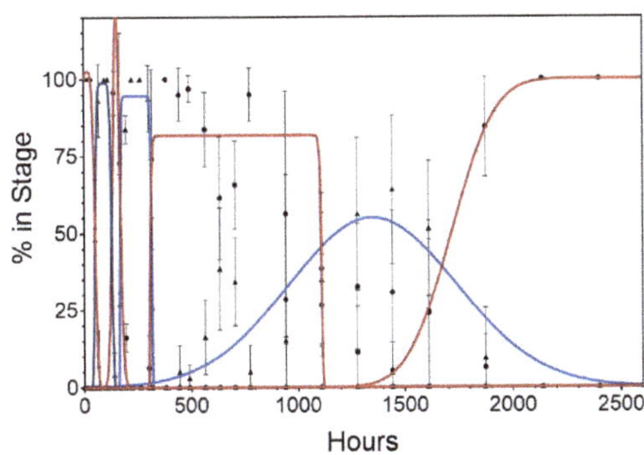

C. 15.0°C *Lucilia sericata* E-A

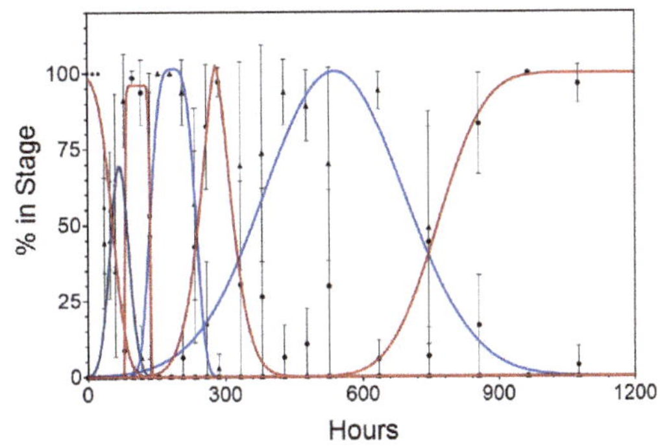

D. 17.5°C *Lucilia sericata* E-A

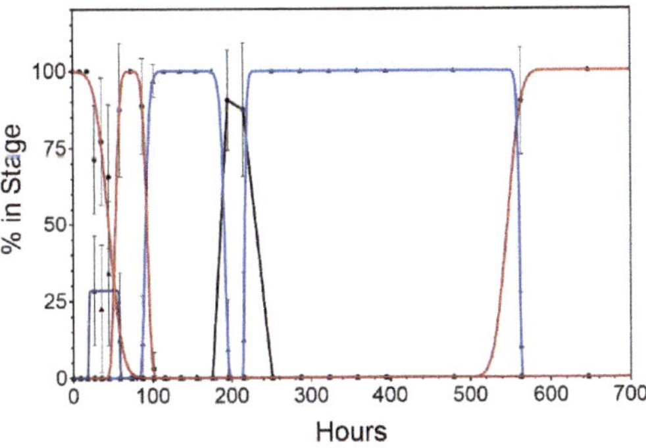

E. 20.0°C *Lucilia sericata* C E-A

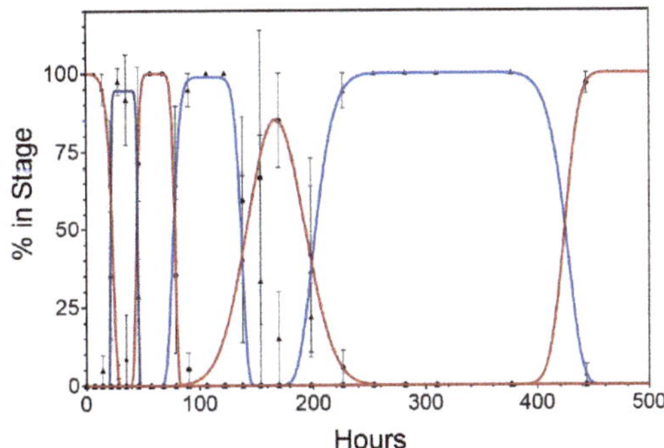

F. 22.5°C *Lucilia sericata* E-A

Figure 1. *Cont.*

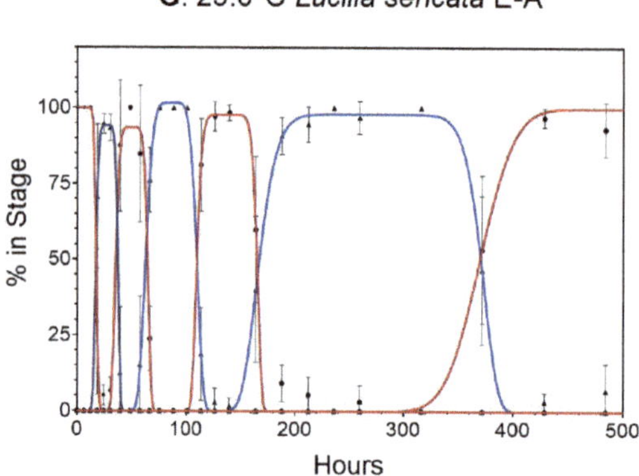

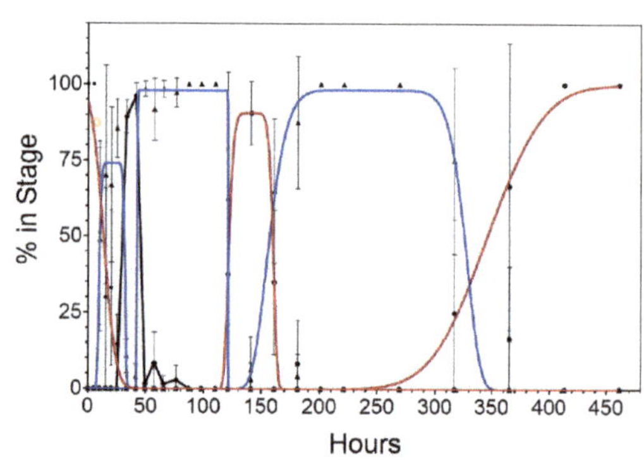

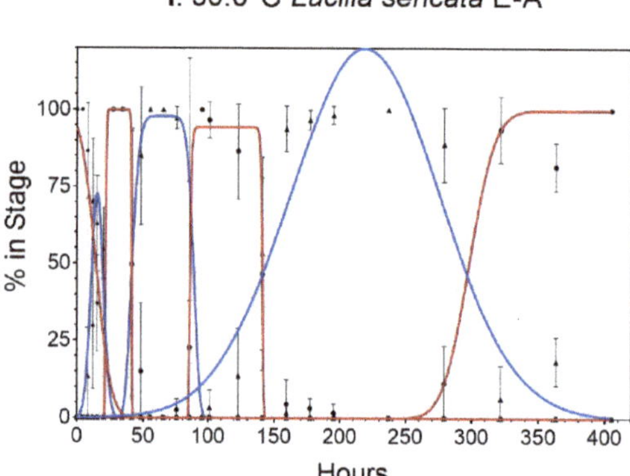

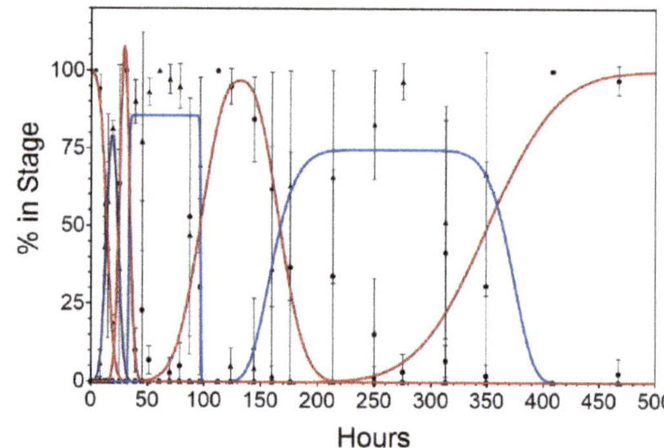

Figure 1. (**A–F**) Gaussian and modified Gaussian curves (those curves with plateau) fit to all life stages (egg to adult) of *Lucilia sericata* for temperatures 10.0 to 32.5 °C. (**A–J**), respectively. Models were not constrained, so occasionally the best fit curve can exceed 100%. Different line colors and black dot and triangles are used to distinguish adjacent curves. (**G–J**). Gaussian and modified Gaussian curves (those curves with plateau) fit to all life stages (egg to adult) of *Lucilia sericata* for temperatures 10.0 to 32.5 °C. (**A–J**), respectively. Models were not constrained, so occasionally the best fit curve can exceed 100%. Different line colors and black dot and triangles are used to distinguish adjacent curves.

Table 4. Mean time for *Phormia regina* to reach 50% of the population for each life stage and temperature (hours).

	Temperature (°C)										
Life Stage	7.5	10.0	12.5	15.0	17.5	20.0	22.5	25.0	27.5	30.0	32.5
Egg–1st	N/A	173	70	61	40	24	16	15	14	10	10
1st–2nd	N/A	N/A	354	154	127	75	50	43	34	27	25
2nd–3f	N/A	N/A	564	231	218	123	92	71	57	47	42
3f–3m	N/A	N/A	612	521	277	202	134	123	95	76	78
3m–Pupal	N/A	N/A	843	469	319	245	178	155	122	107	106
Pupal–Adult	N/A	N/A	1450	966	622	447	356	283	239	218	206

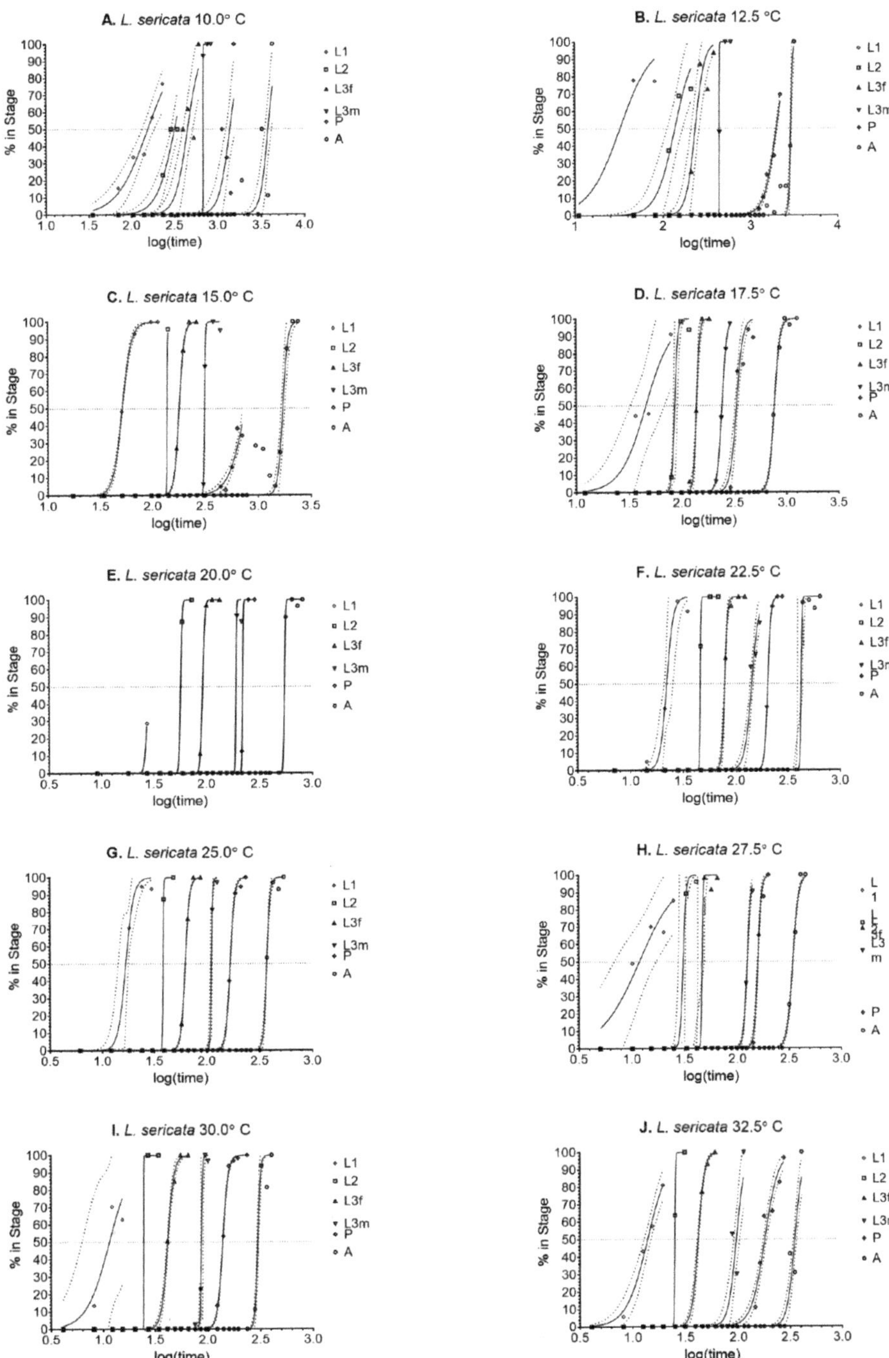

Figure 2. Probit analysis curves used to determine 50% of the transition population for *Lucilia sericata* at 10.0–32.5 °C (Figure 2**A**–**J**, respectively). Confidence intervals are represented by dotted lines.

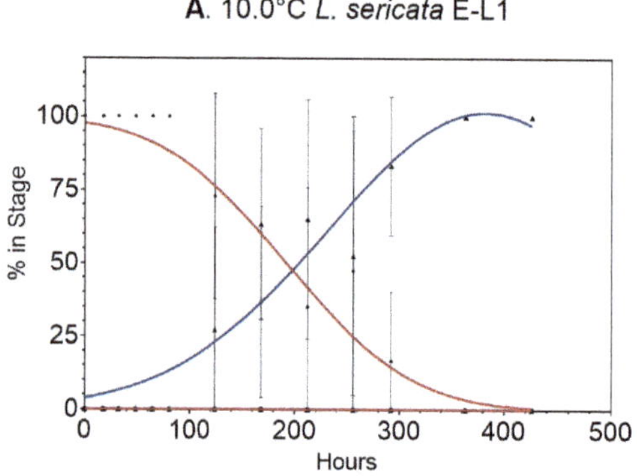

A. 10.0°C *L. sericata* E-L1

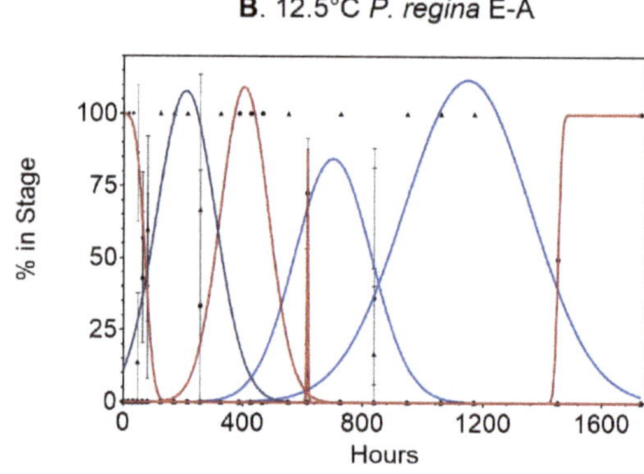

B. 12.5°C *P. regina* E-A

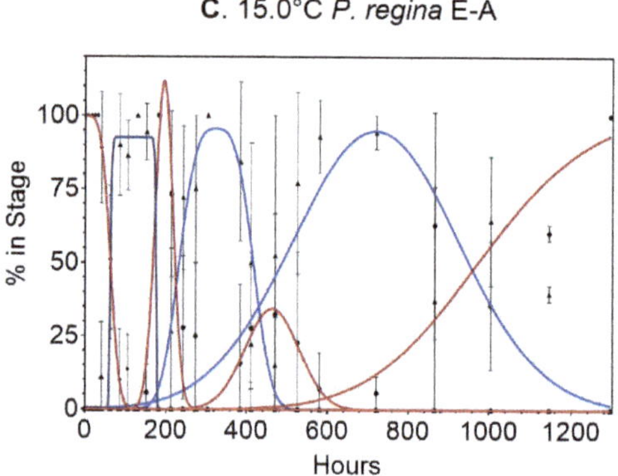

C. 15.0°C *P. regina* E-A

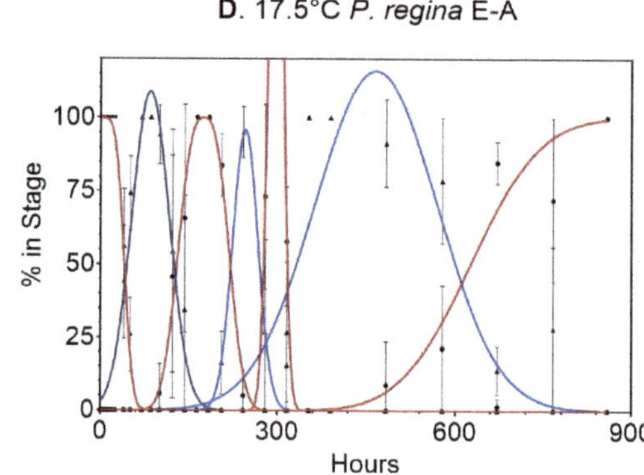

D. 17.5°C *P. regina* E-A

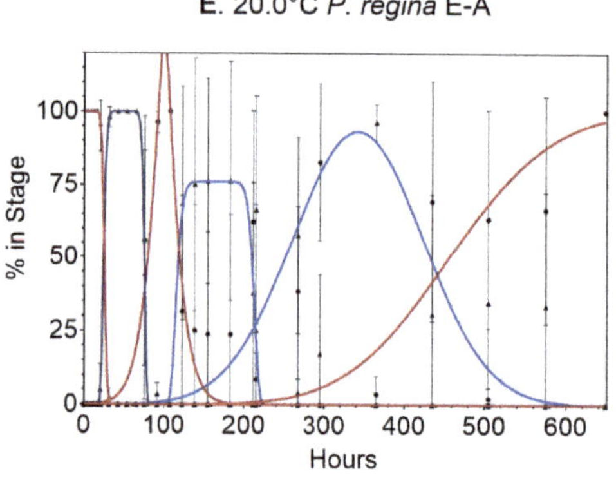

E. 20.0°C *P. regina* E-A

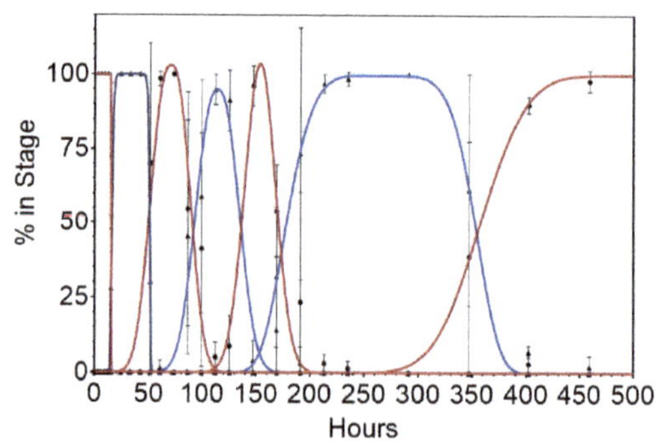

F. 22.5°C *P. regina* E-A

Figure 3. *Cont.*

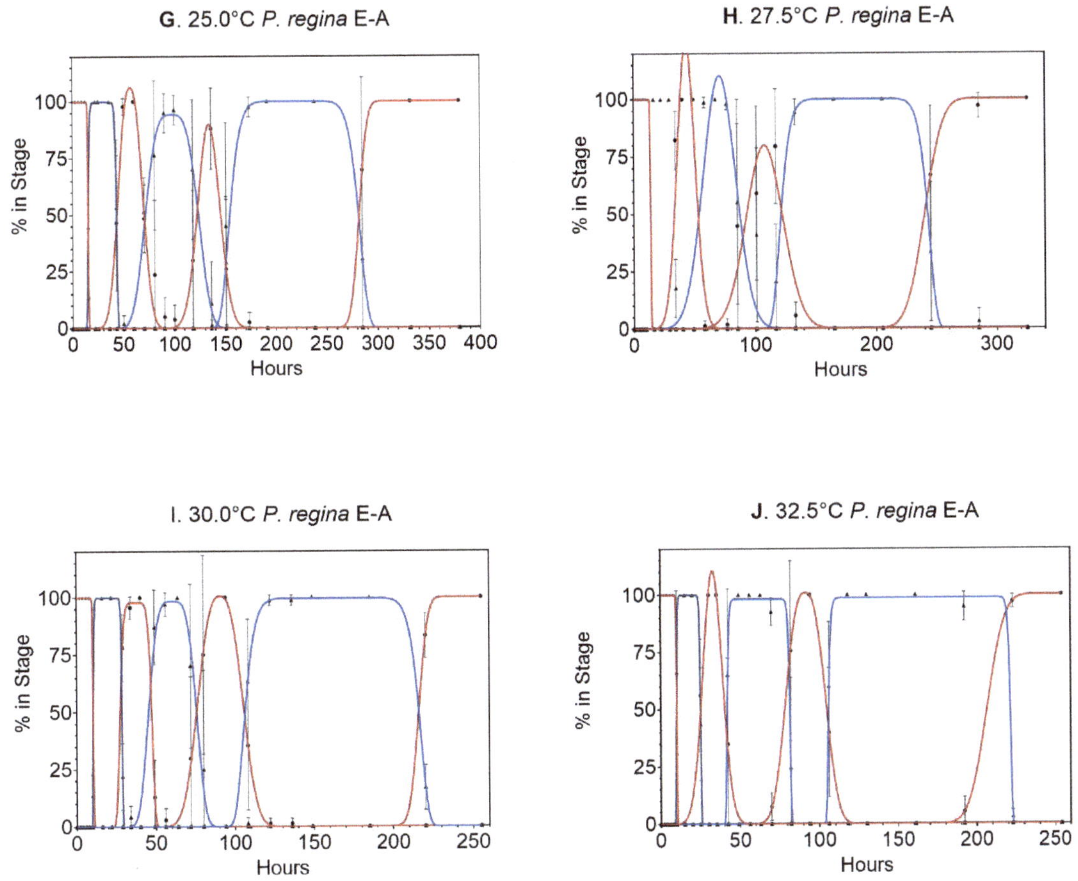

Figure 3. (**A–F**) Gaussian and modified Gaussian curves (those curves with plateau) fit to all life stages (egg to adult) of *Phormia regina* for temperatures 10.0 to 32.5 °C (Figure 1A–J, respectively). Models were not constrained, so occasionally the best fit curve can exceed 100%. Different line colors and black dot and triangles are used to distinguish adjacent curves. (**G–J**) Gaussian and modified Gaussian curves (those curves with plateau) fit to all life stages (egg to adult) of *Phormia regina* for temperatures 10.0 to 32.5 °C (Figure 1A–J, respectively). Models were not constrained, so occasionally the best fit curve can exceed 100%. Different line colors and black dot and triangles are used to distinguish adjacent curves.

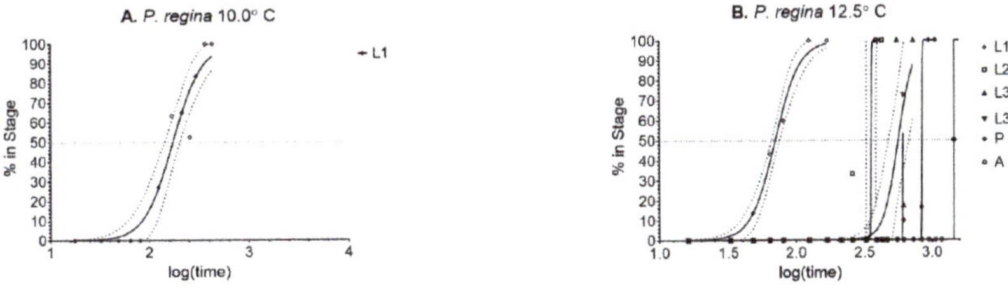

Figure 4. *Cont.*

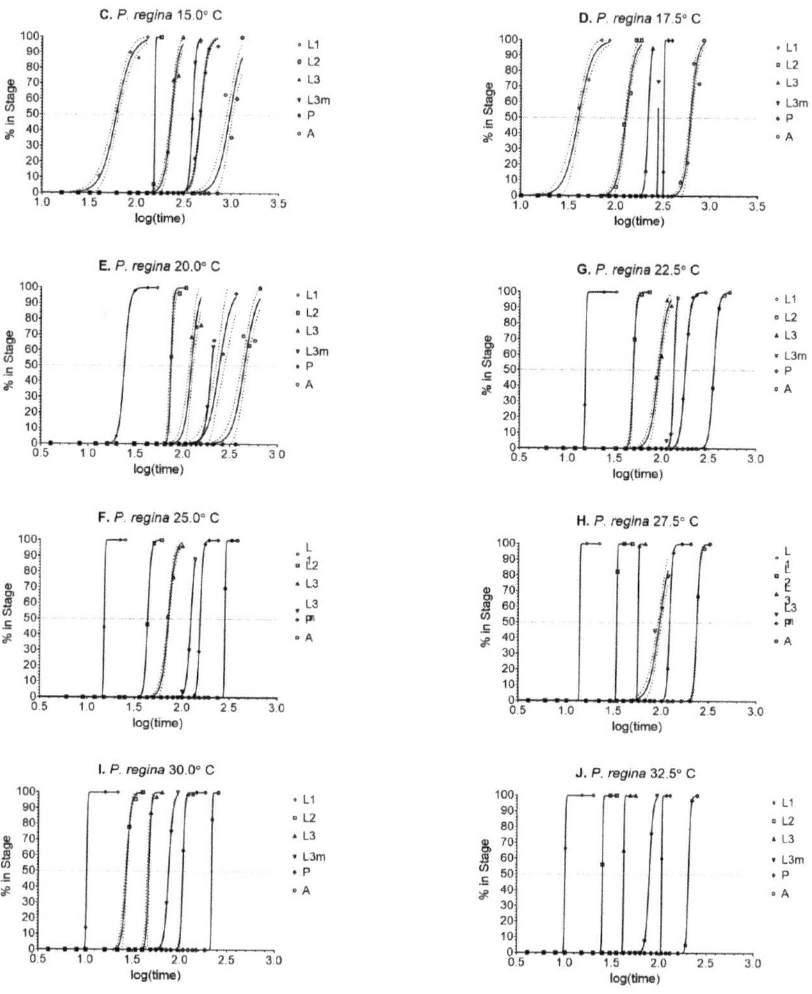

Figure 4. (**A–E**). Probit analysis curves used to determine 50% of the transition population for *Phormia regina* at 10.0–32.5 °C (**A–J**, respectively). Confidence intervals are represented by dotted lines.

4. Discussion

4.1. Lucilia sericata

Just as cooler temperatures yield longer development times [1,4,8,9], cooler temperatures also produce longer transitions times, as well as larger confidence intervals. Since blow flies are poikilothermic, extreme temperatures interfere with biological processes including metabolism, movement, and the regulation of growth hormones. This leads to developmental variation at these temperatures, which was observed at 10.0 and 12.5 °C and 30.0 and 32.5 °C. Additionally, there also appears to be an inherent variability in development, as shown by the wide transition times. The insects used in these experiments had been inbred through 100+ generations, making it improbable the transition times were due to underlying genetic variability. Long transition times could be associated with suboptimal rearing conditions; however, other evidence (e.g., total development time, larval size, and survivorship) is not consistent with this explanation. Consequently, the observed variation in stage transition by individual maggots, which gives rise to long transition times, seems to be an intrinsic trait in *L. sericata*. Possibly, this underlying variability is a means for individuals to survive in an ephemeral environment. Carcasses are rarely in the environment for very long and can be found in a wide expanse of temperatures, humidity, and locations. The ability to complete a life cycle over a broad range of conditions could reduce intraspecific competition, making the survival of the species more likely.

When we examine the data that can be most strongly compared (since we are not using the exact temperatures), our results are similar to those in the modes reported in Kamal [7] (Table 5), with a temperature difference of 26.7 °C (Kamal [7]) versus 27.5 °C. We can compare more directly as percent of larvae within a stage, which eliminates trying to compare mode to mean measurements (Table 6). Most variability is observed in the later life stages (third migratory and pupation), which is where most transition variability is found. There is also a strong comparison between our data and the limited data from Ash and Greenberg [9], with 27.0 °C (A & G) versus 27.5 °C (here). When we compare our calculated means, six of nine are within 2 SD and the remaining three are within 3 SD (Table 6), with the differences shown in Table 7.

Table 5. Comparison between Kamal [7] and Roe and Higley of *Lucilia sericata* as percent in stage.

	Temp	Egg	L1	L2	L3f	L3m	P
R and H	27.5	5.2%	3.7%	5.4%	22.0%	9.7%	54.1%
Kamal	26.7	5.2%	5.7%	3.4%	11.5%	25.9%	48.3%
Difference		0.0%	−2.1%	1.9%	10.5%	−16.2%	5.8%

Table 6. Comparison between Ash and Greenberg [9] and Roe and Higley of *Lucilia sericata* mean transition times.

Transition Stages	Mean of Transition Time in Hours					
	A & G 1975	R & H	A & G 1975	R & H	A & G 1975	R & H
	19.0 C	20.0 C	27.0 C	27.5 C	35.0 C	32.5 C
E-L1	29.4	28.1	14.4	17.9	10.2	13.8
L3m-P	691.2	216.8	194.4	157.6	333.6	180.0
P-A	1312.8	553.7	384.0	343.5	295.2	352.5

Table 7. Difference in transition means observed here for *Lucilia sericata* as proportion of Ash and Greenberg [9] standard deviations.

Transition Stages	Difference in Transition Means (A&G (1975)—Presented Data) as Proportion of A&G Standard Deviations		
	19 and 20 C	27 and 27.5 C	35 and 32.5 C
E-L1	0.6	2.7	1.9
L3m-P	3.9	0.5	0.9
P-A	2.5	0.3	1.0

Both Kamal [7] and Ash and Greenberg [9] used continuous lighting during their development studies, which could account for some of the variation, since it has been shown light regimes can affect development [16]. Interestingly, Kamal's population of *L. sericata* was collected from Pullman, WA, Ash and Greenberg's population were collected from Chicago, IL, and ours were collected from Morgantown, WV. There was also a large time difference in population examinations: 56 and 39 years, respectively. The fact that these data sets produce similar results raises a question about geographic variation and its impact on development. If geographic variation caused a considerable difference in development times, the transitions among the three data sets should have been significantly different, but they were not.

4.2. Phormia regina

Phormia regina is known for its preference for cooler temperatures [12]. However, this species did not mature past the first stage at 10.0 °C in this study and appeared to have just as much difficulty maturing to adults at 12.5 °C and 15.0 °C as *L. sericata*. Unlike data in Byrd and Allen [11], there was an egg hatch at 10.0 °C, but since there was no development past the L1 stage, *P. regina*'s biological minimum temperature is between 10.0 °C and 12.5 °C.

While there are some blow fly species, specifically *Calliphora vicina*, that can complete their life cycle at temperatures below 10.0 °C [17], *P. regina* does not appear to be one of them. Surprisingly, the transition rates at 30.0 °C and 32.5 °C were not adversely affected by temperature. This could be partially explained by mortality rates. There was not an increase in mortality as the temperatures increased, unlike with *L. sericata*, where mortality did increase at the higher temperatures (less mortality equals more data to analyze).

There is similar variation between the two species during the later life stages transitions. Both *P. regina* and *L. sericata* have large variation during the L3m and pupation stages. Unlike *L. sericata*, however, *P. regina*'s pupal stage does not begin to plateau until 22.5 °C and 6 out 10 curves (Figure 4) calculated for L3m did not reach 100%. This could be an artifact of the sampling protocol, where the method of calculating sampling times was not as accurate in this species. Therefore, although the sampling times were divided into five equal times, those times did not align with the transitions during the later life stages. Regardless, as with *L. sericata*, the observed variation in stage transition by individual maggots seems to be intrinsic in *P. regina*. Because all necrophagous species rely on ephemeral resources, it could be expected that those with more intrinsic variation are most likely to survive over a broad range of environmental factors and the trait is shared between blow fly subfamilies.

Because of the vast differences in methodologies and temperatures studied, it is difficult to directly compare data sets. Our results do compare favorably with those of Kamal [7] when we compare percent of larvae within a stage (Table 8) at similar temperatures (26.7 °C (Kamal [7]) and 27.5 °C). Converting transition times to percentages eliminates the need to attempt to compare mode to mean measurements (Table 9). Most variability is observed in the L3m stage in both data sets, which coincides with where we detected the majority of transition variation. Our data also favorably compare to those of Byrd and Allen [11] at the higher temperatures (25.0 and 30.0 °C), with our transition times for all life stages fitting within their given ranges. Our transition times are considerably greater than their reported ranges at the cooler temperatures (15.0 and 20.0 °C).

Table 8. Comparison between Kamal [4] and Roe and Higley of *Phormia regina* development means and modes.

		E-L1	E-L2	E-L3f	E-L3m	E-P	E-A
R and H	mean	13.8	33.9	56.6	95.3	122.1	239.2
Kamal	mode	16	34	45	81	165	309
Difference		−2.2	−0.1	11.6	14.3	−42.9	−69.8

Table 9. Comparison between Kamal [4] and Roe and Higley of *Phormia regina* as percent in stage.

	Temp	Egg	L1	L2	L3f	L3m	P
R and H	27.5	6.1%	8.9%	10.1%	17.2%	11.9%	52.0%
Kamal	26.7	5.2%	5.8%	3.6%	11.7%	27.2%	46.6%
Difference		0.9%	−3.1%	6.5%	5.5%	−15.3%	5.4%

Differences between these data and those of Byrd and Allen [11] could be a result of methodology. Byrd and Allen used 400 eggs per subsample (with a total of three samples). It has been proposed that maggots in large masses feed more efficiently due to external digestive enzymes [3]. In this case, faster feeding and subsequent digestion could push development forward and could explain the faster development times reported by Byrd and Allen [11].

As with *L. sericata*, the question of whether geographical variation significantly impacts development rates is raised. Kamal's [7] population of *P. regina* was obtained from Pullman, WA; Byrd and Allen's [11] population was obtained from Florida, and ours was collected from Lincoln, NE. Although there are differences between the data sets, it is unlikely that the variation was caused by the geographic differences in the fly populations due to them being similar despite methodological differences between the experiments and the high intrinsic variation seen within our highly inbred population.

While this study was very similar to the one conducted with *L. sericata*, the result is the same: accurate transition data leads to more accurate developmental data, which leads to more accurate development models. Models can accommodate the linear and curvilinear areas of development and can be used over a series of temperatures. As discussed with *L. sericata*, collecting development data for as many forensic species as possible is imperative. Comparisons between multiple, comprehensive data sets allows similarities, differences, and patterns to be recognized, increasing our knowledge of basic development biology and the variables that affect it.

4.3. Forensic Implications of Findings

Because transitions were normally distributed, the assumption that stage transitions are rapid with a long fall off is disproved. Consequently, the current practice/recommendation of collecting the largest maggots from a mass (e.g., [18]) is incorrect. By using a small sample of the largest maggots, an underrepresentation of the actual cohort age is initiated, both through sampling error and failure to properly represent transition distribution. While maggot size has often been used as an age determinate, there is a vast difference in maggots reared in a laboratory (where food and/or competition are not limiting factors) versus a carcass (where food and competition are limiting factors). Since nutrition has a direct impact on larval size, it is difficult to control for size among "unknown" age samples, such as those commonly found in death/myiasis investigations. Even with controlled populations, during the earlier life stages, stage is "more effective than size for estimating the [larval] age" [6].

Because the calculation of degree day requirements by stage depends on the determination of when one stage transitions into another, our results highlight a potential source of error in determining degree-day requirements by stage. Additionally, the large range of transition values for the L3-L3m, L3m-P, and P-A stage transitions suggest that degree-day determinations for these stages (i.e., L3, L3m, and P) are particularly given to high variability. One approach for addressing this variation would be to explicitly consider the proportion of individuals in multiple stages and using these to determine where the population is on a stage transition curve. Obviously, for such an approach to be viable, detailed transition curves must be available and samples of maggot must accurately represent the underlying population.

A final implication we have previously mentioned is the controversial question of ways to measure geographic variation between populations of a given species. Given the high degree of variation we see in transition times in this study (with species in different subfamilies), unless the variation in transition time is explicitly considered, it is easy to misconstrue variation associated with transitions with geographic variation.

Supplementary Materials: The following supporting information can be downloaded at: https://www.mdpi.com/article/10.3390/insects14040315/s1, Table S1: Development data for Lucilia sericata (from Excel); Table S2: Datasets and statistical analysis summary of Lucilia sericata nonlinear regressions for percent in stage by temperature; Table S3: Datasets and statistical analysis summary of Lucilia sericata nonlinear regressions for 50% stage transition by temperature; Table S4: Development data for Phormia regina; Table S5: Datasets and statistical analysis summary of Phormia regina nonlinear regressions for percent in stage by temperature; Table S6: Datasets and statistical analysis summary of Phormia regina nonlinear regressions for 50% stage transition by temperature.

Author Contributions: Conceptualization, A.L.R. and L.G.H.; methodology, A.L.R. and L.G.H.; formal analysis, L.G.H.; investigation, A.L.R.; writing—original draft preparation, A.L.R.; writing—review and editing, A.L.R. and L.G.H.; visualization, A.L.R. and L.G.H.; funding acquisition, A.L.R. and L.G.H. All authors have read and agreed to the published version of the manuscript.

Funding: Financial support for this research was provided by the National Institute of Justice, Office of Justice Programs, U.S. Department of Justice (Grant No. 2010-DN-BX-K231). The funders had no role in study design, data collection and analysis, decision to publish, or preparation of the manuscript.

Data Availability Statement: Data for this study are available in Supplementary Materials.

Acknowledgments: We thank Jeff Wells and Anne Perez for the initial *L. sericata* used to start our colonies and Christian Elowsky for his assistance in the lab.

Conflicts of Interest: The authors declare no conflict of interest. The funders had no role in the design of the study; in the collection, analyses, or interpretation of data; in the writing of the manuscript; or in the decision to publish the results.

References

1. Higley, L.G.; Haskell, N.H. Insect development and forensic entomology. In *Forensic Entomology: The Utility of Arthropods in Legal Investigations*; Byrd, J.H., Castner, J.L., Eds.; CRC Press: Boca Raton, FL, USA, 2010; pp. 389–405.
2. Greenberg, B. Flies as forensic indicators. *J. Med. Entomol.* **1991**, *28*, 565–577. [CrossRef]
3. Rivers, D.B.; Dahlem, G.A. *The Science of Forensic Entomology*; Wiley: Oxford, UK, 2014; pp. 131–149.
4. Hall, D.G. *The Blowflies of North America*; Thomas Say Foundation: Annapolis, MD, USA, 1948.
5. Wells, J.D.; LaMotte, L.R. Estimating postmortem interval. In *Forensic Entomology: The Utility of Arthropods in Legal Investigations*; Byrd, J.H., Castner, J.L., Eds.; CRC Press: Boca Raton, FL, USA, 2010; pp. 367–388.
6. Wells, J.D.; Kurahashi, H. *Chrysomya megacephala* (Fabricius) (Diptera: Calliphoridae) development: Rate, variation and the implications for forensic entomology. *Japan. J. Sanit. Zool.* **1994**, *45*, 303–309. [CrossRef]
7. Kamal, A.S. Comparative study of thirteen species of sarcosaprophagous Calliphoridae and Sarcophagidae (Diptera) I. Bionomics. *Ann. Entomol. Soc. Am.* **1958**, *51*, 261–271. [CrossRef]
8. Richards, C.S.; Villet, M.H. Factors affecting accuracy and precision of thermal summation models of insect development used to estimate post-mortem intervals. *Int. J. Leg. Med.* **2008**, *122*, 401–408. [CrossRef] [PubMed]
9. Ash, N.; Greenberg, B. Developmental temperature responses of the sibling species *Phaenicia sericata* and *Phaenicia pallescens*. *Ann. Entomol. Soc. Am.* **1975**, *68*, 197–200. [CrossRef]
10. Anderson, G.S. Minimum and maximum developmental rates of some forensically important Calliphoridae (Diptera). *J. Forensic Sci.* **2000**, *45*, 824–832. [CrossRef] [PubMed]
11. Byrd, J.H.; Allen, J.C. The development of the black blow fly, *Phormia regina* (Meigen). *Forensic Sci. Inter.* **2001**, *120*, 79–88. [CrossRef] [PubMed]
12. Byrd, J.H.; Castner, J.L. Insects of forensic importance. In *Forensic Entomology: The Utility of Arthropods in Legal Investigations*, 2nd ed.; Byrd, J.H., Castner, J.L., Eds.; CRC Press: Boca Raton, FL, USA, 2010; pp. 39–126.
13. Nabity, P.D.; Higley, L.G.; Heng-Moss, T.M. Effects of temperature on development of *Phormia regina* (Diptera: Calliphoridae) and use of development data in determining time intervals in forensic entomology. *J. Med. Entomol.* **2006**, *43*, 1276–1286. [CrossRef] [PubMed]
14. Roe, A.; Higley, L.G. Development modeling of *Lucilia sericata* (Diptera: Calliphoridae). *PeerJ* **2015**, *3*, e803. [CrossRef] [PubMed]
15. Authement, M.L.; Higley, L.G.; Hoback, W.W. Anoxia tolerance in four forensically important calliphorid species. *Forensic Sci.* **2023**, *3*, 1. [CrossRef]
16. Nabity, P.; Higley, L.; Heng-Moss, T. Light-induced variability in development of forensically important blow fly *Phormia regina* (Diptera: Calliphoridae). *J. Med. Entomol.* **2007**, *44*, 351–358. [CrossRef] [PubMed]
17. Davies, L.; Ratcliffe, G.G. Development rates of some pre-adult stages in blowflies with reference to low temperatures. *Med. Vet. Entomol.* **1994**, *8*, 245–254. [CrossRef] [PubMed]
18. Grassberger, M.; Reiter, C. Effect of temperature on *Lucilia sericata* (Diptera: Calliphoridae) development with special reference to the isomegalen- and isomorphen-diagram. *Forensic Sci. Int.* **2001**, *120*, 32–36. [CrossRef] [PubMed]

Article

Implications of the Niche Partitioning and Coexistence of Two Resident Parasitoids for *Drosophila suzukii* Management in Non-Crop Areas

María Josefina Buonocore Biancheri [1], Segundo Ricardo Núñez-Campero [2,3], Lorena Suárez [4,5], Marcos Darío Ponssa [1], Daniel Santiago Kirschbaum [6,7], Flávio Roberto Mello Garcia [8] and Sergio Marcelo Ovruski [1,*]

[1] Planta Piloto de Procesos Industriales Microbiológicos y Biotecnología (PROIMI-CONICET), División Control Biológico, Avda. Belgrano y Pje. Caseros, San Miguel de Tucumán 4000, Tucumán, Argentina
[2] Centro Regional de Investigaciones Científicas y Transferencia Tecnológica de La Rioja (CRILAR), Provincia de La Rioja, UNLaR, SEGEMAR, UNCa, CONICET, Entre Ríos y Mendoza s/n, Anillaco 5301, La Rioja, Argentina
[3] Departamento de Ciencias Exactas, Físicas y Naturales, Instituto de Biología de la Conservación y Paleobiología, Universidad Nacional de La Rioja (UNLaR), Av. Luis M. de la Fuente s/n., Ciudad de La Rioja 5300, La Rioja, Argentina
[4] Dirección de Sanidad Vegetal, Animal y Alimentos de San Juan (DSVAA)—Gobierno de la Provincia de San Juan, CONICET, Nazario Benavides 8000 Oeste, Rivadavia 5413, San Juan, Argentina
[5] CCT CONICET San Juan, Argentina Av. Libertador Gral. San Martín 1109, Capital 5400, San Juan, Argentina
[6] INTA Estación Experimental Agropecuaria Famaillá, Tucumán Ruta Prov. 301, km 32, Famaillá 4132, Tucumán, Argentina
[7] Cátedra Horticultura, Facultad de Agronomía y Zootecnia, Universidad Nacional de Tucumán, San Miguel de Tucumán 4000, Tucumán, Argentina
[8] Departamento de Ecologia, Zoologia e Genética, Instituto de Biologia, Universidade Federal de Pelotas, Pelotas 96000, Rio Grande do Sul, Brazil
[*] Correspondence: sovruski@conicet.gov.ar

Citation: Buonocore Biancheri, M.J.; Núñez-Campero, S.R.; Suárez, L.; Ponssa, M.D.; Kirschbaum, D.S.; Garcia, F.R.M.; Ovruski, S.M. Implications of the Niche Partitioning and Coexistence of Two Resident Parasitoids for *Drosophila suzukii* Management in Non-Crop Areas. *Insects* **2023**, *14*, 222. https://doi.org/10.3390/insects14030222

Academic Editor: Claus P. W. Zebitz

Received: 1 February 2023
Revised: 18 February 2023
Accepted: 20 February 2023
Published: 23 February 2023

Simple Summary: *Drosophila suzukii*, internationally known as the spotted-wing drosophila (SWD), is an invasive insect pest that mainly causes economic damage to fresh and healthy, as well as soft and stone, fruit crops. The SWD has quickly spread throughout all the Argentinean fruit-growing regions. Natural enemies, such as parasitoids, can be an important environmental friendly tool within an SWD management strategy. However, understanding the biological mechanisms that enable the coexistence of different parasitoid species in a particular environment is essential to improve their use as biocontrol agents. Therefore, this study assessed the coexistence of two resident pupal parasitoids, *Trichopria anastrephae* (*Ta*) and *Pachycrepoideus vindemiae* (*Pv*), on SWD-infested guava and peach in non-crop areas of northwestern Argentina, based on spatial (microhabitat) and/or resource (host flies) differentiation. Results revealed that both biological mechanisms might mediate the coexistence of these two pupal parasitoid species. *Ta* showed a preference for resident saprophytic drosophilid puparia located mainly inside fruit flesh, whereas *Pv* searched for the host in less competitive habitats, such as in the soil or outside fruit flesh, where SWD puparia prevailed. Such a differential exploitation of host microhabitats influenced parasitoid efficiency in suppressing SWD populations. The combined use of both parasitoid species may be advisable for local SWD management.

Abstract: Understanding the mechanisms associated with the coexistence of competing parasitoid species is critical in approaching any biological control strategy against the globally invasive pest spotted-wing drosophila (=SWD), *Drosophila suzukii* (Matsumura). This study assessed the coexistence of two resident pupal parasitoids, *Trichopria anastrephae* Lima and *Pachycrepoideus vindemiae* Rondani, in SWD-infested fruit, in disturbed wild vegetation areas of Tucumán, northwestern Argentina, based on niche segregation. Drosophilid puparia were collected between December/2016 and April/2017 from three different pupation microhabitats in fallen feral peach and guava. These microhabitats were "inside flesh (mesocarp)", "outside flesh", but associated with the fruit, and "soil", i.e., puparia buried close to fruit. Saprophytic drosophilid puparia (=SD) belonging to the *Drosophila melanogaster* group and SWD were found in all tested microhabitats. SD predominated in both inside and outside flesh, whereas SWD in soil. Both parasitoids attacked SWD puparia. However, *T. anastrephae* emerged mainly from SD puparia primarily in the inside flesh, whereas *P. vindemiae* mostly foraged SWD puparia in less competitive microhabitats, such as in the soil or outside the flesh. Divergence in host choice and spatial patterns of same-resource preferences between both parasitoids may mediate their coexistence in non-crop environments. Given this scenario, both parasitoids have potential as SWD biocontrol agents.

Keywords: spotted-wing drosophila; drosophilid abundance; pupal parasitoid coexistence; ecological profiles; feral fruit host; non-crop environment

1. Introduction

In nature, resident and introduced parasitoid species may be able to coexist in the same host species by niche partitioning, i.e., the process by which competing species move into different patterns of resource use or different niches [1], or through different ecological profiles and life histories [2–4]. Among various mechanisms enabling the coexistence of competing parasitoids, the temporal and spatial partitioning of resources may be highlighted [5]. As a result, the co-occurrence of competing parasitoid species may depend on the occupation of competitor-free spaces [6]). Thus, niche differences may imply divergence and a competitive avoidance history [7]. Consequently, any information related to the mechanisms associated with the coexistence of competing parasitoid species is essential in addressing any biological control strategy against invasive insect pests [8,9].

The globally invasive pest spotted-wing drosophila (=SWD), *Drosophila suzukii* (Matsumura) (Diptera: Drosophilidae), a native of South East Asia [10] and currently occurring on all continents [11], has quickly spread throughout all the fruit-growing regions of Argentina since it was first recorded in 2014 [12]. The SWD is an economically important pest of small, soft, and stone fruits worldwide, because females lay eggs in fresh, healthy, ripening fruit [13]. Curiously, SWD is one of the few *Drosophila* species that has evolved into a serrated ovipositor, which allows females to drill into the skin of healthy fruits to oviposit inside the fleshy mesocarp [14].

Fly larvae feed deep into the fruit's fleshy mesocarp, resulting in fruit rot. Mature larvae emerge from the fruit to pupate mainly in the soil, although larvae usually also pupate inside fallen fruit or beneath the fruit without burying themselves, remaining attached to the fruit skin [15]. Although SWD is mainly a pest of berry and cherry crops, this dipteran is highly polyphagous, as it has a broad host fruit range, mainly throughout Asia, Europe, and America [10]. In addition to crop host species, mainly Rosaceae, the SWD larva can develop in both native and exotic fruit of ornamental and wild non-crop hosts [15].

The SWD is found in 64 host plants in 25 families in Latin America. Although most hosts are exotic in this region, about 39% are native plants that can become alternative hosts and reservoirs of the pest in the intercrop period [11]. In Argentina, 15 fruit species have been recorded as hosts of SWD, including both feral guava (*Psidium guajava* L.) (Myrtaceae) and feral peach [*Prunus persica* (L.) Stokes] (Rosaceae) [11,16]. These SWD host plants are among the most common and widespread exotic feral fruit species growing in wild vegetation patches, adjacent to commercial fruit crops in northwestern Argentina. Natural infestation levels by SWD in feral guava and feral peach range between 5 and 10% per kg of sampled fruit [12,16]. Guava is not commercially grown in Argentina. It can only be found as an ornamental plant in gardens or as a backyard fruit tree or scattered in wilderness areas with high levels of human disturbance. Peach is cultivated in northwestern Argentina at a very low scale, with no influence on the local or national supply. Peaches are mainly grown in the central–western and south–northeastern regions of Argentina. However, cultivated peaches were not reported to be infested by the SWD, although there are some records of SWD adults caught in liquid traps placed inside the commercial peach orchard (SWD). In spite of the SWD infesting cultivated peaches in some Asian, European and North American countries, in fruits mainly with previous wounds, it is not a natural host of this pest [15].

As SWD spreads almost worldwide, many countries have immediately adopted preventive and intervention measures to minimize economic losses. Thus, SWD mitigation strategies, including exclusion netting, mass trapping supported on attractant-based traps, crop sanitation, and chemical and biological controls, were implemented [13], while the sterile insect technique is currently being evaluated [17]. Concerning biological control, natural enemies may be particularly important as an eco-friendly tool in a network of SWD management strategies, maximizing ecosystem services' benefits. In this regard, information on wild host fruit status, on which SWD populations may increase, is critical to support management strategies, particularly in wilderness environments surrounding commercial fruit crops [18–20]. Therefore, it is imperative to understand better the trophic interactions between SWD and the components of newly invaded landscapes regarding the available hosts, other frugivorous dipterans, and natural enemies [15]. Among SWD's biological controllers, parasitoid hymenopterans are the best studied and most likely to be successful [21–23]. An assemblage of resident koinobiont larval and idiobiont pupal parasitoids has been associated with SWD in crop and non-crop areas of northwestern Argentina [16]. Among all these species, two pupal parasitoids, *Trichopria anastrephae* Lima (Hymenoptera: Diapriidae) and *Pachycrepoideus vindemiae* Rondani (Hymenoptera: Ptromalidae) are commonly abundant, and are often found foraging in search of host puparia on the same fallen fruit [16]. The diapriid *T. anastrephae*, native to South America [24], is a pupal endoparasitoid whose female lays the egg into the hemocoel of the host fly pupa [25]. In contrast, the cosmopolitan *P. vindemiae* not only attacks a wide range of dipteran species but is also a pupal ectoparasitoid, because the female lays the egg inside the space between the puparium shell and host pupal body [26]. Both parasitoids were recorded from tephritid puparia, particularly from *Anastrepha* spp. and *C. capitata*, in Argentina and Brazil [27]. Furthermore, *P. vindemiae* was associated with *D. suzukii* in berry and cherry crops of different Argentinian regions [28]. Although both are idiobiont pupal parasitoids, *T. anastrephae* is an endo-parasitoid, and *P. vindemiae* is an ecto-parasitoid, so they belong to different guilds [29]. A parasitoid guild can be acceptably defined as two or more sympatric species that equally exploit a particular developmental stage of the host [30].

Both *P. vindemiae* [31,32] and *T. anastrephae* [33,34] can be successfully lab-reared on SWD puparia and have shown high potentials as *D. suzukii* biocontrol agents [21,22]. However, competitive tests, undertaken under lab conditions, between *P. vindemiae* and *T. anastrephae* [25], and also with the cosmopolitan *Trichopria drosophilae* Perkins [31], showed that both diaprid species out-competed the pteromalid. The studies above revealed the superiority of the two *Trichopria* species over *P. vindemiae* in intrinsic competitiveness and foraging efficiency. In addition, SWD's resident parasitoid surveys in Brazil recorded a predominance of *T. anastrephae* on *P. vindemiae* [35]. In contrast, two interesting shreds of evidence have been revealed in a recent survey carried out at a non-crop habitat overgrown by feral peach trees in Tucumán, northwestern Argentina [16]: (1) *P. vindemiae* mostly parasitized SWD puparia, and (2) *Trichopria* sp., identified later as *T. anastrephae*, predominantly parasitized puparia from drosophilids of the *Drosophila melanogaster* group. In light of the preceding information, a question arises: how do these competing pupal parasitoids coexist, attacking both SWD and saprophytic *Drosophila* spp. Puparia, in the same fruit at the same time? Therefore, it was hypothesized that the coexistence of both pupal parasitoids on drosophilid-infested fruits in wild vegetation areas of Tucumán results from niche segregation, including spatial partitioning or resource partitioning, or both. In the first scenario, it is assumed that *P. vindemiae* occupies the *T. anastrephae*-free space provided by SWD puparia from microhabitats poorly exploited by the diaprid species. For the second option, it is postulated that in a shared niche situation, *P. vindemiae* is more specialized to inhabit the newly introduced host species, taking into account its cosmopolitan status and because it is far more generalist than *T. anastrephae*. A third situation involves a combination of both parasitoid species. Based on these assumptions, it is predicted that: (1) the distribution of drosophilid puparia in several microhabitats associated with the fruit will reveal differences in the space-use pattern between the SWD and resident drosophilid species; (2) *P. vindemiae* females will focus their search for, and attack, SWD puparia located outside the internal part of the fallen fruit; (3) parasitism on SWD puparia by *P. vindemiae* will increase when the density of resident drosophilid puparia highly exceeds that of SWD; (4) should such a spatial or a resource partitioning occur, *P. vindemiae* females will reduce their competitive interactions with *T. anastrephae* and thus avoid endangering their offspring. To test these predictions, a survey of drosophilid puparia scattered in different microhabitats associated with the fruit, e.g., inside the mesocarp, outside it but attached to the fruit, and buried beneath the fruit of non-crop hosts, such as feral guava and peach, in two disturbed wild vegetation sites of Tucumán was performed. Microhabitat differentiation was addressed for host puparia sampling based on the strong influence of the microhabitat type, e.g., soil or canopy, on the parasitoid assemblage, associated with saprophytic drosophilids consuming decaying organic matter, rather than habitat type [36].

The findings of this study will be useful for planning SWD biological control strategies within an area-wide integrated pest management approach [21] in Argentinian fruit-growing regions and elsewhere around the Americas.

2. Materials and Methods

2.1. Study Area

The study was carried out in an area characterized by a mosaic of environments, such as suburban sectors occupied by housing within a secondary rainforest matrix, with a predominance of exotic plants, citrus crops, and mountain slopes, slightly disturbed with a high presence of indigenous plant species. This area, located in Horco Molle, Yerba Buena district, Tucumán province, northwestern Argentina, belongs originally to the first vegetation level of the Yungas rainforest eco-region, called "Premontane Forest" [37]. The Yungas is a narrow strip of South American subtropical montane rainforest located along the eastern slopes of the Andes mountain range, starting from Peru and extending into northwestern Argentina [38]. The study area belongs to the southernmost extension of the Yungas. Two sampling sites were chosen within the study area (Figure 1). Site #1, located at 26°48′ S latitude and 65°19′ W longitude, and 520 m altitude, was a 2 ha patch of secondary structure rainforest with feral guava trees predominating. This site borders a road to the east, with a suburban sector within a secondary forest dominated by *Ligustrum lucidum* W. T. Aiton ("Evergreen tree") to the west, south, and north. Site #2 was located at 26°43′ S latitude and 65°22′ W longitude, and 660 m altitude, within the Sierra de San Javier park, a protected wildlife area belonging to Universidad Nacional de Tucumán (UNT). The site is surrounded by buildings belonging to UNT, mixed with disturbed wild vegetation patches. Both sampling sites were 4.7 km apart and located at the foothills of the San Javier Mountain (Figure 1), where the climate is subtropical, with a dry season from May to October and a humid–warm season from November to March, with 21.5 °C and 900 mm of average annual temperature and rainfall, respectively [37].

2.2. Drosophilid Puparia Sampling

Drosophilid puparia were collected from three different pupation microhabitats (Figure 2): (a) on the fallen fruit, but inside it, i.e., puparia located in the mesocarp (=flesh), (b) on the fallen fruit, but outside it, i.e., puparia attached to the fruit rind, into shallow external fruit fissures, and largely or partially protruding from the fruit skin, and (c) in the soil, i.e., puparia buried either underneath the fruit or close to it. These three pupation microhabitats will hence forth be named "inside fruit flesh", "outside fruit flesh", and "soil", respectively.

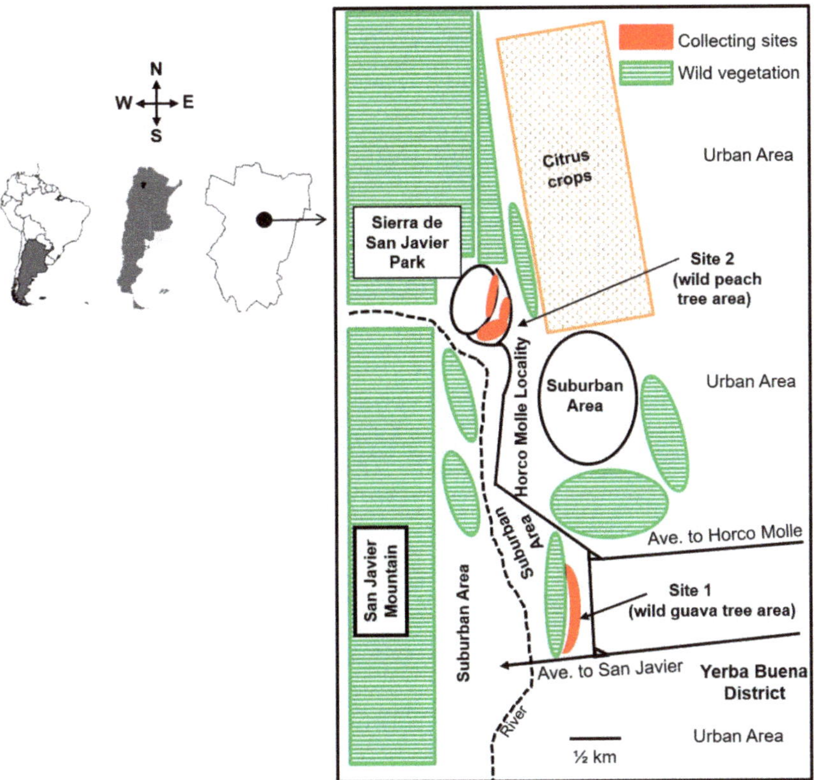

Figure 1. Study area showing the two sampling sites: Site 1, a guava tree-dominated habitat; Site 2, a peach tree-dominated habitat. The area is located in Horco Molle, Yerba Buena district, Tucumán province, northwestern Argentina.

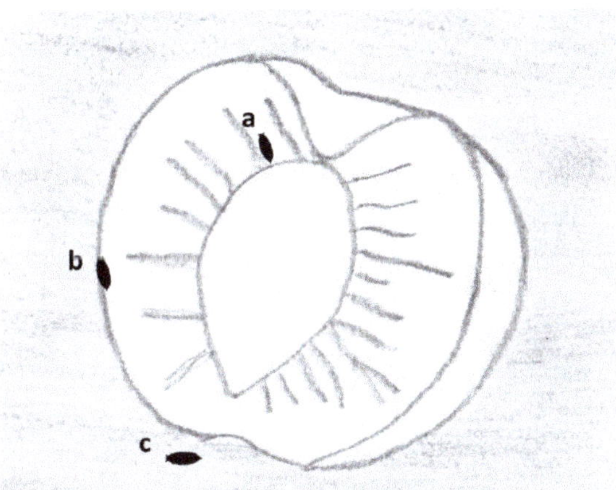

Figure 2. Scheme of a peach fruit fallen on the soil showing three microhabitats, in which the larvae of both saprophytic *Drosophila* spp. and *Drosophila suzukii* can pupate: (**a**) "inside fruit flesh", (**b**) "outside fruit flesh" and (**c**) "soil".

The puparia-collecting procedure involved randomly selecting 20 fallen fruits per sampling date during feral peach and guava fruiting seasons. In this regard, six surveys of drosophilid puparia of peach and six of guava were carried out between December 2016 and January 2017 (early summer) and March 2017 and April 2017 (late summer and early autumn), respectively. Each selected fruit (peach or guava) was removed and examined using a hand-held magnifier with an X20 glass lens at the study site. All drosophilid puparia found inside pulp fruit or attached to the fruit rind were extracted with either a blunt-tip tweezer or a soft-bristled paintbrush. Then, puparia were placed separately into 8 × 5 cm (diameter × height) plastic cups according to the pupation habitat where they were found. Each cup had a thin layer of sterilized, moistened vermiculite Intersum® (Aislater S.R.L., Cordoba, Argentina) on the bottom to avoid desiccation during transport to the lab. Cups were covered with plastic lids with pinholes. In addition, the soil underneath each fruit and the soil sector around the fruit in a 3 cm radius were dug with a hand shovel up to ~2 cm deep to find buried puparia. The extracted soil of each sampled fruit was placed individually in a plastic bag, and its top was closed with a rubber band. Both cups and bags were placed in plastic crates (32 × 24 × 12 cm) and taken to the Pest Biological Control Department (DCBP, Spanish acronym) from the Planta Piloto de Procesos Industriales Microbiológicos y Biotecnología (PROIMI,) in San Miguel de Tucumán, the capital city of the Tucumán province. PROIMI is ~6 and ~10 km away from study sites #1 and #2, respectively.

2.3. Drosophilid Puparia Processing and Identification

Each soil sample was sieved through a 1 mm metal-mesh sieve at the DCBP-PROIMI's laboratory. Puparia retained in the sieve were removed and then identified, as were puparia from the fallen fruit. *Drosophila suzukii* puparia were differentiated from those of other drosophilids by the external shape of the characteristic anterior spiracles, composed of two tubes with plumose-shaped tips on the top [23,39]. Identified *D. suzukii* puparia were separated from the remaining drosophilid puparia and placed into 5 × 6-cm (diameter × height) disposable clear plastic cups. These cups had sterilized 2 mm-thick vermiculite in the bottom and a plastic lid with pinholes to facilitate internal oxygenation. The vermiculite inside the cups was sprinkled every three days with purified water. Puparia were differentiated according to the habitat from which they were recovered and placed in individualized cups. The same procedure was carried out with other *Drosophila* Fallén puparia, identified as belonging to the *Drosophila melanogaster* species group [40], but not differentiated at the species level. These saprophytic drosophilids will henceforth be referred to as *Drosophila* spp. in the text. All cups were conditioned in a room at 26 ± 1 °C, 80 ± 5% RH, and 10:14 h L:D until adult flies and parasitoids emerged.

2.4. Adult Parasitoid and Fly Identification

Drosophilid flies were identified by M.J.B.B., and parasitoid specimens by S.M.O. and Fabiana Gallardo (Facultad de Ciencias Naturales y Museo, Universidad Nacional de La Plata, La Plata, Argentina). Gibson's [41] and Risbec's [42] keys were used to identify the pteromalid and the diaprid, respectively. Voucher fly and parasitoid adult specimens were stored at the entomological collection of the Fundación Miguel Lillo, in San Miguel de Tucumán. Parasitoid specimens were also deposited into the entomological collection of the Museo de Ciencias Naturales de la Plata, Buenos Aires, Argentina.

2.5. Data Analysis

The response variables analyzed were the drosophilid and parasitoid relative abundance per microhabitat, as well as the parasitism. The drosophilid relative abundance was calculated as the total number of *D. suzukii* or *Drosophila* spp. puparia recovered per microhabitat over the total number of *Drosophila* puparia. Parasitoid relative abundance was calculated as the total number of *P. vindemiae*, or *T. anastrephae* adults that emerged from *D. suzukii* or *Drosophila* spp. puparia per microhabitat over the total number of parasitoid individuals recovered from each drosophilid species. The parasitism percentage was estimated as the number of emerged parasitoids over the number of *D. suzukii* or *Drosophila* spp. puparia recovered from each microhabitat per 100.

The statistical analysis was performed using the software R [43]. For the analysis of drosophilids' habitat usage, and parasitoid attack, the factorial model for nonparametric analysis of variance, Aligned Rang Transformation ANOVA in the packages '*ARTool*' [44], was performed. First, the algorithm aligned the fixed effects and classified them using the model function, then generated linear models from the transformed data and analyzed the variance using the "*anova.art*" function. A post hoc pairwise comparison (Fisher's least significant difference = LSD) was conducted to show differences between factor levels using a Bonferroni–Holm adjustment method using the 'art.con' function [45]. Kruskal–Wallis tests from the "*Agricolae*" package [46] were performed to determine microhabitat preferences for pupation between saprophytic drosophilids and *D. suzukii*. The library '*rcompanion*' function was used to obtain letters that display the significant difference in figures. Violin box plots are used to show the resulting data from the study. Aside from displaying the summary statistics, using violin box plots of the package "ggplot2" [47] to plot numerical data, the entire data distribution (raw data) is shown (Supplementary Files S1–S4).

3. Results

3.1. Drosophilid Fly Abundance and Relationship with Microhabitats Tested

The abundance of saprophytic drosophilid puparia belonging to the *Drosophila melano gaster* group was two- and four-fold higher than that of *D. suzukii* puparia found on peach and guava, respectively (Figure 3A,B).

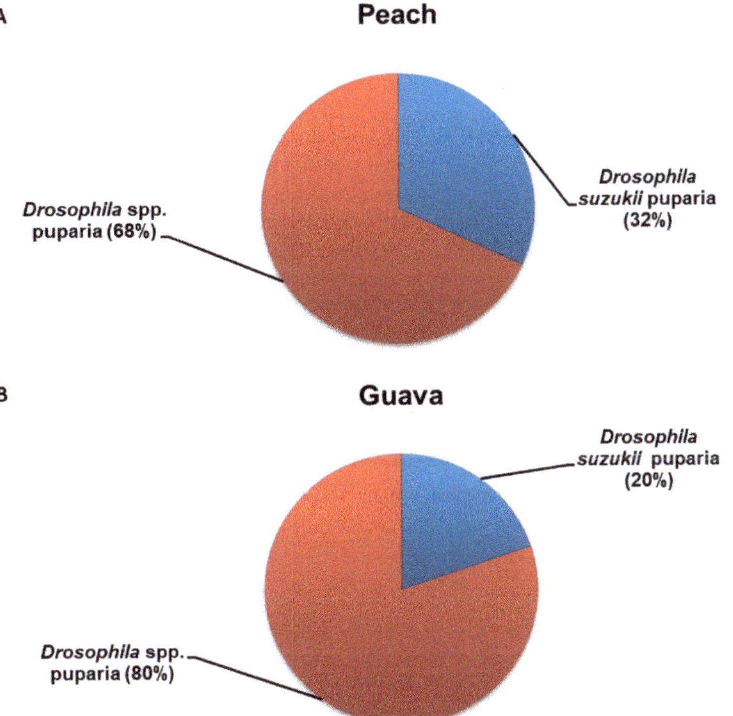

Figure 3. Relative abundance of both *Drosophila suzukii* and saprophytic *Drosophila* spp. puparia recovered from (**A**) peach and (**B**) guava.

Both saprophytic drosophilids and *D. suzukii* were found in all three microhabitats tested, but with remarkable abundance differences. In both host fruit species, *Drosophila* spp. puparia were significantly predominant in the "inside fruit flesh" habitat, while the lowest number of saprophytic drosophilid puparia was found in the soil (Peach, $X^2 = 295.66$, $df = 2$, $p < 0.0001$; Guava, $X^2 = 295.61$, $df = 2$, $p < 0.0001$) (Figure 4A,B). In peach, SWD puparia were slightly more abundant in both "inside fruit flesh" and "outside fruit flesh" microhabitats than in "soil" ($X^2 = 40.51$, $df = 2$, $p < 0.0001$), while in guava, there were no significant differences between the three microhabitats ($X^2 = 5.11$, $df = 2$, $p = 0.0770$) (Figure 4A,B). A comparative analysis of the abundance of saprophytic *Drosophila* spp. and SWD puparia by microhabitat and fruit species showed significant differences between all tested conditions (Table 1). Numbers of *Drosophila* spp. puparia were 4-, 2-, 6-, and 3.5-fold considerably higher than those recorded for SWD from both "inside fruit flesh" and "outside fruit flesh" microhabitats, in both peach and guava, respectively (Figure 4A,B).

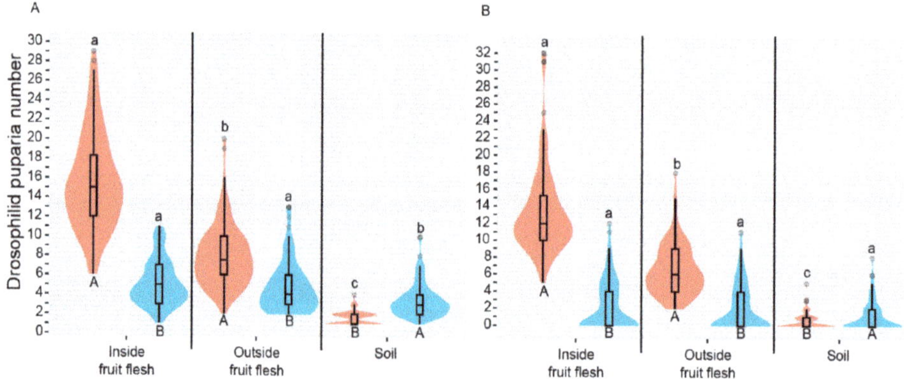

Figure 4. Violin-box plots showing the kernel probability density of a drosophilid (saprophytic *Drosophila* spp. from *D. melanogaster* group = red, and *Drosophila suzukii* = blue) choosing a specific microhabitat in peach and in guava. Violin-box plots showing the kernel probability density. Different letters show significant differences at $p = 0.05$ (LSD test with the Bonferroni–Holm adjustment method). To compare all microhabitats, lowercase, and uppercase letters are used, respectively, for *Drosophila* spp. and *Drosophila suzukii*. The rectangular white bar in the center of the violin box and the black horizontal line inside the bar show the interquartile range and the median, respectively; the black vertical lines stretched from the bar show the lower/upper adjacent values, while black dots display the outlier data.

Table 1. Summary of the Kruskal–Wallis test on puparia abundance comparison between saprophytic *Drosophila* spp. and *D. suzukii* recorded by microhabitat and host fruit species.

Microhabitat	Host Fruit	Statistical Analysis Outcome		
		df	X^2	*p*
Inside fruit flesh	Peach	1	166.94	<0.0001 *
	Guava	1	173.95	<0.0001 *
Outside fruit flesh	Peach	1	66.78	<0.0001 *
	Guava	1	108.91	<0.0001 *
Soil	Peach	1	122.78	<0.0001 *
	Guava	1	0.02	=0.0270 *

* Significant variables.

Similarly, the number of SWD puparia found in the soil beneath or near fruit increased by four- and two-fold, significantly higher than that recorded for *Drosophila* spp. in peach and guava tree-dominated environments, respectively (Figure 4A,B).

3.2. Pupal Parasitoid Abundance and Relationship with Microhabitats Tested

The only pupal parasitoid species associated with drosophilids in both host fruits were *T. anastrephae* and *P. vindemiae*. The former species was 1.8-fold more abundant than the second one. The number of *T. anastrephae* specimens recovered from saprophytic *Drosophila* spp. was three- and seven-fold higher than that recorded from SWD in peach and guava, respectively (Figure 5A,B). The number of *P. vindemiae* specimens recorded from SWD puparia was slightly higher, 1.3- and 1.4-fold, than that obtained from saprophytic *Drosophila* spp. puparia in peach and guava, respectively (Figure 5A,B). When the success of both *T. anastrephae* and *P. vindemiae* in the parasitizing puparia of both *Drosophila* spp. and *D. suzukii* in peach and guava was tested, significant differences were recorded for both categorical factors, such as the type of microhabitat used for host parasitization and the parasitized drosophilid species, as well as their interaction (Table 2).

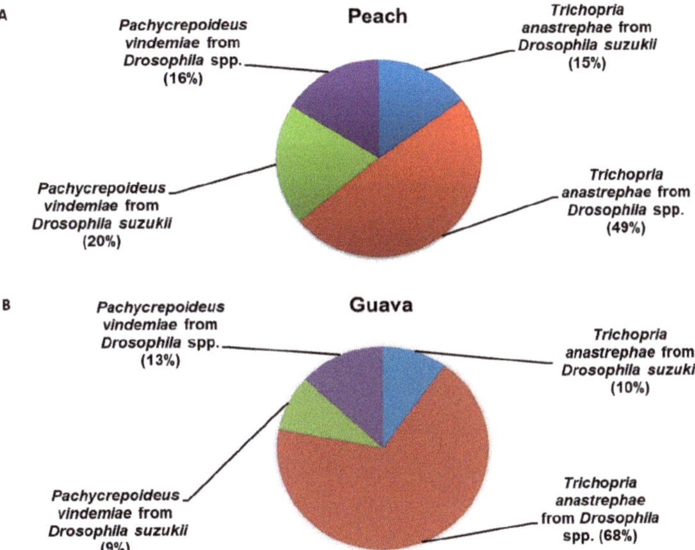

Figure 5. Relative abundance of *Pachycrepoideus vindemiae* and *Trichopria anastrephae* adults recovered from *Drosophila suzukii* and saprophytic *Drosophila* spp. Puparia, associated with (**A**) peach and (**B**) guava.

Trichopria anastrephae was remarkably successful in parasitizing saprophytic *Drosophila* spp. puparia located inside peach and guava flesh, followed, in decreasing order, by puparia found "outside fruit flesh" and in the "soil" (Figure 6A,B). The above pattern was also recorded in SWD puparia (Figure 6A,B). However, *T. anastrephae* was considerably more successful in parasitizing *Drosophila* spp. puparia than SWD puparia in both "inside fruit flesh" and "outside fruit flesh" microhabitats; this was not the case for puparia located in the soil, as there was no significant difference in parasitism between drosophilids (Figure 6A,B). *Pachycrepoideus vindemiae* significantly parasitized more saprophytic *Drosophila* spp. puparia outside fruit flesh when compared to the other tested microhabitats on peach and guava (Figure 6C,D). Nevertheless, the parasitism success of *P. vindemiae* on SWD puparia located in both "outside fruit flesh" and "soil" was statistically similar to that recorded from *Drosophila* spp. puparia outside fruit flesh when only peach was evaluated (Figure 6C). In guava, significantly more SWD puparia were also parasitized by *P. vindemiae* in both "outside fruit flesh" and "soil" microhabitats, but the success of such parasitism was statistically lower than that recorded from *Drosophila* spp. puparia located outside the flesh (Figure 6D). *Pachycrepoideus vindemiae* parasitized a significantly higher number of *Drosophila* spp. puparia located "inside fruit flesh" than SWD puparia found in the same microhabitat in peach, although there were no statistical differences for guava (Figure 6C,D).

Table 2. Summary of Aligned Rank Transform ANOVA on the effect of the type of microhabitat used for host parasitism (=THU), the parasitized drosophilid species (=PDS), and their interaction on the adult emergence of both *Trichopria anastrephae* and *Pachycrepoideus vindemiae*, with data recorded from saprophytic *Drosophila* spp. and *D. suzukii* puparia recovered from peach and guava.

Host Fruit	Parasitoid Species	Source of Variation	df	Residuals df	F	p
Peach	*T. anastrephae*	THU	2	714	566.19	<0.0001 *
		PDS	1	714	389.58	<0.0001 *
		THU × PDS	2	714	204.23	<0.0001 *
Guava	*T. anastrephae*	THU	2	714	599.24	<0.0001 *
		PDS	1	714	733.73	<0.0001 *
		THU × PDS	2	714	298.26	<0.0001 *
Peach	*P. vindemiae*	THU	2	714	85.728	<0.0001 *
		PDS	1	714	84.386	<0.0001 *
		THU × PDS	2	714	113.08	<0.0001 *
Guava	*P. vindemiae*	THU	2	714	52.673	<0.0001 *
		PDS	1	714	102.91	<0.0001 *
		THU × PDS	2	714	42.630	<0.0001 *

* Significant variables.

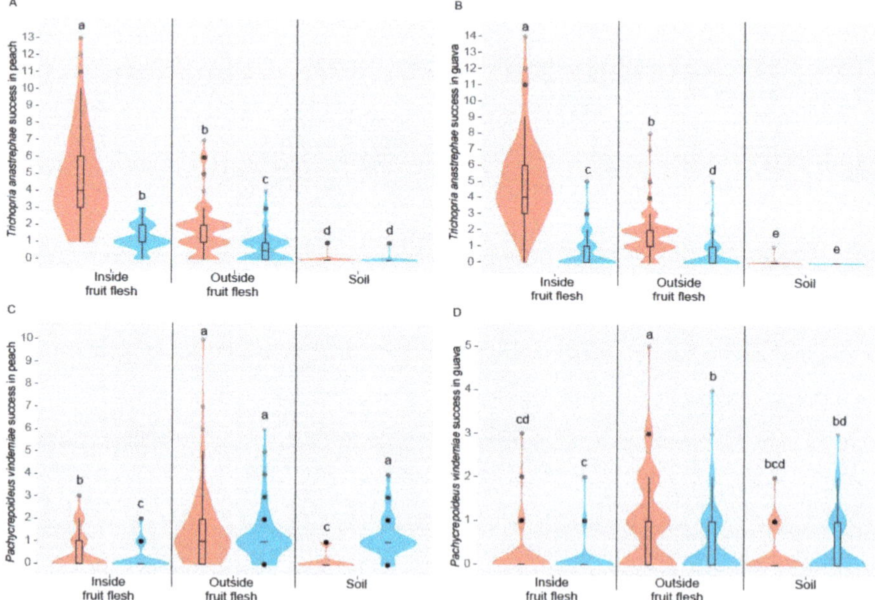

Figure 6. Violin-box plots showing the kernel probability density that drosophilids (saprophytic *Drosophila* spp. from *D. melanogaster* group = red, and *Drosophila suzukii* = blue) are parasitized in a different microhabitat by (**A**) *Trichopria anastrephae* in peach, (**B**) *Trichopria anastrephae* in guava, (**C**) *Pachycrepoideus vindemiae* in peach, and (**D**) *Pachycrepoideus vindemiae* in guava. Different letters show significant differences at *p* = 0.05 (LSD test with the Bonferroni–Holm adjustment method). The rectangular white bar in the center of the violin box and the black horizontal line inside the bar show the interquartile range and the median, respectively; the black vertical lines stretched from the bar show the lower/upper adjacent values, while black dots display the outlier data.

4. Discussion

The success of biological control programs involving parasitoids relies, among many factors, on the knowledge of the resident parasitoid assemblage associated with the invading pest and, crucially, on understanding the mechanisms that allow the coexistence of competing parasitoid species. This sort of ecological insight provides a better understanding of the impact exerted by each resident parasitoid species on the target pest population. Furthermore, all such information is critical for developing and implementing biological control strategies, including exotic or local parasitoid species. In this framework, results of the field study carried out in the fruit-growing province of Tucumán, northwestern Argentina, evidenced niche partitioning as a mechanism involved in facilitating the coexistence of two resident generalist parasitoids, *T. anastrephae* and *P. vindemiae*, attacking puparia of both the invasive fruit pest *D. suzukii* and local saprophytic drosophilid species in the same non-crop fallen fruit. In particular, results revealed interesting aspects of the fruit–drosophilid–parasitoid trophic relationship: (1) differentiated patterns of drosophilid puparia distribution in microhabitats; (2) proportions of *P. vindemiae* adults recovered from SWD puparia that are higher than or similar to those found for *Drosophila* spp. puparia from the *D. melanogaster* group; (3) a discernible trend of *P. vindemiae* females to focus their attack on the puparia of both SWD and resident saprophytic *Drosophila* spp. outside the fleshy inner of the dropped fruit; (4) a strong preference of *T. anastrephae* females for targeting puparia of resident saprophytic drosophilid species, particularly those located inside the fruit.

Two issues should be emphasized concerning the first finding. Firstly, a markedly higher abundance of resident drosophilid puparia over SWD puparia in both host fruits surveyed was observed, but this difference was more evident in guava. Secondly, there was increased resident drosophilid puparia in both inside and outside peach and guava flesh, whereas SWD puparia prevailed in the soil relative to all the other drosophilids. The above is consistent with the prediction based on space-use patterns between the invasive drosophilid species and the local ones. These differences may be related to the fruit ripeness stage preferred by SWD females for egg laying. Rather than overripe, fallen and damaged fruit, these females choose ripe, fresh fruit still on the plant [15]. Therefore, the SWD female exploits mostly fruit in the ripening stages, due to their availability for other *Drosophila* species [31]. In turn, the mature larvae usually tend to migrate from the fruit hanging on the branch, in order to pupate in the soil [48,49]. Thus, the preference for healthy fruits enables the SWD female to exploit novel niches by avoiding competition with other drosophilids [14]. However, SWD females may sometimes lay their eggs in fallen, wounded, and/or fermenting fruit, in situations involving a shortage or non-availability of suitable hosts [50,51]. The females of the *Drosophila melanogaster* group (e.g., *D. melanogaster* and *D. simulans*) are saprophytic flies; they feed and oviposit on damaged, decaying, or fermenting fruits [52], and their larvae usually pupate in the dropped fruit, covering a significant part of their biological cycle in the same microenvironment, in contrast to the standard SWD female oviposition behavior. Given these differences in fruit ripeness preference, the female of resident drosophilids usually oviposits on the host at a later stage than the SWD female. Thus, SWD larvae may complete their development first, and mature larvae usually drop out of the fruit. Therefore, another critical factor to consider is the interspecific larval competition for resources between SWD and saprophytic drosophilids on damaged or already rotten fruit. Such competition may cause a decrease in SWD population growth [50], which might also influence the lower density of SWD puparia inside the host fallen fruit versus resident drosophilid puparia. However, a higher proportion of SWD puparia were found in the fruit than buried around the fallen fruit. Although the SWD mature larva tends to leave the fruit, it can pupate inside the ripe fruit, as [31] demonstrated using infested cherries under lab conditions. These authors found more SWD larvae pupated in the cherry fruit than in the soil. However, recent laboratory trials, carried out by the senior author of the current study, revealed that 60–70% of the total SWD larvae pupated in an artificial pupation medium close to infested peaches, while 75–80% of total *D. melanogaster* puparia remained on fruits. In those lab trials, healthy, ripe, soft peaches were offered to 20 mated SWD females for 24 h. The fruit was then kept in the experimental cage, and on the fifth day, when the fruit showed spotted and rotting sectors, the 20 mated *D. melanogaster* females were released and remained in the cage for 24 h. High numbers of drosophilid larvae left the fruit 1–3 days after *D. melanogaster* females oviposited in peaches. It was then verified by puparium identification that all those larvae were *D. suzukii*. Overall, under competitive interspecific conditions, SWD females are specialized to oviposit on healthy fruit, which is highly preferred, although these females can be flexible to use wounded and fermented fruit as well [50].

The second finding showed an apparently closer trophic relationship between *P. vindemiae* and SWD than with saprophytic drosophilid species in the two natural environments studied. This trait was even more striking when puparia recovered from fallen peaches were analyzed. In this regard, SWD puparia were parasitized by *P. vindemiae* 1.4 times more than by *T. anastrephae*, as opposed to saprophytic drosophilid puparia, which were 3.3-fold more parasitized by *T. anastrephae*. In guava, the incidence of *P. vindemiae* affecting Drosophilidae populations was lower than that recorded in peach. The above was evidenced by the lower proportion of *P. vindemiae* adults recovered from puparia taken from guava, compared to *T. anastrephae* adults. The *P. vindemiae* and *T. anastrephae* adult ratios recorded from SWD puparia were close to 1:1, while for resident drosophilid puparia, there were nearly six *T. anastrephae* individuals per *P. vindemiae* adult. This remarkable difference in the proportion of *P. vindemiae* adults recovered from Drosophilidae puparia between both sampling environments may be due to two related events. Firstly, there is a higher predominance of saprophytic drosophilids on SWD on guava than on peach. In this framework, the proportion of resident drosophilid puparia over SWD was twice higher in guava than in peach. Secondly, there is a low natural population of *P. vindemiae* relative to that of *T. anastrephae* in the guava tree-dominated habitat. In this regard, data from direct field observations at the surveying sites recorded an average of 4.2-fold more *P. vindemiae* adults in the peach tree-dominated wild forest compared to the guava-sampling site. Field observations were made every 15 m for 2 h on each sampling date in both natural environments. Regardless of the above two events, *P. vindemiae*, relative to *T. anastrephae*, is a resident parasitoid, mostly predominant on SWD puparia in both tested habitats. These results support data recently published by Buonocore Biancheri et al. [16], which point to *P. vindemiae* as an attractive agent of SWD natural mortality in disturbed wild environments from the province of Tucumán. However, field studies in southern Brazil showed a higher prevalence of *T. anastrephae* in SWD puparia than that of *P. vindemiae* [35,53]. Analogous data were published on *T. drosophilae* naturally attacking SWD puparia in Europe [54]. However, those contrasts between the results of field studies conducted mainly in Tucumán and southern Brazil may be due to a wide range of biotic and abiotic factors. These may include the disturbance characteristics of the study environment, the variation in the population density of the target pest and competitive frugivorous species, weather conditions during the sampling period, and the host fruit species and its abundance.

All the above factors notably influence the composition of the parasitoid assemblage associated with the target pest and the abundance of each resident parasitoid species [55,56]. However, laboratory establishment of the population lines of both pupal parasitoids tested in the current field study is foreseen as a later research step. In this context, studies with Brazilian population lines of *T. anastrephae* and *P. vindemiae* lab-reared on SWD puparia showed a significant prevalence of the former parasitoid species over the latter in terms of parasitism and adult emergence rates [25]. Similar findings were reported by Wang et al. [31], who found a strong predominance of lab-reared *T. drosophilae* over lab-reared *P. vindemiae* as a parasitoid of both *D. suzukii* and *D. melanogaster*. Similarly, Daane et al. [57] and Wang et al. [58] pointed to *T. drosophilae* as a more efficient SWD parasitoid in laboratory tests than other well-known pupal parasitoids, such as *P. vindemiae*. Likewise, Wolf et al. [59] found in the combined release of *T. drosophilae* and *P. vindemiae* under semi-field experiments that almost all the parasitoid offspring that emerged from SWD puparia were *T. drosophilae* adults, despite the two microhabitats tested: soil and foliage.

Under this framework, several authors pointed to *T. drosophilae* as the pupal parasitoid species with the highest potential for SWD control [60–65]. Evidence was provided through augmentative releases of *T. drosophilae* on cherries, in either crop or non-crop areas in southern Trento, Italy, which achieved a 34% reduction in fruit infestation by *D. suzukii* in unmanaged vegetation areas surrounding orchards [20]. Similarly, mass releases of *T. drosophilae* in berry crops at Colima and Jalisco, Mexico, reduced SWD populations by 50 to 55% [66]. Given the contrast between the field findings of the current study and those from both laboratory [25,31] and semi-field [59,61] studies, it is relevant to assess Argentinian population lines of both pupal parasitoids under lab conditions. This would enable a comparative assessment of *P. vindemiae* and *T. anastrephae* as biocontrol agents of *D. suzukii* by determining host preference, regulating offered host densities, and analyzing the interspecific competition.

The third finding showed that *P. vindemiae* was the predominant parasitoid species recovered from Drosophilidae puparia that was externally attached to the fruit skin, enclosed in outer fruit injuries, protruding from the fruit rind, or directly buried under the fallen fruit. On the contrary, results of the fourth finding revealed that the highest levels of *T. anastrephae* adult abundance were mainly recorded from resident drosophilid puparia sampled directly from peach and guava flesh. Such data were reflected when adult proportions of both pupal parasitoid species recorded from the different tested habitats were comparatively assessed. On this issue, around seven and five *P. vindemiae* adults per *T. anastrephae* adult were recovered from SWD puparia collected from peach and guava, respectively, in the two habitats not involving fruit flesh. In turn, about twice as many *P. vindemiae* adults per *T. anastrephae* adult were recorded from resident drosophilid puparia collected from "outside fruit" and "soil" microhabitats in both host fruit species. In connection with the second finding, the above data support *P. vindemiae*'s prevalence over *T. anastrephae* on SWD puparia. Although saprophytic drosophilid puparia found in both "outside fruit" and "soil" microhabitats yielded relatively more *P. vindemiae* than *T. anastrephae* adults, the highest *P. vindemiae* adult abundance was recorded from SWD puparia. This result was even more evident when only SWD puparia sampled from the "soil" were considered. About 92% of the total pupal parasitoid adults recovered from SWD puparia sampled from the "soil" around peach fruit were *P. vindemiae*. In the same way, all SWD puparia collected from soil in the guava-dominated environment yielded only *P. vindemiae* adults. Th soil was the most favorable microhabitat for *P. vindemiae* females to parasitize Drosophilidae puparia. The *P. vindemiae* female may tend to forage in this type of microhabitat rather than inside the fruit. This assertion may also be corroborated by data on the *P. vindemiae*: *T. anastrephae* adult ratio recorded from resident drosophilid puparia found in soil. At least 35- and 11-fold more *P. vindemiae* than *T. anastrephae* adults were recovered from those buried puparia in guava and peach sampling sites, respectively. *Pachycrepoideus vindemiae* either rarely frequented the inside of the fruit or found it difficult to parasitize host puparia in this microhabitat, showing the lowest levels of abundance recorded in both tested fruit species. Thus, *P. vindemiae* adults recovered from either SWD or resident drosophilid puparia found inside guava and peach flesh only accounted for 6–20% of all individuals of this species. Interestingly, previous studies in Switzerland under both semi-field [59] and open-field [36] conditions demonstrated a preference of *P. vindemiae* for parasitizing drosophilid puparia in the foliage, while *T. drosophilae* mostly parasitized host puparia on the ground. The current study focused on sampling drosophila puparia on the soil and those associated with fallen fruit, without considering damaged fruit located in the plant canopy. However, the forthcoming surveys of drosophila parasitoids will cover wounded or rotting fruit still located in the canopy.

Based on the highly contrasting *P. vindemiae* parasitism data between microhabitats tested in the current study, the interference with *T. anastrephae* was more likely a critical factor influencing *P. vindemiae* performance. That is, given a competitive interaction with

T. anastrephae for the resource, *P. vindemiae* probably faces a disadvantageous situation. Laboratory trials previously reported by da Costa Oliveira et al. [25] showed that *T. anastrephae* from the Brazilian population lineage was competitively superior to *P. vindemiae*, and achieved substantially higher levels of parasitism in SWD puparia when the two parasitoids interacted with each other. Similar results were also reported in interspecific competition studies between *T. drosophilae* and *P. vindemiae* under lab conditions [31] or in semi-field trials [59], where *P. vindemiae* only achieved the highest parasitism when released alone. Both *T. anastrephae* and *P. vindemiae* can discriminate hosts previously parasitized by the other species [25], being relevant for their females to oviposit first on the typical host. Usually, the first parasitoid species ovipositing into the host prevails in an intrinsic competition [67]. However, *T. anastrephae* as *T. drosophilae* [31] may have a set of biological features that allow it to out-compete *P. vindemiae*. These *T. anastrephae* traits may include the following: (a) faster embryonic development, (b) first-instar larvae better being equipped (larger mandibles and fast movements) for encountering competitors, and (c) higher foraging efficiency, which involves less time spent handling the host due to a higher mature egg load.

It is also worth noting that *T. anastrephae* was the dominant parasitoid species recovered from host puparia found inside the fruit in the current study. Hence, it is likely that *T. anastrephae* females preferentially foraged in this microhabitat. About 73% of all *T. anastrephae* adults recovered from resident drosophilid puparia in feral peach and guava were from those sampled directly inside the mesocarp. Interestingly, results also revealed that most of the *T. anastrephae* adults associated with *D. suzukii* (over 56%) were from puparia collected from inside guava or peach fruit. This information is consistent with da Costa Oliveira et al. [25], who stated that *T. anastrephae* females of the Brazilian population lineage successfully parasitize SWD puparia inside strawberry fruits. In addition to Drosophilidae puparium survey data, field records through direct inspection inside the fruit evidenced an average proportion of 16.5 *T. anastrephae* adults per each *P. vindemiae* adult, in this microhabitat by testing 36 fruits (18 guavas and 18 peaches) during all six collecting dates (three fruit of each species per sampling date). A comparative laboratory study between *P. vindemiae* and *T. drosophilae* showed that the diaprid was more effective than the pteromalid for attacking SWD, and parasitism by either parasitoid species was higher in puparia located on cherry fruit, rather than in the soil [31]. This study also showed a slight preference for *T. drosophilae*, similar to *T. anastrephae*, for attacking host puparia on fruits, although Wang et al.'s [31] work was, methodology-wise, different from the current study.

Such information would provide evidence of a resource and niche partitioning, probably aimed at reducing or avoiding interspecific competition between resident pupal parasitoid species. Initially, the above-discussed data plus the second finding outcome would reflect a differentiated use of available resources in the surveyed environments, i.e., different drosophilid species as hosts. This background would mainly display *P. vindemiae* females parasitizing *D. suzukii* puparia and *T. anastrephae* females mostly attacking saprophytic drosophilid puparia. These host preference assertions for *P. vindemiae* and *T. anastrephae* might be supported by differences in the co-evolutionary history between the parasitoid and its host. On this basis, *T. anastrephae* is a neotropical-native parasitoid species [29,33] that has co-evolved in sympatry with saprophytic drosophilid species, such as those of *D. melanogaster* group [16]. Thus, a close trophic association occurs between *T. anastrephae* and non-pest saprophytic drosophilids, whereas with *D. suzukii*, a new trophic association has recently been established, which is naturally uncommon, due to the incidence of preferred hosts. In contrast, *P. vindemiae* is a worldwide cyclorrhaphous dipteran parasitoid that was introduced in several Latin American countries as a biocontrol agent against tephritid pests [27]. Although its first record in Argentina dates back to the 1940s, it is most likely an exotic parasitoid species [68]. Therefore, the high level of polyphagy associated with the lack of a common co-evolutionary history with saprophytic drosophilid species in northwestern Argentina supports a closer trophic association between *P. vindemiae* and *D. suzukii*. Subsequently, data from the current study suggest an apparent preference for *P. vindemiae* for parasitizing Drosophilidae puparia in microhabitats mostly exposed to female parasitoid attacks. In line with this statement, the *P. vindemiae* female would exhibit a foraging behavior targeted mainly to host puparia in the soil. Host puparia buried beneath or adjacent to fruit likely provide a *T. anastrephae*-free microhabitat, which would ease *P. vindemiae* females' foraging for the host in the soil, without interference from the closest competitor. This scenario points to probable niche segregation between both pupal parasitoid species at a spatial scale. This means *T. anastrephae* females focuse on foraging mostly inside the fruit for host puparia, whereas *P. vindemiae* females target their host search effort in habitats occasionally frequented by the competitor, such as both "outside fruit flesh" and "soil". As pointed out by Wang et al. [31], when discussing the *T. drosophilae–P. vindemiae* competitive relationship, an alternative host does appear to reduce interspecific competition between such species, although these parasitoids showed no preference for *D. suzukii* or *D. melanogaster* when tested in the laboratory. However, in natural conditions, sympatric species tend to reduce competition by using different resources or habitats [3].

In conclusion, results reveal that both divergence in host choice and spatial patterns of same-resource preference among potential competitors, such as *P. vindemiae* and *T. anastrephae*, may mediate the coexistence of these two pupal parasitoids species in each natural environment tested in the current study. Given the apparent preference of the native *T. anastrephae* for resident saprophytic dipteran puparia, mainly located in guava or peach fruit, *P. vindemiae* might be more suited to forage in less competitive microhabitats, such as in the soil or outside of the fruit flesh, in which puparia of the exotic *D. suzukii* would naturally prevail in these habitats. From a SWD management approach, this scenario suggests that both pupal parasitoids have potential as *D. suzukii* biological control agents. This is because such niche partitioning primarily involves differentiated exploitation of host microhabitats, influencing the efficiency of both parasitoids in suppressing *D. suzukii* populations. Such an approach regarding the use of both pupal parasitoid species, based on a differentiation in host microhabitat preference (soil vs. foliage), was highlighted by Wolf et al. [59] relying on semi-field study results in Switzerland. Likewise, Kruitwagen et al. [69] and Jarret et al. [70], in studies based on experimental adaptation studies of resident parasitoids to the invasive *D. suzukii*, pointed out that both *T. anastrephae* and *P. vindemiae* might offer a greater potential to control SWD natural populations over larval parasitoids. Consequently, combining the two resident parasitoid species in wild non-crop environments may be an advisable alternative for local SWD management, either through augmentative releases [20] or through a conservation biological control program [71]. It is worth analyzing this initiative from an area-wide SWD management approach, as suggested by Garcia et al. [11], Rossi-Stacconi et al. [20], Garcia [21], Wang et al. [22]. In this context, parasitoid releases should mainly be performed in wild areas, where known, non-crop, alternative SWD hosts are abundant and may increase the risk of SWD infestations in surrounding fruit crops [20,22,72]. Furthermore, pupal parasitoids would be more effective if released early in the fruiting season, when SWD numbers are still low, to avoid the pest population increase [20,59,72]. This SWD biological control strategy is particularly relevant for the province of Tucumán, where feral guavas and peaches share the same geographical space with commercial berry orchards, as Tucumán hosts most of the soft fruit crops in fruit-growing regions of northern Argentina [73]. Both feral fruit species allow the sustainability of SWD populations during the season in which commercial berry crops are not in production, representing a high economical risk for the local fruit industry. In this context, the use of both the studied parasitoids is a practical and useful alternative for berry growers in Tucumán; they may release them in areas of wild vegetation adjacent to their crops or in orchards or backyards where there is no phytosanitary control. Finally, it is relevant to examine whether that niche differentiation in both parasitoid species occurs in other fruit host species, such as berries, or in other natural environments, such as berry crops in the outlying areas surrounding crops.

Supplementary Materials: The following supporting information can be downloaded at: https://www.mdpi.com/article/10.3390/insects14030222/s1, File S1: Raw Data_Buonocore Biancheri et al.; File S2: Statistical Analysis_Aligned-Rank-Transformation-ANOVA-Markdown-Guavas; File S3: Statistical Analysis_Aligned-Rank-Transformation-ANOVA-Markdown-Peach; File S4: Statistical Analysis-Habitat-use by Drosophilids.

Author Contributions: Conceptualization, M.J.B.B., S.R.N.-C., L.S., M.D.P., D.S.K., F.R.M.G. and S.M.O.; Data curation, M.J.B.B., S.R.N.-C., L.S., M.D.P. and S.M.O.; funding acquisition, D.S.K. and S.M.O.; Formal Analysis, S.R.N.-C.; Investigation, M.J.B.B., S.R.N.-C., L.S., M.D.P., D.S.K., F.R.M.G. and S.M.O.; Methodology, M.J.B.B., L.S., M.D.P., D.S.K., F.R.M.G. and S.M.O.; project administration, D.S.K.; Resources, D.S.K. and S.M.O.; Software, S.R.N.-C.; Supervision, S.M.O.; Validation, M.J.B.B., S.R.N.-C., L.S., M.D.P., D.S.K., F.R.M.G. and S.M.O.; Visualization, M.J.B.B., S.R.N.-C., L.S., M.D.P. and S.M.O.; writing—original draft preparation; M.J.B.B., S.R.N.-C., L.S., M.D.P., D.S.K., F.R.M.G. and S.M.O.; writing—review and editing, M.J.B.B., S.R.N.-C., L.S., M.D.P., D.S.K., F.R.M.G. and S.M.O. All authors have read and agreed to the published version of the manuscript.

Funding: Fondo para la Investigación Científica y Tecnológica (FONCYT) from Agencia Nacional de Promoción Científica y Tecnológica de Argentina (ANPCyT) (grant PICT 2017–0512).

Data Availability Statement: The data presented in this study are available in Supplementary Material here.

Acknowledgments: The authors would like to thank the authorities of the "Sierra de San Javier" Park (Universidad Nacional de Tucumán, Argentina) for allowing us to collect biological material and Patricia Colombres and Guillermo Borchia (PROIMI—CONICET) for laboratory assistance.

Conflicts of Interest: The authors declare no conflict of interest.

References

1. Reitz, S.R.; Trumble, J.T. Competitive displacement among insects and arachnids. *Annu. Rev. Entomol.* **2002**, *47*, 435–465. [CrossRef] [PubMed]
2. Bonsall, M.B.; Hassell, M.P.; Asefa, G. Ecological trade-offs, resource partitioning, and coexistence in a host–parasitoid assemblage. *Ecology* **2002**, *83*, 925–934.
3. Amarasekare, P. Competitive coexistence in spatially structured environments: A synthesis. *Ecol. Lett.* **2003**, *6*, 1109–1122. [CrossRef]
4. Wang, X.G.; Messing, R.H. Two different life history strategies determine the competitive outcome between *Dirhinus giffardii* and *Pachycrepoideus vindemmiae*, ectoparasitoids of cyclorrhaphous Diptera. *Bull. Entomol. Res.* **2004**, *94*, 473–480. [CrossRef] [PubMed]
5. Pekas, A.; Tena, A.; Harvey, J.A.; Garcia-Marí, F.; Frag, F. Host size and spatiotemporal patterns mediate the coexistence of specialist parasitoids. *Ecology* **2016**, *97*, 1345–1356. [CrossRef] [PubMed]
6. Aluja, M.; Ovruski, S.M.; Sivinski, J.; Córdova-García, G.; Schliserman, P.; Campero, S.R.N.; Ordano, M. Inter-specific competition and competition free space in the tephritid parasitoids *Utetes anastrephae* and *Doryctobracon areolatus* (Hymenoptera: Braconidae: Opiinae). *Ecol. Entomol.* **2013**, *38*, 485–496. [CrossRef]
7. Tscharntke, T. Coexistence, tritrophic interactions and density dependence in a species-rich parasitoid community. *J. Anim. Ecol.* **1992**, *61*, 59–67. [CrossRef]
8. Pedersen, B.S.; Mills, N.J. Single vs. multiple introduction in biological control: The roles of parasitoid efficiency, antagonism and niche overlap. *J. Appl. Ecol.* **2004**, *41*, 973–984. [CrossRef]
9. Duan, J.J.; Van Driesche, R.G.; Schmude, J.M.; Quinn, N.F.; Petrice, T.R.; Rutledge, C.E.; Poland, T.M.; Bauer, L.S.; Elkinton, J.S. Niche partitioning and coexistence of parasitoids of the same feeding guild introduced for biological control of an invasive forest pest. *Biol. Control* **2021**, *160*, 104698. [CrossRef]
10. Asplen, M.K.; Anfora, G.; Biondi, A.; Choi, D.S.; Chu, D.; Daane, K.M.; Gibert, P.; Gutierrez, A.P.; Hoelmer, K.A.; Hutchison, W.D.; et al. Invasion biology of spotted-wing Drosophila (*Drosophila suzukii*): A global perspective and future priorities. *J. Pest Sci.* **2015**, *88*, 469–494. [CrossRef]
11. Garcia, F.R.M.; Lasa, R.; Funes, C.F.; Buzzetti, K. *Drosophila suzukii* Management in Latin America: Current Status and Perspectives. *J. Econ. Entomol.* **2022**, *115*, 1008–1023. [CrossRef]
12. Escobar, L.I.; Ovruski, S.M.; Kirschbaum, D.S. Foreign invasive pests *Drosophila suzukii* (Matsumura) and *Zaprionus indianus* Gupta (Diptera: Drosophilidae) threaten fruit production in northwestern Argentina. *Dros. Info. Serv.* **2018**, *101*, 9–14.
13. De Ros, G.; Grassi, A.; Pantezzi, T. Recent Trends in the Economic Impact of *Drosophila suzukii*. In *Drosophila Suzukii Management*; Garcia, F.R.M., Ed.; Springer Nature: Cham, Switzerland, 2020; pp. 11–28.
14. Atallah, J.; Teixeira, L.; Salazar, R.; Zaragoza, G.; Kopp, A. The making of a pest: The evolution of a fruit-penetrating ovipositor in *Drosophila suzukii* and related species. *Proc. Royal Soc. B* **2014**, *281*, 20132840. [CrossRef]

15. Kirschbaum, D.S.; Funes, C.F.; Buonocore-Biancheri, M.J.; Suárez, L.; Ovruski, S.M. The biology and ecology of *Drosophila suzukii* (Diptera: Drosophilidae) (Chapter 4). In *Drosophila Suzukii Management*; Garcia, F.R.M., Ed.; Springer Nature: Cham, Switzerland, 2020; pp. 41–91.

16. Biancheri, M.J.B.; Suárez, L.; Kirschbaum, D.S.; Garcia, F.R.M.; Funes, C.F.; Ovruski, S.M. Natural parasitism influences biological control strategies against both global invasive pests *Ceratitis capitata* (Diptera: Tephritidae) and *Drosophila suzukii* (Diptera: Drosophilidae), and the Neotropical-native pest *Anastrepha fraterculus* (Diptera: Tephritidae). *Environ. Entomol.* **2022**, *51*, 1120–1135.

17. Sassù, F.; Nikolouli, K.; Stauffer, C.; Bourtzis, K.; Cáceres, C. Insect technique and incompatible insect technique for the integrated *Drosophila suzukii* management (Chapter 9). In *Drosophila Suzukii Management*; Garcia, F.R.M., Ed.; Springer Nature: Cham, Switzerland, 2020; pp. 169–194.

18. Lee, J.C.; Dreves, A.J.; Cave, A.M.; Kawai, S.; Isaacs, R.; Miller, J.C.; Van Timmeren, S.; Bruck, D.J. Infestation of wild and ornamental noncrop fruits by *Drosophila suzukii* (Diptera: Drosophilidae). *Ann. Entomol. Soc. Am.* **2015**, *108*, 117–129. [CrossRef]

19. Kenis, M.; Tonina, L.; Eschen, R.; van der Sluis, B.; Sancassani, M.; Mori, N.; Haye, T.; Helsen, H. Non-crop plants used as hosts by *Drosophila suzukii* in Europe. *J. Pest Sci.* **2016**, *89*, 735–748. [CrossRef]

20. Rossi Stacconi, M.V.; Grassi, A.; Ioriatti, C.; Anfora, G. Augmentative releases of *Trichopria drosophilae* for the suppression of early season *Drosophila suzukii* populations. *BioControl* **2019**, *64*, 9–19. [CrossRef]

21. Garcia, F.R.M. Basis for area-wide management of *Drosophila suzukii* in Latin America (Chapter 5). In *Drosophila Suzukii Management*; Garcia, F.R.M., Ed.; Springer Nature: Cham, Switzerland, 2020; pp. 93–110.

22. Wang, X.G.; Daane, K.M.; Hoelmer, K.A.; Lee, J.C. Biological control of spotted-wing drosophila: An update on promising agents (Chapter 8). In *Drosophila Suzukii Management*; Garcia, F.R.M., Ed.; Springer Nature: Cham, Switzerland, 2020; pp. 143–168.

23. Abram, P.K.; Wang, X.G.; Hueppelsheuser, T.; Franklin, M.T.; Daane, K.M.; Lee, J.C.; Lue, C.H.; Girod, P.; Carrillo, J.; Wong, W.H.L.; et al. A coordinated sampling and identification methodology for larval parasitoids of spotted-wing drosophila. *J. Econ. Entomol.* **2022**, *115*, 922–942. [CrossRef]

24. Garcia, F.R.M.; Corseuil, E. Native hymenopteran parasitoids associated with fruit flies (Diptera: Tephritoidea) in Santa Catarina State, Brazil. *Fla. Entomol.* **2004**, *87*, 517–521. [CrossRef]

25. da Costa Oliveira, D.; Stupp, P.; Martins, L.N.; Wollmann, J.; Geisler, F.C.S.; Cardoso, T.D.N.; Bernardi, D.; Garcia, F.R.M. Interspecific competition in *Trichopria anastrephae* parasitism (Hymenoptera: Diapriidae) and *Pachycrepoideus vindemmiae* (Hymenoptera: Pteromalidae) parasitism on pupae of *Drosophila suzukii* (Diptera: Drosophilidae). *Phytoparasitica* **2021**, *49*, 207–215. [CrossRef]

26. Wang, X.G.; Messing, R.H. The ectoparasitic pupal parasitoid, Pachycrepoideus vindemmiae (Hymenoptera: Pteromalidae), attacks other primary tephritid fruit fly parasitoids: Host expansion and non-target impact. *Biol. Control* **2004**, *31*, 227–236. [CrossRef]

27. Garcia, F.R.M.; Ovruski, S.M.; Suárez, L.; Cancino, J.; Liburd, O.E. Biological control of tephritid fruit flies in the Americas and Hawaii: A review of the use of parasitoids and predators. *Insects* **2020**, *11*, 662. [CrossRef] [PubMed]

28. Funes, C.F.; Garrido, S.A.; Aquino, D.A.; Escobar, L.I.; Gómez Segade, C.; Cichón, L.; Ovruski, S.M.; Kirschbaum, D.S. New records of *Pachycrepoideus vindemmiae* (Hymenoptera: Pteromalidae) associated with *Drosophila suzukii* (Diptera: Drosophilidae) in cherry and berry crops from Argentina. *Rev. Soc. Entomol. Arg.* **2020**, *79*, 39–43. [CrossRef]

29. Ovruski, S.M.; Aluja, M.; Sivinski, J.; Wharton, R. Hymenopteran parasitoids on fruit infesting Tephritidae (Diptera) in Latin America and the Southern United States: Diversity, distribution, taxonomic status and their use in fruit fly biological control. *Integr. Pest Manag. Rev.* **2000**, *5*, 81–107. [CrossRef]

30. Ehler, L.E. Parasitoid communities, parasitoid guilds, and biological control. In *Parasitoid Community Ecology*; Hawkins, B.A., Sheehan, W., Eds.; Oxford University Press: Oxford, UK, 1994; pp. 418–436.

31. Wang, X.G.; Kaçar, G.; Biondi, A.; Daane, K.M. Foraging efficiency and outcomes of interactions of two pupal parasitoids attacking the invasive spotted wing drosophila. *Biol. Control* **2016**, *96*, 64–71. [CrossRef]

32. Hogg, B.N.; Lee, L.C.; Rogers, M.A.; Worth, L.; Nieto, D.J.; Stahl, J.M.; Daane, K.M. Releases of the parasitoid *Pachycrepoideus vindemmiae* for augmentative biological control of spotted wing drosophila, *Drosophila suzukii*. *Biol. Control* **2022**, *168*, 104865. [CrossRef]

33. Krüger, A.P.; Scheunemann, T.; Vieira, J.G.A.; Morais, M.C.; Bernardi, D.; Nava, D.E.; Garcia, F.R.M. Effects of extrinsic, intraspecific competition and host deprivation on the biology of *Trichopria anastrephae* (Hymenoptera: Diapriidae) reared on *Drosophila suzukii* (Diptera: Drosophilidae). *Neotrop. Entomol.* **2019**, *48*, 957–965. [CrossRef]

34. Vieira, J.G.A.; Krüger, A.P.; Scheuneumann, T.; Morais, M.C.; Speriogin, H.S.; Garcia, F.R.M.; Nava, D.E.; Bernardi, D. Some aspects of the biology of *Trichopria anastrephae* (Hymenoptera: Diapriidae), a resident parasitoid attacking *Drosophila suzukii* (Diptera: Drosophilidae) in Brazil. *J. Econ. Entomol.* **2019**, *113*, 81–87. [CrossRef]

35. Wollmann, J.; Schlesener, D.C.H.; Ferreira, M.S.; Garcia, M.S.; Costa, V.A.; Garcia, F.R.M. Parasitoids of drosophilidae with potential for parasitism on *Drosophila suzukii* in Brazil. *Dros. Info. Serv.* **2016**, *99*, 38–42.

36. Trivellone, V.; Meier, M.; Cara, C.; Pollini Paltrinieri, L.; Gugerli, F.; Moretti, M.; Wolf, S.; Collatz, J. Multiscale determinants drive parasitization of Drosophilidae by Hymenopteran parasitoids in agricultural landscapes. *Insects* **2020**, *11*, 334. [CrossRef]

37. Brown, A.D. Las selvas pedemontanas de las Yungas: Manejo sustentable y conservación de la biodiversidad de un ecosistema prioritario del noroeste argentino. In *Selva Pedemontana de las Yungas: Historia Natural, Ecología y Manejo de un Ecosistema en Peligro*; Brown, A.D., Blendinger, P.G., Lomáscolo, T., García Bes, P., Eds.; Ediciones del Subtrópico: Yerba Buena, Argentina, 2009; pp. 13–36.

38. Brown, A.D.; Grau, H.R.; Malizia, L.R.; Grau, A. Argentina. In *Bosques Nublados del Geotrópico*; Kappelle, M., Brown, A.D., Eds.; Instituto Nacional de Biodiversidad: San José, Costa Rica, 2001; pp. 623–659.

39. Hauser, M. A historic account of the invasion of *Drosophila suzukii* (Matsumura) (Diptera: Drosophilidae) in the continental United States, with remarks on their identification. *Pest Manag. Sci.* **2011**, *67*, 1352–1357. [CrossRef]

40. Ranz, J.M.; Maurin, D.; Chan, Y.S.; von Grotthuss, M.; Hillier, L.D.W.; Roote, J.; Ashburner, M.; Bergman, C.M. Principles of Genome Evolution in the *Drosophila melanogaster* Species Group. *PLoS Biol.* **2007**, *5*, e152. [CrossRef]

41. Gibson, G.A.P. Illustrated Key to the Native and Introduced Chalcidoid Parasitoids of Filth Flies in America North of Mexico (Hymenoptera: Chalcidoidea). Available online: http://res2.agr.ca/ecorc/apss/chalkey/keyintro.htm (accessed on 20 October 2000).

42. Risbec, J. La Faune Entomologique des Cultures au Sénégal et au Soudan: II. Contribution à L'étude des Proctotrupidae (Serphiidae). [s.l.], (Travaux du Laboratoire d'Entomologie du Secteur Soudanais de Recherche Agronomique–Station de Bambey); Gouvernement Général de l'Afrique Occidentale Française, 1950; 638p.

43. R Core Team. *R: A Language and Environment for Statistical Computing*; R. Foundat Statistical Computing: Vienna, Austria, 2022; Available online: https://www.R-project.org/ (accessed on 10 October 2022).

44. Wobbrock, J.O.; Findlater, L.; Gergle, D.; Higgins, J.J. The Aligned Rank Transform for nonparametric factorial analyses using only Anova Procedures. In Proceedings of the SIGCHI Conference on Human Factors in Computing Systems, Vancouver, BC, Canada, 7–12 May 2011. [CrossRef]

45. Elkin, L.A.; Kay, M.; Higgins, J.J.; Wobbrock, J.O. An Aligned Rank Transform procedure for multifactor contrast tests. In Proceedings of the 34th Annual ACM Symposium on User Interface Software and Technology, 10–14 October 2021. [CrossRef]

46. de Mendiburu, F. Agricolae: Statistical Procedures for Agricultural Research. R. Package Version 1.3-5. 2021. Available online: https://CRAN.R-project.org/package=agricolae (accessed on 10 October 2022).

47. Wickham, H. *ggplot2: Elegant Graphics for Data Analysis*; Springer: New York, NY, USA, 2016.

48. Woltz, J.M.; Lee, J.C. Pupation behavior and larval and pupal biocontrol of *Drosophila suzukii* in the field. *Biol. Control* **2017**, *110*, 62–69. [CrossRef]

49. Lee, J.C.; Wang, X.G.; Daane, K.M.; Hoelmer, K.A.; Isaacs, R.; Sial, A.A.; Walton, V.M. Biological control of spotted-wing drosophila (Diptera: Drosophilidae): Current and pending tactics. *J. Integr. Pest Manag.* **2019**, *10*, 13. [CrossRef]

50. Kienzle, R.; Groß, L.B.; Caughman, S.; Rohlfs, M. Resource use by individual *Drosophila suzukii* reveals a flexible preference for oviposition into healthy fruits. *Sci. Rep.* **2020**, *10*, 3132. [CrossRef]

51. Mazzetto, F.; Lessio, F.; Giacosa, S.; Rolle, L.; Alma, A. Relationships between *Drosophila suzukii* and grapevine in Northwestern Italy: Seasonal presence and cultivar susceptibility. *Bull. Insectol.* **2020**, *73*, 29–38.

52. Scheidler, N.H.; Liu, C.; Hamby, K.A.; Zalom, F.G.; Syed, Z. Volatile codes: Correlation of olfactory signals and reception in *Drosophila*-yeast chemical communication. *Sci. Rep.* **2015**, *5*, 14059. [CrossRef]

53. Schlesener, D.C.H.; Wollmann, J.J.; de Bastos Pazini, J.; Padilha, A.C.; Grützmacher, A.D.; Garcia, F.R.M. Insecticide Toxicity to *Drosophila suzukii* (Diptera: Drosophilidae) parasitoids: *Trichopria anastrephae* (Hymenoptera: Diapriidae) and *Pachycrepoideus vindemmiae* (Hymenoptera: Pteromalidae). *J. Econ. Entomol.* **2019**, *112*, 1197–1206. [CrossRef]

54. Stacconi, M.V.R.; Panel, A.; Baser, N.; Ioriatti, C.; Pantezzi, T.; Anfora, G. Comparative life history traits of indigenous Italian parasitoids of *Drosophila suzukii* and their effectiveness at different temperatures. *Biol. Control* **2017**, *112*, 20–27. [CrossRef]

55. Schliserman, P.; Aluja, M.; Rull, J.; Ovruski, S.M. Habitat degradation and introduction of exotic plants favor persistence of invasive species and population growth of native polyphagous fruit fly pests in a Northwestern Argentinean mosaic. *Biol. Invasions* **2014**, *16*, 2599–2613. [CrossRef]

56. Schliserman, P.; Aluja, M.; Rull, J.; Ovruski, S.M. Temporal diversity and abundance patterns of parasitoids of fruit-infesting Tephritidae (Diptera) in the Argentinean Yungas: Implications for biological control. *Environ. Entomol.* **2016**, *45*, 1184–1198. [CrossRef] [PubMed]

57. Daane, K.M.; Wang, X.G.; Biondi, A.; Miller, B.; Miller, J.C.; Riedl, H.; Shearer, P.W.; Guerrieri, E.; Giorgini, M.; Buffington, M. First exploration of parasitoids of *Drosophila suzukii* in South Korea as potential classical biological agents. *J. Pest Sci.* **2016**, *89*, 823–835. [CrossRef]

58. Wang, X.G.; Kaçar, G.; Biondi, A.; Daane, K.M. Life-history and host preference of the pupal parasitoid *Trichopria drosophilae* of spotted wing drosophila. *BioControl* **2016**, *61*, 387–397. [CrossRef]

59. Wolf, S.; Barmettler, E.; Eisenring, M.; Romeis, J.; Collatz, J. Host searching and host preference of resident pupal parasitoids of *Drosophila suzukii* in the invaded regions. *Pest Manag. Sci.* **2021**, *77*, 243–252. [CrossRef]

60. Mazzetto, F.; Marchetti, E.; Amiresmaeili, N.; Sacco, D.; Francati, S.; Jucker, C.; Dindo, M.L.; Lupi, D.; Tavell, L. Drosophila parasitoids in northern Italy and their potential to attack the exotic pest *Drosophila suzukii*. *J. Pest Sci.* **2016**, *89*, 837–850. [CrossRef]

61. Kaçar, G.; Wang, X.G.; Biondi, A.; Daane, K.M. Linear functional response by two pupal Drosophila parasitoids foraging within single or multiple patch environments. *PLoS ONE* **2017**, *12*, e0183525. [CrossRef]

62. Rossi-Stacconi, M.V.; Amiresmaeili, N.; Biondi, A.; Carli, C.; Caruso, S.; Dindo, M.L.; Francati, S.; Gottardello, A.; Grassi, A.; Lupi, D.; et al. Host location and dispersal ability of the cosmopolitan parasitoid *Trichopria drosophilae* released to control the invasive spotted wing Drosophila. *Biol. Control* **2018**, *117*, 188–196. [CrossRef]
63. Pfab, F.; Stacconi, M.V.R.; Anfora, G.; Grassi, A.; Walton, V.M.; Pugliese, A. Optimized timing of parasitoid release: A mathematical model for biological control of *Drosophila suzukii*. *Theor. Ecol.* **2018**, *11*, 450–489. [CrossRef]
64. Wang, X.G.; Serrato, M.A.; Son, Y.; Walton, V.M.; Hogg, B.N.; Daane, K.M. Thermal performance of two indigenous pupal parasitoids attacking the invasive *Drosophila suzukii* (Diptera: Drosophilidae). *Environ. Entomol.* **2018**, *47*, 764–772. [CrossRef]
65. Yi, C.; Cai, P.; Lin, J.; Liu, X.; Ao, G.; Zhang, Q.; Xia, H.; Yang, J.; Ji, Q. Life History and Host Preference of *Trichopria drosophilae* from Southern China, One of the Effective Pupal Parasitoids on the *Drosophila* Species. *Insects* **2020**, *11*, 103. [CrossRef]
66. Gonzalez-Cabrera, J.; Moreno-Carrillo, G.; Sanchez-Gonzalez, J.A. Single and combined release of *Trichopria drosophilae* (Hymenoptera: Diapriidae) to control *Drosophila suzukii* (Diptera: Drosophilidae). *Neotrop. Entomol.* **2019**, *48*, 949–956. [CrossRef]
67. Harvey, J.A.; Tanaka, T.; Poelman, E.H. Intrinsic inter- and intraspecific competition in parasitoid wasps. *Annu. Rev. Entomol.* **2013**, *58*, 333–351. [CrossRef]
68. Ovruski, S.M.; Schliserman, P. Biological control of tephritid fruit flies in Argentina: Historical review, current status, and future trends for developing a parasitoid mass-release program. *Insects* **2012**, *3*, 870–888. [CrossRef]
69. Kruitwagen, A.; Beukeboom, L.W.; Wertheim, B. Optimization of native biocontrol agents, with parasitoids of the invasive pest *D. suzukii* as an example. *Evol. Appl.* **2018**, *11*, 1473–1497. [CrossRef]
70. Jarrett, B.J.M.; Linder, S.; Fanning, P.D.; Isaacs, R.; Szucs, M. Experimental adaptation of native parasitoids to the invasive insect pest, *Drosophila suzukii*. *Biol. Control* **2022**, *167*, 104843. [CrossRef]
71. Vargas, G.; Rivera-Pedroza, L.-F.; Cracía, L.F.; Mundstock Jahnke, S. Conservation Biological Control as an Important Tool in the Neotropical Region. *Neotrop. Entomol.* **2022**. [CrossRef]
72. Wolf, S.; Boycheva-Woltering, S.; Romeis, J.; Collatz, J. *Trichopria drosophilae* parasitizes *Drosophila suzukii* in seven common non-crop fruits. *J. Pest Sci.* **2020**, *93*, 627–638. [CrossRef]
73. Funes, C.F.; Kirschbaum, D.S.; Escobar, L.I.; Heredia, A.M. *La Mosca de las alas Manchadas, Drosophila suzukii (Matsumura), Nueva Plaga de las Frutas Finas en Argentina*; PDF, Digital Book; Ediciones INTA: Famaillá, Argentina, 2018; 28p.

MDPI AG
Grosspeteranlage 5
4052 Basel
Switzerland
Tel.: +41 61 683 77 34

Insects Editorial Office
E-mail: insects@mdpi.com
www.mdpi.com/journal/insects